Mechanisms and Machine Science

Volume 100

This book series establishes a well-defined forum for monographs, edited Books, and proceedings on mechanical engineering with particular emphasis on MMS (Mechanism and Machine Science). The final goal is the publication of research that shows the development of mechanical engineering and particularly MMS in all technical aspects, even in very recent assessments. Published works share an approach by which technical details and formulation are discussed, and discuss modern formalisms with the aim to circulate research and technical achievements for use in professional, research, academic, and teaching activities.

This technical approach is an essential characteristic of the series. By discussing technical details and formulations in terms of modern formalisms, the possibility is created not only to show technical developments but also to explain achievements for technical teaching and research activity today and for the future.

The book series is intended to collect technical views on developments of the broad field of MMS in a unique frame that can be seen in its totality as an Encyclopaedia of MMS but with the additional purpose of archiving and teaching MMS achievements. Therefore, the book series will be of use not only for researchers and teachers in Mechanical Engineering but also for professionals and students for their formation and future work.

The series is promoted under the auspices of International Federation for the Promotion of Mechanism and Machine Science (IFToMM).

Prospective authors and editors can contact Mr. Pierpaolo Riva (publishing editor, Springer) at: pierpaolo.riva@springer.com

Indexed by SCOPUS and Google Scholar.

More information about this series at http://www.springer.com/series/8779

Jinyang Xu · Krishna Murari Pandey
Editors

Mechanical Engineering and Materials

Proceedings of ICMEM 2020

IFToMM

Springer

Editors
Jinyang Xu
School of Mechanical Engineering
Shanghai Jiao Tong University
Shanghai, China

Krishna Murari Pandey
Department of Mechanical Engineering
National Institute Of Technology Silchar
Silchar, Assam, India

ISSN 2211-0984 ISSN 2211-0992 (electronic)
Mechanisms and Machine Science
ISBN 978-3-030-68302-3 ISBN 978-3-030-68303-0 (eBook)
https://doi.org/10.1007/978-3-030-68303-0

This Springer imprint is published by the registered company Springer Nature Switzerland AG
The registered company address is: Gewerbestrasse 11, 6330 Cham, Switzerland

Preface

The 2020 International Conference on Mechanical Engineering and Materials (ICMEM 2020) has been held on October 16th–17th, 2020 in Beijing, China. The ICMEM 2020 organized by Hubei Zhongke Institute of Geology and Environment Technology intends to invite worldwide famous scientists, experts, scholars, and researchers for academic presentations. About 80 participants took part in the ICMEM 2020, and they are mainly from China, USA, France, and South Africa. Three keynote speeches, thirteen oral presentations, and thirty-nine poster presentations are provided at the conference. The proceedings of ICMEM 2020 contain twenty-three papers related to recent developments in mechanical engineering and manufacturing. All of the accepted papers have been thoroughly peer-reviewed by experts in related fields to meet the scientific publication standards.

We would like to thank all the members of the Technical Program Committee for the success of ICMEM 2020. And we wish to express our sincere gratitude to the authors and reviewers for their valuable scientific contributions and useful feedbacks to ensure a high scientific quality of ICMEM 2020.

Finally, we thank all the staff members at Springer for their valuable editorial support.

Shanghai, China — Assoc. Prof. Jinyang Xu
Silchar, India — Prof. Krishna Murari Pandey

Contents

About the Author

Dr. Jinyang Xu is an associate professor and a doctoral supervisor of Mechanical Engineering at Shanghai Jiao Tong University, China. He was awarded the Shanghai Pujiang Scholar by the Shanghai Municipality in 2017. He received his M.Sc. (2013) in Mechanical Manufacturing & Automation from Shanghai Jiao Tong University, China, and his Ph.D. (2016) in Mechanical Engineering & Materials from Arts et Métiers ParisTech, France. His research interests focus on composites machining, numerical modeling, micro/nano cutting, and surface texturing. He is now serving as an associate editor of Simulation-Transactions of the Society for Modeling and Simulation International (indexed by SCIE/EI). He is also on the editorial board of several JCR-ranked journals, including Green Materials, International Journal of Aerospace Engineering, Advances in Materials Science and Engineering, *etc*. Moreover, Dr. Xu is also serving as a regular reviewer for over 40 international journals indexed by SCIE/EI, such as IJMTM, JMPT, IJAMT, *etc*. He has been invited as Keynote/Invited Speaker, Section Chair, and TPC member for a number of international conferences with themes in materials science and manufacturing processes. For his notable research in the field of Materials Processing and Manufacturing, he was honored with the IAAM Scientist Medal of the year 2020, and received the Best Paper Awards at the CJUMP2017 and the ISGMA2015 conferences.

Chapter 1
Controlled and Localized Electrochemical Microfabrication with Moving Sub-Micron Rounded Tip as Tool Electrode

Yufeng Wang and Yongbin Zeng

Abstract Electrochemical reaction has been extensively applied to fabricate microstructures. Electrochemical micromachining (EMM) with ultrashort voltage pulses holds great potential in micromachining with sub-micro or nano precision. In this paper, a moving sub-micro spherical tip was adopted to EMM in processing microchannels with the controlled profiles and sub-micro accuracy. Both the mechanisms and advantages of EMM using the moving spherical tip were analyzed. The electric field density distribution on the workpiece surface determined the machined profile of the microchannels, which has been verified by simulation and experimental methods. An experimental setup for electrochemical micromachining was developed with a workpiece moving resolution of nanometer scale. Experiments were conducted to investigate the influences of the applied voltage, initial inter-electrode gap, and spherical tip diameter on the machined depth and precision of EMM. Results showed that microchannels with smaller gap could be processed with the lower voltage, proper smaller inter-electrode gap, and smaller spherical tip diameter. Microchannels with the width 0.9 μm and side gap 0.12 μm were fabricated by using applied voltage 4 V, spherical tip diameter 0.8 μm, inter-electrode gap 1.0 μm. Microchannels with the controlled profile and precision could be fabricated by EMM potentially. Preliminary experiments also demonstrated the possibility of high-aspect-ratio processing by layer-by-layer EMM with moving spherical tip as the tool electrode.

Keywords Electrochemical micromachining · Microchannels · Spherical tip · Moving path · High aspect ratio

---

Y. Wang (✉)
Ningbo Institute of Materials Technology and Engineering, Chinese Academy of Sciences, Ningbo 315201, Zhejiang, China
e-mail: wangyufeng@nimte.ac.cn

Y. Zeng
College of Mechanical and Electrical Engineering, Nanjing University of Aeronautics and Astronautics, Nanjing 210016, China

J. Xu and K. M. Pandey (eds.), *Mechanical Engineering and Materials*, Mechanisms and Machine Science 100,
https://doi.org/10.1007/978-3-030-68303-0_1

1.1 Introduction

Micro/sub-micron fabrication techniques have become critical issues as the miniaturization trend in various areas, such as microelectronics, micro electromechanical system (MEMS), lab-on-a-chip, optics, and bioengineering [1]. The fabrication of microchannels, micro holes, and controlled microstructures on various components is attracting much attention in manufacturing discipline in recent years. Lithography based processes are widely applied to fabricate microstructures on semiconductor and metal materials. However, the lithography-based micromachining processes are high in cost and require a vacuum or ultra-clean working environment [2].

Micromachining techniques have been extensively studied in recent years. A variety of processes have been applied to machining microstructures such as laser beam machining, micro electric discharge machining, electrochemical micromachining, electrochemical discharge machining, plasma etching. Thermal based micromachining technique could introduce the recast layer and heat affected zone to the machined surface. The lithography based micromachining processsed always suffered from a higher cost. Electrochemistry-based micromachining, referred to as electrochemical micromachining (EMM), has received much attention in micro and nano surface patterning and fabrication of metal materials. It is a non-contact, heat and residual stresses free process, regardless of the mechanical properties of the anodic materials. Thus, the processed surface by EMM has a higher surface integrity compared with that processed by other thermal based machining processes. Micro and even nanostructures could be fabricated by various electrochemistry-based micromachining processes [3]. In electrochemistry-based micromachining processes, materials are removed with redox reaction at the anode/electrolyte interface. In creating microstructures with high localization, electrochemical reactions should be restricted in a small and controlled area. Many technologies taking the advantages of electrochemistry principles have been developed for microfabrication, such as LIGA (Lithography, electroplating, and molding) [4, 5], electrochemical nanolithography [6], confined etchant layer technique (CELT) [7], three-dimensional electrodeposition [8], and EMM with ultrashort voltage pulses [9], etc. In order to process microstructures, electrochemical dissolution or deposition should be confined in an area of a micrometer or sub-micrometer scale. The electrochemical nanolithography combines the electrochemistry and scanning probe microscope using a nanometer-scale tip as the tool electrode, which is capable of fabricating profiles with nanometer and atomic resolution [10, 11]. CELT could replicate microstructures by confining the outer boundary of the diffusion layer of the etchant and can retain the 3-dimensional fine micro-pattern of the mold electrode [12]. High aspect ratio microstructures obtained by electrochemical deposition could be fabricated with the LIGA technique [13]. Electrochemical nanoimprint lithography provided a nanoscale accuracy method for producing semiconductor microdevices [14]. With the introduction of ultrashort voltage pulses of nanosecond duration to electrochemical micromachining, the dissolution area of the materials is highly confined in a small area with sub-micrometer-scale precision [15]. Microstructures of various contours could be replicated by applying tool cathode, which is prepared with contours that is complementary to the fabricated microstructures to EMM [16]. In addition, the machining accuracy could be further improved by decreasing pulse duration [17]. Even spatial resolutions below 100 nm could be available while decreasing pulse duration to the picosecond range [15]. However, the pulse generator with the capacity of producing picosecond pulse voltage is rarely applied to EMM, due to the limit of pulse power. It also improves the cost of EMM. Moreover, EMM is a multi-disciplinary process containing electromagnetic, flow field, and thermal effects. To improve the controllability and precision of EMM has become an urgent issue. Few studies attempted to apply the sub-micron spherical tip to the EMM process.

This paper has applied a moving sub-micro spherical stylus as the tool electrode in EMM. Both the mechanisms and advantages of EMM using a sub-micron spherical tip were analyzed. The EMM processed profiles considering various parameters, including applied voltage, inter-electrode gap, and spherical tip diameter, has been studied by simulation and experiments. The influences of various controlled parameters on the side gap and depth were experimentally studied with the developed setup. Furthermore, layer-by-layer EMM has also been adopted to fabricate high aspect ratio microfluidic channels.

1.2 Method and Materials

1.2.1 Method

In EMM, the workpiece serves as the anode, and the tool electrode serves as the cathode. When a spherical tip and workpiece are separated at a distance in the electrolyte environment, electrochemical reactions occur at the electrode/electrolyte interface, while applying a voltage source between the two counter polarities. The principle of electrochemistry-based micromachining relies on the faradic processes occurring at the tip and the workpiece surface. The conducting workpiece materials (M) are dissolved by anodic dissolution, i.e., $M^{-}ne^{-} \rightarrow Mn^{+}$. And hydrogens are generated on the surface of the spherical end with the electrochemical reaction of $2H^{+} + 2e^{-} \rightarrow H_2(g)$, that is, the cathode surface, as illustrated in Fig. 1.1. Thus, the anodic workpiece materials are removed. Based on Faraday's law, the processed contour is mainly concerned with the distribution of the electric current density on the workpiece surface, which could be expressed by Eq. (1.1) [18]:

$$j(r) = \frac{U}{D}\left[1 + \left(\frac{r}{R + D}\right)^2\right]^{-\frac{3}{2}} \cdot \kappa \tag{1.1}$$

where U is the tip-workpiece pulsed voltage, D is the inter-electrode gap, r is the distance between a point on the workpiece surface, and the point under the rounded

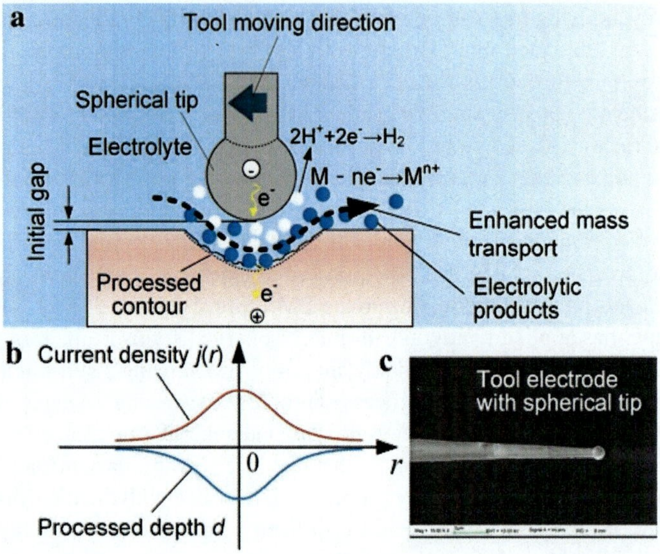

Fig. 1.1 Mechanisms of EMM with a moving spherical tip as the tool electrode, **a** principles of EMM removing process, **b** Schematic diagram of electric current density distribution and machined profile by EMM, **c** SEM images of the tool electrode with a spherical tip

tip apex, R is the radius of the spherical tip, and k is electrolyte conductivity. The distribution of electric current density on the workpiece is schematically illustrated in Fig. 1.1.

Based on the Faraday's law and Ohm's law, the materials removal rate (v_a) is proportional to the current density [19], as expressed in Eq. (1.2),

$$v_a = \eta \omega j(r) \tag{1.2}$$

where η is the electric current efficiency, ω is the electrochemical equivalent volume of the workpiece materials. The area under the spherical tip is removed with a contour affected by the electric current density distribution. Hence, the processed contour by EMM with a spherical tip could be regulated by controlling the electric current density distribution on the workpiece. According to Eqs. (1.1) and (1.2), the machined contour could be controlled by adjusting the available parameters, such as applied voltage, inter-electrode gap, and spherical tip diameter, in principle. By moving the spherical tip along with a programmed trajectory, microstructures with the controlled profiles could be directly machined on the workpiece.

In EMM with the dissolution of anodic workpiece materials, electrolytic products on the workpiece/electrolyte interface are formed. The electrolytic products contain metal ions (Mn^+) and small quantities of metal hydroxide ($M(OH)_n$), as shown in Fig. 1.1. The accumulation of the electrolytic products would decrease the electrolytic conductivity, which would then reduce the electrochemical machining rate, based on Eq. (1.1). By moving the spherical tip, the flow field of the electrolyte in the machining gap could be improved, thus enhancing the diffusion efficiency of electrolytic products out of the machining area, as illustrated in Fig. 1.1a. Also, the electrolyte could be refreshed effectively in the machining area. The EMM rate could be less restricted by the diffusion rate by moving the spherical tip.

Moreover, the application of spherical tip in EMM could relieve the electric current density concentration at a region with sharp angles, with is beneficial to obtain a more homogeneous tip/workpiece gap, improving the efficiency of renewing of electrolyte in the machining area. EMM is a non-contact process without wearing of the electrode tip. Thus sub-micrometer scale rounded tip could be utilized for reducing the dimension of the areas affected by electrochemical reactions and then decreasing the sizes of the processed microstructures.

1.2.2 Experimental Setup and Materials

As shown in Fig. 1.2, an experimental setup for EMM was developed, which consisted of three-axis linear precision stage, three-dimensional piezo transducer, computer control system, pulse generator, electrolyte transfer system, and data acquisition system. A sub-micron spherical tip made of tungsten was employed as the tool cathode and was fixed to the spindle end. The tool electrode with a spherical tip in the controlled diameter, as shown in the scanning electron microscopy (SEM) image of Fig. 1.1c has been fabricated using a liquid membrane pulsed electrochemical etching process [20]. A tungsten microrod was penetrated through the center of the cathode ring on which a liquid membrane of electrolyte was suspended. The anodic tungsten rod was electrochemically etched to form a neck region with the smaller diameter under the pulsed voltage source. While the maximum electric current density at the necking region is larger than a threshold value, the tungsten necking is melted, and then the melt rounds into a sub-micron spherical tip due to the liquid surface tension after the tungsten rod breaks. The workpiece of superalloy

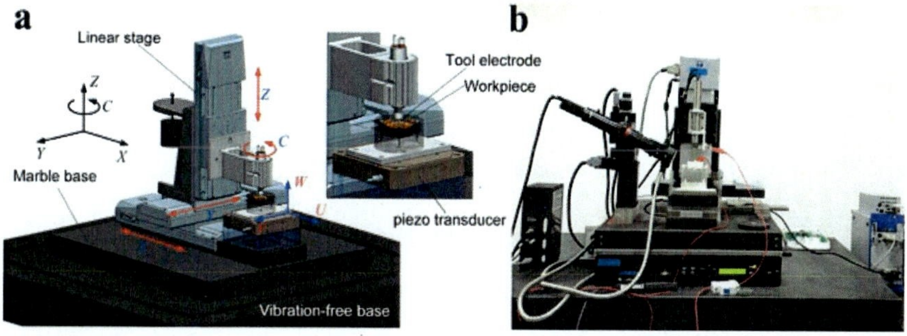

Fig. 1.2 Experimental setup developed for electrochemical micromachining, **a** 3D model, **b** photograph of the experimental setup

GH4169 (Shanghai INESA Scientific Instrument Co., Ltd., thickness: 1 mm) is fixed into the electrochemical cell mounted on the piezo transducer (P561.3CD, PI Co.). The workpiece was cleaned by acetone, ethanol, and deionized water in sequence. The piezo transducer allows the workpiece to move in three-dimensional directions with a resolution of 0.8 nm and a range of 100 μm.

To set the initial inter-electrode gap, the tip approached the workpiece by feeding the Z-axis linear stage (M511.HD, PI Co.), while applying a small tip-workpiece voltage. The electric short circuit response has been used for determining the tip-workpiece distance [21]. The tip advances to the workpiece at a rate of 0.1 μm/step under a small potential 2 V. The position where an electric short circuit occurs was defined as zero point. Afterward, the tip retreated at the controlled initial tip-workpiece gap. An electrolyte solution of 0.1 M H2SO_4 served as the electrolyte. During EMM, a pulse voltage with the period 500 ns, pulse duration 45 ns, and varying voltage was provided by a high-frequency pulse generator (8110A, KEYSIGHT Co.), in order to localize the machining area in the proximity of the spherical tip, and thus reduce the effect of stray current corrosion. An atomic force microscope (AFM, Dimensional Edge, Bruker Co.) was employed to characterize the three-dimensional profile of the machined microchannels. The precision of EMM using the moving spherical tip was represented by the side gap (Δ), which could be calculated by the Eq. (1.3):

$$\Delta = (W - 2R)/2 \tag{1.3}$$

where W is the width of the fabricated microchannels, R is the spherical tip radius of the tool electrode.

1.2.3 *Electric Field Simulation*

As the analysis above, the EMM processed contour was determined by the electric current density. The machined contour was in accordance with the electric field density distribution in the inter-electrode gap. To explore the characteristics of electric current density distribution precisely, the electric field simulation was conducted, as shown in Fig. 1.3a. The assumptions were as follows:

(1) The process of EMM is in the equilibrium state.
(2) The electric parameters are constant, such as electrolyte conductivity, temperature, et al.
(3) The concentration gradient in the bulk electrolyte is negligible.

The electric potential φ in the inter-electrode gap domain could be calculated by Laplace's equation (Eq. (1.4)):

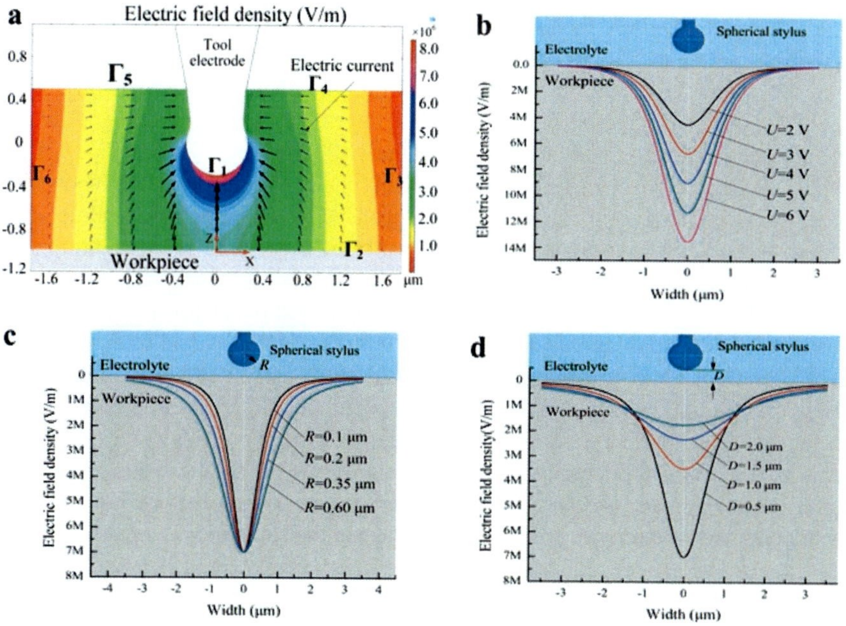

Fig. 1.3 Electric field simulation of EMM using a spherical tip, **a** Modelling of the electric field in the inter-electrode gap, **b** Variation of electric field density distribution with voltage, **c** Variation of electric field density distribution with spherical tip diameter, **d** Variation of electric field density distribution with the inter-electrode gap

$$\frac{\partial^2 \varphi}{\partial x^2} + \frac{\partial^2 \varphi}{\partial y^2} = 0 \tag{1.4}$$

Boundary conditions are given in Eqs. (1.5), (1.6) and (1.7):

$$\varphi|\Gamma_2 = U \text{ (Anode)} \tag{1.5}$$

$$\varphi|\Gamma_1 = 0 \text{ (Cathode)} \tag{1.6}$$

$$\frac{\partial \varphi}{\partial n}|\Gamma_{3,4,5,6} = 0 \tag{1.7}$$

where U is the applied voltage. The electric current density j is then given by Ohm's law as the normal derivative of the potential, as is shown in Eq. (1.8)

$$j = -\kappa \cdot \left(\frac{\partial \varphi}{\partial x} + \frac{\partial \varphi}{\partial y} \right) \tag{1.8}$$

By using the MEMS module of the COMSOL Multiphysics, the electric field density value on each point of the anode surface has been calculated. Figure 1.3b shows the variation of electric field density with applied voltage. Both the machined depth and side gap increased with increasing voltage. Thus, a higher voltage enhanced the materials removal rate but deteriorated the machining localization. Figure 1.3c shows the variation of electric field density with spherical stylus diameter. It could be noted that the EMM localization could be enhanced by using the spherical tip with a smaller diameter. Microstructures with smaller sizes could be processed utilizing smaller rounded tip as the tool electrode. Figure 1.3d shows the variation of electric field density with the initial inter-electrode gap. Micro contours with higher depth and narrower width could be obtained by EMM with the smaller initial inter-electrode gap. Therefore, the electric current density distribution and the machined contour on the workpiece surface could be controlled by regulating the various parameters. As the analysis above, microfeatures with higher precision and smaller sizes could be fabricated by EMM utilizing moving spherical tip, with the smaller voltage, smaller spherical tip diameter, and narrower inter-electrode gap. The following experiments were conducted to verify the proposed method and the simulation results.

1.3 Results and Discussions

1.3.1 Influences of the Applied Voltage

Based on the principle of EMM with a moving spherical tip as the tool electrode, microchannels were fabricated with the workpiece feeding rate of 0.1 μm/s in the lateral directions (XY plane). Voltage ranging from 4 to 6 V was applied between the workpiece and the tool electrode. A tool electrode with a spherical tip of 650 nm in diameter was employed. The initial inter-electrode has been set to 1 μm. Afterward, the contour of the fabricated microchannels was measured by AFM. The width of the microchannel has been defined as the with at the half depth.

Figure 1.4a shows the variation of depth and side gap of the machined microchannels, and the three-dimensional profiles of the microchannels at each corresponding applied voltage. Figure 1.4b shows the cross-sectional contours of the microgrooves under different pulse voltage. It was demonstrated that both the width and the depth of the microchannels increased with increasing applied voltage. The experimental results were in accordance with the simulation results. The workpiece materials removal rate increases with increasing voltage. As illustrated in Eq. (1.1), the electric current density increases with a bigger voltage amplitude, which makes the dissolution rate increase. During the electrochemical reaction, the dissolution current (i_{diss}) could be expressed as Eq. (1.9) [2]:

$$i_{diss} = i_0 \cdot \exp\left(\frac{\alpha \Delta \varphi}{KT}\right) \tag{1.9}$$

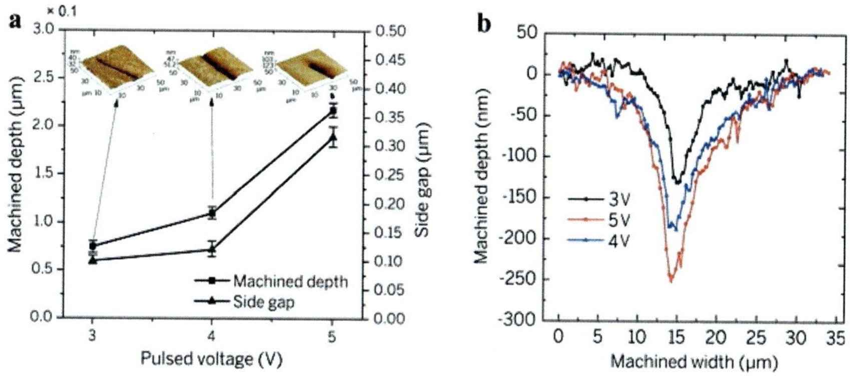

Fig. 1.4 **a** Variation of the depth and side gap of the machined microchannels, and the three-dimensional profiles of the microchannels at each corresponding applied voltage, **b** The cross-sectional profiles of the microchannels under voltage of 4, 5, and 6 V

where i_0 is the exchange current density, α is the transfer coefficient (α = 0.5) [2], Δφ is the over-potential of the double layer on the workpiece surface, K is a gas constant, T is the temperature. The increase of over-potential resulting from increasing tip-workpiece applied voltage contributes to the improvement of the dissolution rate. As shown in Fig. 1.4b, the cross-sectional contours of the microgrooves under different voltage were in accordance with the simulation results (Fig. 1.3b). Both the width and the depth of the microchannels have been increased by increasing the applied voltage. Micro channels with the samller width could be fabricated as the electric current density is much localized in proximity to the spherical end, as the simulation results. Therefore, microchannels with smaller sizes and higher localization could be obtained using the smaller voltage.

1.3.2 *Influences of Inter-Electrode Gap*

The initial inter-electrode gap affects not only the distribution of electric current density but also the electrolyte flow field and the electrolytic products transfer rate in the machining area [22]. To investigate the influences of the inter-electrode gap on EMM, the applied voltage of 4 V and a spherical tip diameter of 0.65 μm was applied. The inter-electrode gap has been set to 0.5 μm, 1.0 μm, and 1.5 μm, respectively.

Figure 1.5 shows the variation of depth and width of the fabricated microchannels by EMM under different inter-electrode gaps. The machined depth of the microchannels decreased with an increasing inter-electrode gap with an inter-electrode gap smaller than 1.0 μm, which is in accordance with the simulation results, as shown in Fig. 1.3d. When the inter-electrode gap was larger than 1.0 μm, the machining depth kept constant due to the high diffusion rate of the electrolytic products with the larger inter-electrode gap. Thus, the electrolyte in the machining area could be

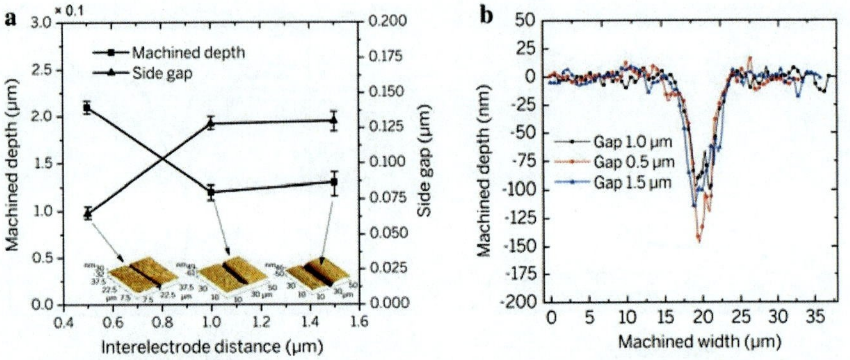

Fig. 1.5 **a** Variation of depth and side gap of the machined microchannels, and the three-dimensional profiles of the microchannels at each inter-electrode gap, **b** The cross-sectional profiles of the microchannels under the inter-electrode gap of 0.5 μm, 1.0 μm, and 1.5 μm

refreshed efficiently. The actual electric conductivity (κ') of the electrolyte in the machining area could be expressed as Eq. (1.10):

$$\kappa' = \frac{2(1-\beta)}{2+\beta} \cdot \kappa \tag{1.10}$$

where κ is electrolyte conductivity, β is the volume percent of electrolytic products in the electrolyte in the inter-electrode gap. Under a larger inter-electrode gap, the percent of electrolytic products were smaller. Thus, the machined depth could not decrease further, contrary to the simulation results. Figure 1.5b shows the cross-sectional profile of the microchannels under the different inter-electrode gap. Additionally, as shown in Fig. 1.5a, the machining side gap increased with the increasing inter-electrode gap, as the spread of the electric current density becomes wider under a larger inter-electrode gap. Experiments also revealed that an inter-electrode gap smaller than 0.3 μm could not be employed to EMM due to the depletion of the electrolyte reaction ions. The width of the microchannel has been decreased with increasing inter-electrode gap, as the electric current density was much restricted around the spherical tip illustrated by the simulation results. Microchannels could not be fabricated steadily under such a small inter-electrode gap. Therefore, a proper smaller inter-electrode gap was preferred to fabricate microchannels with smaller sizes with a higher resolution.

1.3.3 Influences of Spherical Tip Diameter of Tool Electrode

Microchannels were fabricated by the tool electrode with a spherical tip diameter of 0.40 μm, 0.65 μm, and 0.90 μm. The applied voltage was 4 V, and the initial inter-electrode was set as 1 μm. As the simulation results, the width of the spread of the electric current density increased with increasing spherical tip diameter. It leads to an increase in the width of the processed microchannels, which was in accordance with the experimental results shown in Fig. 1.6. The electric current density is confined under the tip with a smaller scale spherical tip diameter. Figure 1.6b shows the cross-sectional profiles of the microchannels under the spherical tip diameter of 0.40 μm, 0.65 μm, and 0.90 μm, respectively. However, different from the simulation results, with the decrease of the spherical tip size, the machined depth decreased, as shown in Fig. 1.6a. This is may due to the difficulty in electrolytic product diffusion in the smaller machining area. As illustrated by Eq. (1.9), the accumulation of the electrolytic products could decrease the electrolyte conductivity, and thus reduce the machining depth of the microchannels. Therefore, a proper smaller spherical tip diameter is favorable to confine the electrochemical micromachining in a much smaller area, thus reducing the sizes of the fabricated microchannels.

Experimental results showed that microchannels could be fabricated with the controlled sizes and precision by EMM while using moving spherical tip as the tool electrode. Microchannels with smaller gap could be processed with the lower voltage, proper smaller inter-electrode gap, and smaller spherical tip diameter. Thus, microchannels with the controlled profile and precision could be fabricated by EMM. Figure 1.7 shows the microchannels fabricated by EMM, in which the voltage 4 V, spherical tip diameter 0.75 μm, inter-electrode gap 1.0 μm were employed. Microchannels with the width 0.96 μm and side gap 0.12 μm were fabricated. The side was 0.11 μm. As shown in Fig. 1.7b, d, microchannels with the programmed shape could be fabricated via numerical control of the spherical tip moving path.

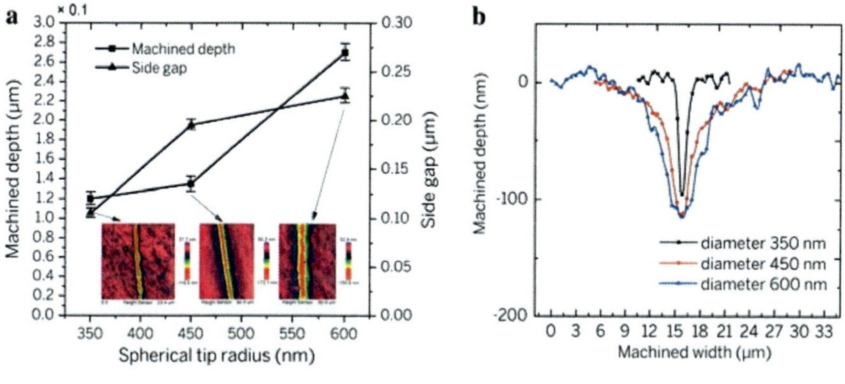

Fig. 1.6 **a** Variation of the depth and side gap of the machined microchannels, and the three-dimensional profiles of the microchannels at each spherical tip diameter, **b** The cross-sectional profiles of the microchannels under the spherical tip diameter of 0.40 μm, 0.65 μm, and 0.90 μ

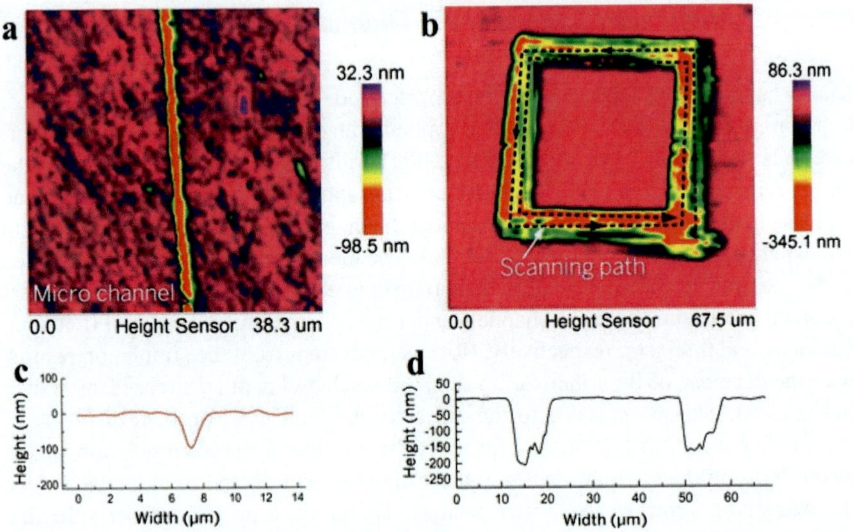

Fig. 1.7 **a**, **b** Two-dimensional AFM image of the fabricated microchannels with the programmed shape by EMM while using the moving spherical tip as the tool electrode, **c**, **d** The corresponding cross-sectional profiles of the microchannels

1.3.4 Fabrication of High Aspect Ratio Microchannels

To fabricate microchannels with higher depth, layer-by-layer EMM has been studied, as illustrated in Fig. 1.8a. The aspect ratio has been specified as the ratio between the

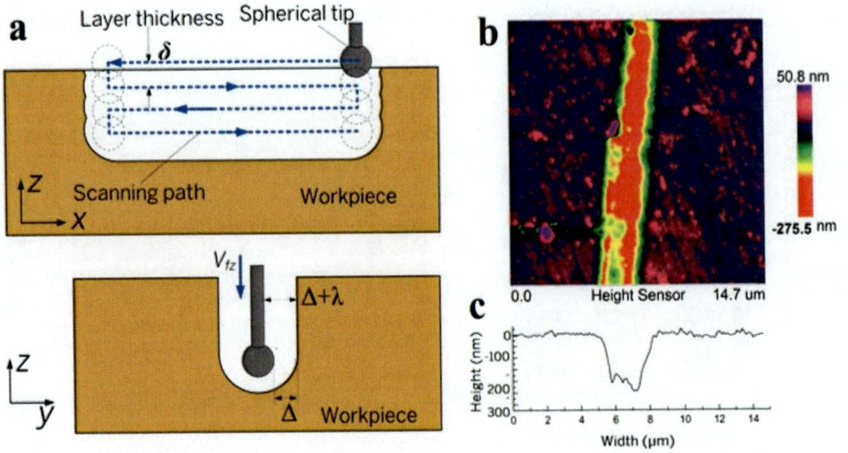

Fig. 1.8 Schematic illustration of layer-by-layer EMM with moving spherical tip as the tool electrode, **b** Two-dimensional AFM image of the fabricated high aspect ratio microchannel, **c** Cross-sectional profile of the microchannels

depth and width of the processes microchannels. When a moving period is completed, the spherical tip feeds downward with a controlled layer thickness δ. Then the moving path is repeated. Thus, microchannels with a higher depth could be fabricated. The electrolyte electric resistance $\left(R_{electrolyte}\right)$ could be expressed as $\left(R_{electrolyte}\right) = \rho \cdot d$, where $\rho\rho$ is the specific electrolyte resistivity, and d is the local separation between the two electrodes [2]. Because the side gap at spherical end (Δ) is smaller than that face to the shank ($\Delta + \lambda$), the electric resistance ($\rho \cdot \Delta$) at the spherical tip is smaller than that around the electrode shank ($\rho(\Delta + \lambda)$). Thus, the EMM rate at the spherical tip is larger than that around the electrode shank. With the feeding of the spherical tip towards the workpiece, the taper of the fabricated microchannels could be reduced. Therefore, the aspect ratio of the microchannel could be improved.

Figure 1.8b shows the microchannel with a depth of 0.25 μm fabricated by layer-by-layer EMM using the moving spherical tip as the tool electrode. A layer thickness of 0.1 μm, applied voltage of 4 V, the inter-electrode gap of 1 μm were used. EMM was implemented at room temperature. As the cross-sectional profile shown in Fig. 1.8c, a microchannel with an aspect ratio of 0.125 has been fabricated, while the processed microchannel had an aspect ratio of smaller than 0.1 in the previous study [22]. Experimental results preliminary verified the possibility of high-aspect-ratio processing by layer-by-layer EMM with moving spherical tip as the tool electrode. The following studies would be conducted to investigate the characteristics of layer-by-layer EMM while using moving spherical tip as the tool electrode. However, the achieved maximum machining depth by EMM may be restricted by the spherical diameter, due to the conical shape of the tool electrode shank, as illustrated in Fig. 1.1c. The taper angle could be deteriorated by the shank part induced stray corrosion. Thus, spherical tool electrode with cylindrical shank of the smaller diameter is urgently needed to process micro channels with a higher depth, which would be studied in the following study.

1.4 Conclusions

In this paper, electrochemical micromachining using a moving spherical tip as the tool electrode has been implemented to fabricate microchannels. Both the mechanisms and advantages of EMM using the moving spherical tip were analyzed. The electric field density distribution on the workpiece surface determined the machined profile of the microchannels, which was verified by the simulation and experimental methods. An experimental setup for electrochemical micromachining was developed with a workpiece moving resolution of nanometer scale. Experiments were conducted to investigate the influences of the applied voltage, initial inter-electrode gap, and spherical tip diameter on the machined depth and precision of EMM. Results showed that microchannels with smaller gap could be processed with the lower voltage, proper smaller inter-electrode gap, and smaller spherical tip diameter. Microchannels with the controlled profile and precision could be fabricated by EMM potentially. Preliminary experiments also demonstrated the possibility of high-aspect-ratio processing by layer-by-layer EMM with moving spherical tip as the tool electrode.

Acknowledgements This work was supported by the National Natural Science Foundation of China (51905525), the Zhejiang Provincial Natural Science Foundation (LQ19E050003), and the Natural Science Foundation of Ningbo (2019A610152).

References

1. Ghoshal, B., Bhattacharyya, B.: Shape control in micro borehole generation by EMM with the assistance of vibration of tool. Precis. Eng. **38**(1), 127–317 (2014)
2. Schuster, R., Kirchner, V., Allongue, P., Ertl, G.: Electrochemical micromachining. Science **289**(5476), 98–101 (2000)
3. Li, Y., Maynor, B.W., Liu, J.: Electrochemical AFM "Dip-Pen" Nanolithography. J. Am. Chem. Soc. **123**(9), 2105–2106 (2001)
4. Xie, Z.X., Kolb, D.M.: Spatially confined copper dissolution by an STM tip: a new type of electrochemical reaction. J. Electroanal. Chem. **481**(2), 177–182 (2000)
5. Sheffer, M., Mandler, D.: Scanning electrochemical imprinting: a tool for surface patterning. J. Electrochem. Soc. **155**(3), D203–D208 (2008)
6. Abril, O., Gündel, A., Maroun, F., Allongue, P., Schuster, R.: Single-step electrochemical nanolithography of metal thin films by localized etching with an AFM tip. Nanotechnology **19**(32), 325301 (2008)
7. Hu, J., Yu, M.F.: Meniscus-confined three-dimensional electrodeposition for direct writing of wire bonds. Science **329**(5989), 313–316 (2010)
8. Lai, L.J., Zhou, H., Du, Y.J., Zhang, J., Jia, J.C., Jiang, L.M., Zhu, L.M., Tian, Z.W., Tian, Z.Q., Zhan, D.P.: High precision electrochemical micromachining based on confined etchant layer technique. Electrochem. Commun. **28**, 135–138 (2013)
9. Schuster, R., Kirchner, V., Xia, X.H., Bittner, A.M., Ertl, G.: Nanoscale electrochemistry. Phys. Rev. Lett. **80**, 5599–5602 (1998)
10. Xie, X.H., Chung, H.J., Sow, C.H., Wee, A.T.S.: Nanoscale materials patterning and engineering by atomic force microscopy nanolithography. Mat. Sci. Eng. **54**(1–2), 1–48 (2006)
11. Garcia, R., Martinez, R.V., Martinez, J.: Nano-chemistry and scanning probe nanolithographies. Chem. Soc. Rev. **35**(1), 29–38 (2006)
12. Jiang, L.M., Du, Y.J., Jia, J., Lai, L.J., Zhou, H., Zhu, L.M., Tian, Z.W., Tian, Z.Q., Zhan, D.: Three dimensional micromachining on aluminum surface by electrochemical wet stamping technique. Electrochem. Commun. **33**, 119–122 (2013)
13. Zhan, D., Han, L., Zhang, J., Shi, K., Zhou, J., Tian, Z.: Confined chemical etching for electrochemical machining with nanoscale accuracy. Accounts. Chem. Res. **49**(11), 2596–2604 (2016)
14. Zhang, L., Zhang, J., Yuan, D., Han, L., Zhou, J., Tian, Z., Tian, Z., Zhan, D.: Electrochemical nanoimprint lithography directly on n-type crystalline silicon (111) wafer. Electrochem. Commun. **75**, 1–4 (2017)
15. Kock, M., Kirchner, V., Schuster, R.: Electrochemical micromachining with ultrashort voltage pulses-a versatile method with lithographical precision. Electrochim. Acta **48**(20), 3213–3219 (2003)
16. Bi, X., Zeng, Y., Qu, N.: Wire electrochemical micromachining of high-quality pure-nickel microstructures focusing on different machining indicators. Precis. Eng. **61**, 14–22 (2019)
17. Ma, X., Schuster, R.: Locally enhanced cathodoluminescence of electrochemically fabricated gold nanostructures. J. Electroanal. Chem. **662**(1), 12–16 (2011)

18. Gomer, R.: Extensions of the field-emission fluctuation method for the determination of surface diffusion coefficients. Appl. Phys. A **39**, 1–8 (1986)
19. Chen, X., Qu, N., Li, H., Guo, Z.: Removal of islands from micro-dimple arrays prepared by through-mask electrochemical micromachining. Precis. Eng. **39**, 204–211 (2015)
20. Wang, Y., Zeng, Y., Qu, N., Zhu, D.: Fabrication of sub-micro spherical probes by liquid membrane pulsed electrochemical etching. J. Mater. Process. Tech. **231**, 171–178 (2016)
21. Wang, Y., Zeng, Y., Zhang, W.: Improving the machining efficiency of electrochemical micromachining with oscillating workpiece. Int. J. Adv. Manuf. Technol. **102**, 2695–2708 (2019)
22. Yuan, Y., Han, L., Huang, D., Su, J., Tian, Z., Tian, Z., Zhan, D.: Electrochemical micromachining under mechanical motion mode. Electrochim. Acta **183**, 3–7 (2015)

Chapter 2
Peridynamic Simulation for Thermal and Mechanical Behavior of Manufactured Structures with MIG Welding

Hui Li, Liping Zhang, Zhenyu Liu, Lijuan Li, Yixiong Zhang, and Xuejiao Shao

Abstract A new numerical simulation method is developed for the evaluation of thermal and mechanical analyses of manufactured structures with Metal Inert Gas (MIG) welding on the basis of the peridynamics. The peridynamic formulations of the weak coupled thermo-mechanical problems are derived based on a heat transfer model and a microplastic model. To simulate the MIG welding process, a birth–death material point method in the peridynamic framework is proposed to model the addition of weld metal to the workpiece. To simplify the calculation process and improve the computational efficiency due to no need of updating the neighbourhoods of material points, an additional scalar field is introduced in the equations to describe the birth and death states of the material points. In addition, the present method is applied to simulate a classical model of two plates being joined together by the MIG welding. The results indicate that the residual tensile deformations are formed in the welded area while the residual compressive deformations are formed far from the wild line after the release of the clamps on account of the plastic deformation. These results also demonstrate the validity of the present method for the simulation of the MIG welding by comparing with the reference solution.

Keywords Peridynamics · Welded structures · Birth–death material point · Residual deformation · Numerical simulation

2.1 Introduction

With the development of industry, the welding plays a more and more important role in a lot of manufacturing fields. The welded structures can achieve preferably the functions and properties of products to satisfy the needs of automobile, aerospace,

H. Li (✉) · L. Zhang · Z. Liu · L. Li · Y. Zhang · X. Shao
Science and Technology On Reactor System Design Technology Laboratory, Nuclear Power Institute of China, No. 328 Changshun Avenue, Chengdu 610200, Sichuan, P. R. China
e-mail: lhcnpi@163.com

J. Xu and K. M. Pandey (eds.), *Mechanical Engineering and Materials*, Mechanisms and Machine Science 100,
https://doi.org/10.1007/978-3-030-68303-0_2

and shipbuilding industries [1–3], etc. However, the welding is a challenge because of the different material properties, the surface contact, the residual stresses and the brittle intermetallic compounds [4, 5]. So far, all kinds of welding techniques have been developed and used in the metal welding, such as the laser welding, friction stir welding (FSW), resistance spot welding (RSW), metal inert gas (MIG) welding, [6–8], etc. At the same time, many numerical simulation methods have been proposed to analyze and improve these welding techniques. For instance, Chao et al. [9] and Riahi and Nazari [10] investigated the FSW for aluminum alloy 6061-T6 based on a thermal and thermo-mechanical modeling. Kumaresan et al. [11] analyzed the heat transfer and residual stress in a butt welded plate. Tang et al. [12] studied the heat input and temperature distribution of FSW. Zhang and Zhang [13] presented a fully coupled thermos-mechanical model for FSW. Gannon et al. [14] revealed the effect of welding sequence on residual stress and distortion in flat-bar stiffened plates. Moreover, Pardo and Weckman [15] predicted the welding pool and reinforcement dimensions of GMA welds using a finite element method with the element birth and death technique. In summary, these works mentioned above have made a great contribution to evaluating and improving the welding processes, welding technologies and welded structures. However, a novel and efficient computational method is still grateful for the simulation of thermal and mechanical analyses of welded structures.

In the last twenty years, a nonlocal peridynamics proposed by Silling [16] has attracted a lot of researchers' interests due to its natural advantages to deal with the discontinuous and contact problems. The peridynamic method is expressed by an integral motion equation and mainly contains two kinds of forms, i.e., the bond-based peridynamics and the state-based peridynamics. Since the proposal of the peridynamic method, its basic theories have been deeply developed from several perspectives [17–22]. Meanwhile, the peridynamic method has been widely extended to many fields. For example, an improved peridynamic fracture model for the analyses of the brittle fracture problems was developed by Huang et al. [23]. A peridynamic formulation for transient heat conduction was presented by Oerkus et al. [24] and Bobaru and Duangpanya [25] to simulate the heat conduction problems in bodies with evolving discontinuities. Further, Ouchi et al. [26] and Zhang et al. [27] proposed a coupling peridynamic approach for the consolidation and dynamic analyses of saturated porous media. Li et al. [28] developed a peridynamic model for the nonlinear analyses of bimodular truss structures, tensegrity structures and membranes. In addition, a peridynamic modeling of pitting corrosion was presented by Chen et al. [29] for the simulation of the chemical corrosion and damage in metal. According to the successful applications of the peridynamic method mentioned above, it is revealed that the peridynamic method could commendably deal with the heat conduction problems and the discontinuous problems. Therefore, it is promising to propose a peridynamic method for the simulation of the thermal and mechanical analyses of MIG welded structures.

2.2 Description of Physical Mechanism

The physical phenomena in MIG welding process of a manufactured structure, contains essentially the thermal transfer and mechanical distortion problems. Those related physical mechanisms are usually supposed to solve on the basis of a weak coupled thermo-mechanical modeling, in which the change of temperature field will generate the linear expansion of materials, however, the material deformation does not influence temperature field in converse. In addition, the geometrical discontinuities and boundaries are changed in the wake of welding process. To overcome the problems mentioned above, the peridynamic method is applied to simulate the MIG welding process due to its capacity of solving thermo-mechanical problems and advantage of dealing with geometrical discontinuities. The corresponding expressions of the thermo-mechanical problems in peridynamics are expounded in the following.

2.2.1 *Heat Transfer Mechanism*

On the basis of the non-local peridynamic theory [16] and the corresponding spatial discretization of material points with associated mass and volume, the peridynamic equation of heat transfer problems for the material point X^i can be expressed as Eq. (2.1) [25]:

$$\rho c\frac{\partial T\left(x^i,t\right)}{\partial t}=\int\limits_{\mathcal{H}_{x^i}} K[\delta]\omega\left(\left\|\xi^{ij}\right\|\right)\frac{T\left(x^i,t\right)}{\left\|\xi^{ij}\right\|^2}dV_{x^i}+Q_{x^i}, \tag{2.1}$$

where ρ is the mass density, c is the thermal capacity. The bond $\xi^{ij}=x^j-x^i$, in which x^i and x^j are the coordinates of the ith and jth material points. $\mathcal{H}_{x^i}\mathcal{H}_{x^i}$ is the neighborhood of the material point x^i with a certain horizon.V_{x^i} is the volume and $\omega\left(\left\|\xi^{ij}\right\|\right)$ is the weight function. T presents the temperature of the material point. Q_{x^i} is the heat source per unit volume. $K[\delta]K[\delta]$ is the micro-conductivity related to the local thermal conductivity k and the horizon δ. The micro-conductivity $K[\delta]$ for different weight functions can be defined by Eqs. (2.2) and (2.3):

$$K[\delta]=\begin{cases}\frac{k}{\delta}, & \omega\left(\left\|\xi^{ij}\right\|\right)=1,\ 1D\\ \frac{4k}{\pi\delta^2} & \omega\left(\left\|\xi^{ij}\right\|\right)=1,\ 2D\end{cases}, \tag{2.2}$$

$$K[\delta]=\begin{cases}\frac{k}{\delta}, & \omega\left(\left\|\xi^{ij}\right\|\right)=1-\frac{\left\|\xi^{ij}\right\|}{\delta},\ 1D,\\ \frac{12k}{\pi\delta^2}, & \omega\left(\left\|\xi^{ij}\right\|\right)=1-\frac{\left\|\xi^{ij}\right\|}{\delta},\ 2D.\end{cases} \tag{2.3}$$

In addition, on the basis of the method to impose the thermal boundary conditions listed in literature [25], both the temperature and heat-flux boundary conditions are considered in this paper by adding the fictitious material points outside the corresponding boundaries.

2.2.2 Mechanical Distortion Mechanism with Thermal Effect

On the basis of the bond-based peridynamic theory and the thermal expansion, the peridynamic equation of motion for the material point x^i can be expressed as Eq. (2.4) [18]:

$$\rho\ddot{\mathbf{u}}\left(\mathbf{x}^i, t\right) = \int\limits_{\mathcal{H}_{\mathbf{x}^i}} \mathbf{f}(\boldsymbol{\xi}, \boldsymbol{\eta}, T, t) dV_{\mathbf{x}^j} + \mathbf{b}^i, \tag{2.4}$$

in which, $\ddot{u}$ is the accelerated velocity, f is the pairwise force and b^i is the density force. Investigating the MIG welding process, the high temperature will cause the regions near the welding seam to plastically deform, and thereby develop a significant residual tensive stresses in the welded structures. Consequently, the plastic deformation is considered in the simulation of welding process and the coresspoding expression of f based on the microplastic model can be defined by Eq. (2.5) [18]:

$$\mathbf{f}(\boldsymbol{\xi}, \boldsymbol{\eta}, T, t) = f\frac{\boldsymbol{\eta} - \boldsymbol{\xi}}{\|\boldsymbol{\eta} - \boldsymbol{\xi}\|}, \tag{2.5}$$

in which, $\dot{f}$ can be defined as Eq. (2.6):

$$\dot{f} = \begin{cases} c\varpi(\|\boldsymbol{\xi}\|)\left(\dot{s} - \alpha_{\Theta}\dot{\underline{T}}\right), \text{ if } |f| < f_y \text{or } f\left(\dot{s} - \alpha_{\Theta}\dot{\underline{T}}\right) < 0, \\ 0, \qquad\qquad\qquad\qquad \text{otherwise} \end{cases} \tag{2.6}$$

where c is the micro-modulus, $s = \frac{\|\eta+\xi\| - \|\xi\|}{\|\eta+\xi\|}$ is the bond strain, α_{Θ} is the coefficient of thermal expansion, $\underline{T}$ is the average temperature of the related two material points, and f_y is a prescribed density force at yield that can be calibrated according to Eq. (2.7):

$$f_y = \begin{cases} \dfrac{2\sigma_y}{A\delta^2}, & 1\text{D}, \\ \dfrac{12\sigma_y}{\pi h\delta^2}, & 2\text{D}, \end{cases}. \tag{2.7}$$

where σ_y is the normal stress at yield in the continuum mechanics, A is the cross-section area for 1D and h is the thickness for 2D.

In addition, the both the displacement and force boundary conditions are demonstrated with the fictitious material points consistent with the thermal boundary conditions.

2.3 Numerical Simulation of MIG Welding

The MIG welding process is modelled in an alternative algorithm of decoupled thermal and mechanical problems in this paper. As shown in Fig. 2.1, the interest is discretized into uniform material points, in which the blue region is the base metal while the green region is the weld metal. In the peridynamic analysis of MIG welding process, the material properities of the material points in the heat-affected zone are changed while the corresponding temperatures achive the critical temperatures of the material. To evaluate this prolem, a birth–death material point method in the peridynamic framework is developed. The birth and death states of the material points are presented by a scalar field, as Eq. (2.8),

$$\varphi\left(\mathbf{x}^i, t\right) = \begin{cases} 1, & T_{\max}\left(\mathbf{x}^i, t\right) \geq T_s \text{ and } T\left(\mathbf{x}^i, t\right) < T_s, \\ 0, & \text{otherwise,} \end{cases} \tag{2.8}$$

in which, T_{max} is the maximum temperature in process of welding and T_s is solidification temperature of the material, respectively. It is noted that the birth–death material point method can simplify the calculation process and improve the computational efficiency because it has no need to update the neighbourhoods of the material points during the simulation.

Next, the interaction of the material points x^i and x^j can be given by Eq. (2.9):

$$\varphi\left(\mathbf{x}^i, \mathbf{x}^j, t\right) = \min\left\{\varphi\left(\mathbf{x}^i, t\right), \varphi\left(\mathbf{x}^j, t\right)\right\}. \tag{2.9}$$

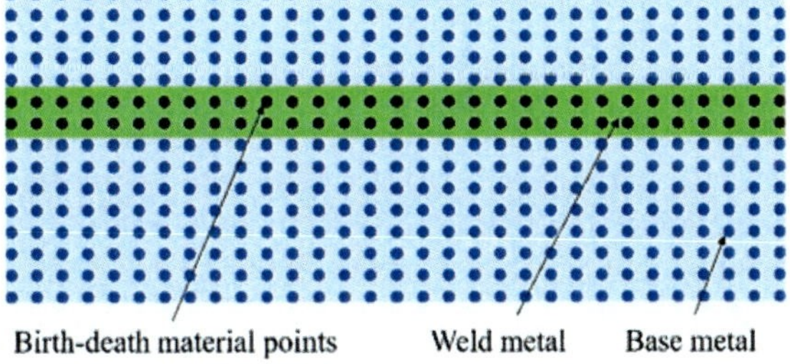

Fig. 2.1 Illustration of the discretization and birth–death material points

Substituting Eq. (2.9) into Eqs. (2.1) and (2.4) yields Eqs. (2.10) and (2.11):

$$\rho c \frac{\partial T(\mathbf{x}^i, t)}{\partial t} = \int\limits_{\mathcal{H}_{\mathbf{x}^i}} K[\delta]\omega(\|\boldsymbol{\xi}^{ij}\|)\varphi(\mathbf{x}^i, \mathbf{x}^j, t)\frac{T(\mathbf{x}^j, t) - T(\mathbf{x}^i, t)}{\|\boldsymbol{\xi}^{ij}\|^2} dV_{\mathbf{x}^i} + Q_{\mathbf{x}^i}, \tag{2.10}$$

$$\rho\ddot{\mathbf{u}}(\mathbf{x}^i, t) = \int\limits_{\mathcal{H}_{\mathbf{x}^i}} \varphi(\mathbf{x}^i, \mathbf{x}^j, t)\mathbf{f}(\boldsymbol{\xi}, \boldsymbol{\eta}, T, t) dV_{\mathbf{x}^j} + \mathbf{b}^i. \tag{2.11}$$

It is noted that Eq. (2.11) is a dynamic equation. However, the mechanical distortion caused by the changes of temperature should be under a quasi-static process. Therefore, in order to weaken the dynamic effects of the structure, an artificial damping method is utilized for the peridynamic equation of motion, which yields Eq. (2.12):

$$\rho\ddot{\mathbf{u}}(\mathbf{x}^i, t) + \lambda\dot{\mathbf{u}} = \int\limits_{\mathcal{H}_{\mathbf{x}^i}} \varphi(\mathbf{x}^i, \mathbf{x}^j, t)\mathbf{f}(\boldsymbol{\xi}, \boldsymbol{\eta}, T, t) dV_{\mathbf{x}^j} + \mathbf{b}^i, \tag{2.12}$$

where λ is a damping coefficient obtained by an adaptive dynamic relaxation technique [28].

On the other hand, it is assumed that 60% of the total heat input of MIG welding is transferred to the work piece via the molten metal droplets [13], and the total heat input is calculated by Eq. (2.13):

$$Q_{\mathbf{x}^i} = \eta V I / V_{\mathbf{x}^i}, \tag{2.13}$$

where η is the arc efficiency, V is the arc voltage and I is the arc current. In addition, the heat flux of the molten metal droplets in the Gaussian distribution [9] is adapted in this paper.

So far, Eqs. (2.10) and (2.12) show the decoupled peridynamic formulations for the numerical simulation of MIG welding.

2.4 Results and Discussions

To verify the present method for MIG welding process, a classical model of two plates being joined together is considered in the following situations. The geometry and welding configuration of the test are shown in Fig. 2.2. The plate is measured 500 mm long, 250 mm wide and 4 mm thick, respectively. The plate is made of aluminium alloy 6061-T6 and the temperature dependent physical properties of the material are shown in Tables 2.1 and 2.2. The solidification temperature of aluminium 6061-T6

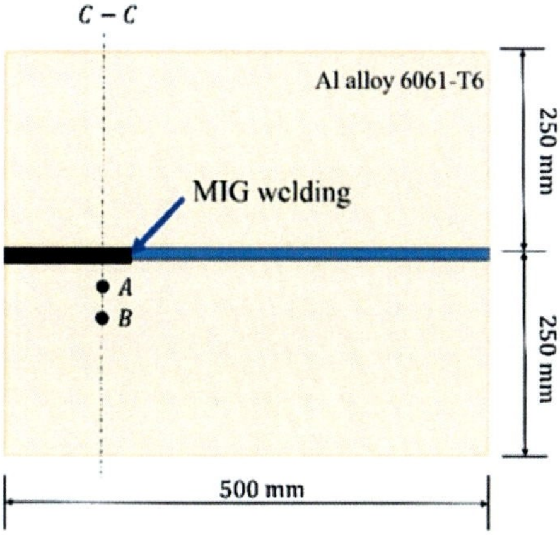

Fig. 2.2 Illustration of the model for MIG welding

Table 2.1 Temperature dependent physical properties of Al 6061-T6

T[°C]	25	37.8	93.3	148.9	204.4	260	315.5	371.1	426.7
$k[\mathrm{W}/(\mathrm{m}\cdot{}^{\circ}\mathrm{C})]$	167	170	177	184	192	201	207	217	223
$c[\mathrm{J}/(\mathrm{Kg}\cdot{}^{\circ}\mathrm{C})]$	896	920	978	1004	1028	1052	1078	1104	1133
$\alpha_{\Theta}[\mu\mathrm{m}/(\mathrm{m}\cdot{}^{\circ}\mathrm{C})]$	22.0	23.5	24.6	25.7	26.6	27.6	28.5	29.6	30.7
$\rho[(\mathrm{Kg}/\mathrm{m}^3)]$	2700	2685	2685	2667	2657	2657	2630	2620	2602

Table 2.2 Temperature dependent mechanical properties of Al 6061-T6

T[°C]	25	100	148	204.4	260	315.5	371.1	426.7	482.2
E[GPa]	66.9	63.2	61.3	56.8	51.2	47.2	43.5	28.8	20.2
σ_y[MPa]	278.1	260.7	251.2	221.0	152.3	73.9	36.8	21.6	10.5

is 610 °C. The two plates are supported on a backing plate and clamped by L-shaped steel trips through their length at the distance of 125 mm from the weld line. For the numerical simulation, the domain is discreitized into 250 × 250 uniform material points. The horizon size is $\delta = 6\,\mathrm{mm}$. The MIG welding parameters are as follows: $\mathrm{V} = 26\,\mathrm{V}$, $\mathrm{I} = 140\,\mathrm{I}$ and Speed $= 10\,\mathrm{mm/s}$.

Figure 2.3 depicts the temperature history profile of the material points measured 6 and 10 mm from the weld line. It reveals that the temperatures of the material points which is 6 mm far from the weld line are lower than the melting point of the material. Figures 2.4 and 2.5 show the comparison of temperatue and displacement distributions along the C–C cross section for different welding lengths of $\mathrm{l_w} = 100$, 250 and 500 mm, here $\mathrm{l_w}$ is the welding length. It can be seen from Fig. 2.4 that the maximum temperature caused by MIG welding and plastic deformation reached up to 233% higher than the melting point of the material. Figure 2.5 presents that the tensile and compressive zones are formed by the plastic deformations. These results are consistent with the solution obtained by Farajkhah et al. [6]. Moreover,

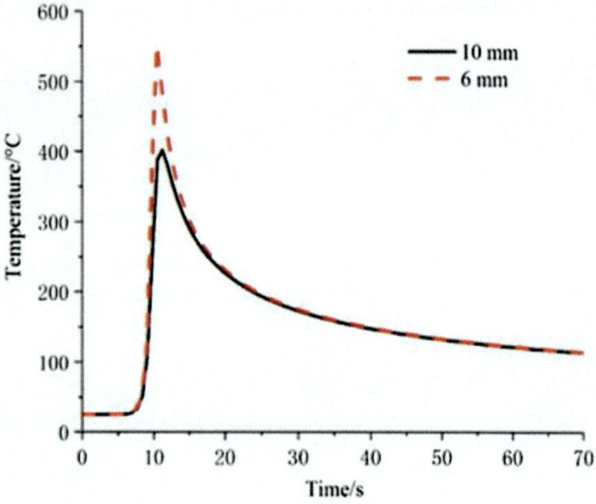

Fig. 2.3 Temperature history profile at 6 and 10 mm away from the weld line

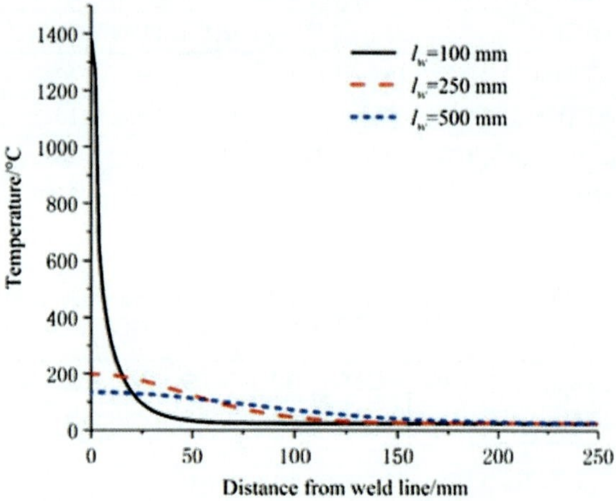

Fig. 2.4 Comparison of temperature distributions along C–C cross section for different welding lengths

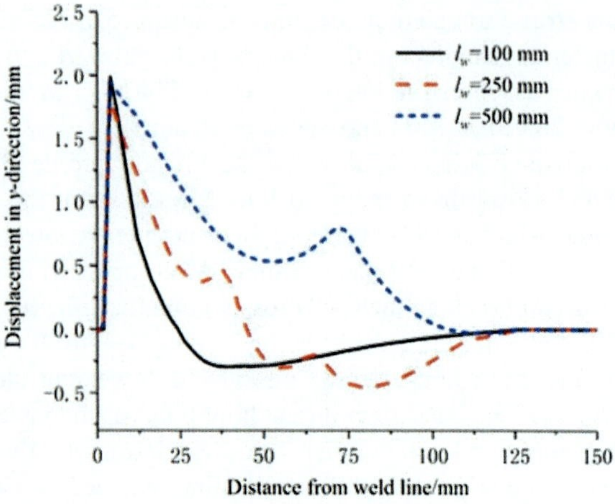

Fig. 2.5 Comparison of displacement distributions along C–C cross section for different welding lengths

Table 2.3 MIG welding parameters

Case	Voltage (V)	Current (A)	Speed (mm/s)
1	26	140	10
2	28	200	10
3	30	260	8

another two kinds of MIG welding parameters are taken into account, see Table 2.3. Figure 2.6 shows the temperature history profile of the material point at 6 mm away from the weld line with different welding parameters. The temperature and displacement distributions along the C–C cross section for these three cases are given in Figs. 2.7 and 2.8. These curves indicate that the maximum temperature

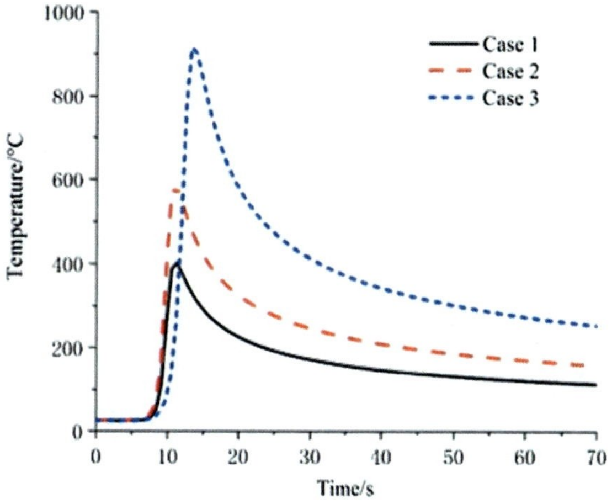

Fig. 2.6 Temperature history profile at 6 mm away from the weld line for different MIG welding parameters

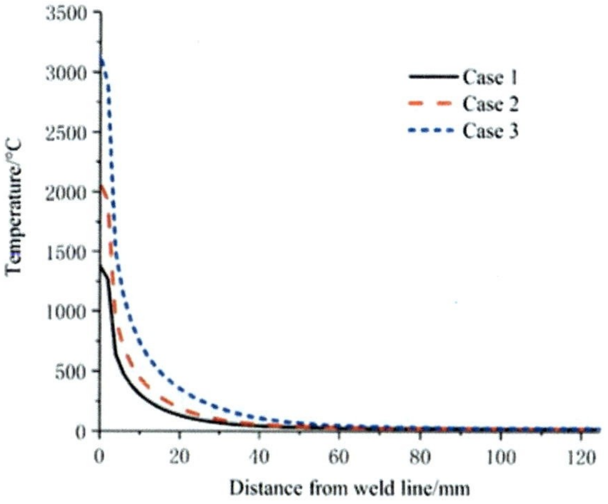

Fig. 2.7 Temperature distributions along C–C cross section for different MIG welding parameters

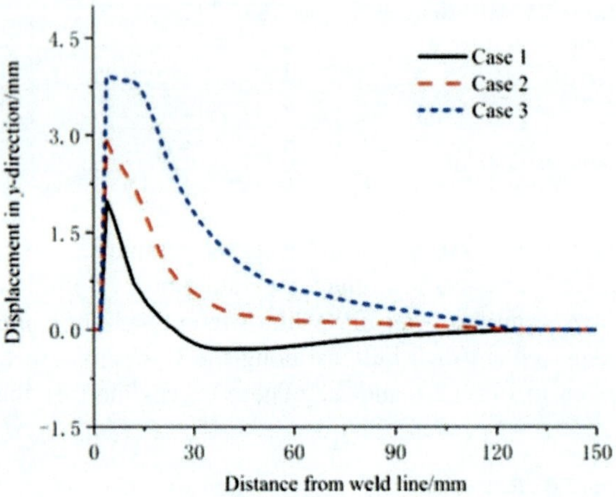

Fig. 2.8 Displacement distributions along C–C cross section for different MIG welding parameters

during the MIG is higher while the value of heat input is larger or the welding speed is slower.

In addition, to evaluate the residual deformation, the clamps for securing the plates in place are released after wleding, i.e., when t ≥ 65 s. Figure 2.9 plots the displacement fields along the C–C cross section after the release of the clamps. It can be seen from these curves that the tensile residual deformations are formed in the welded area while the compressive residual deformations are formed far from the wild line. Furthermore, Fig. 2.10 shows the contours of the temperature and displacement at time t = 28 s, 42 s, 56 s and 70 s, respectively. Consequently, these results demonstrate the validity of the present method for the simulation of the MIG welding.

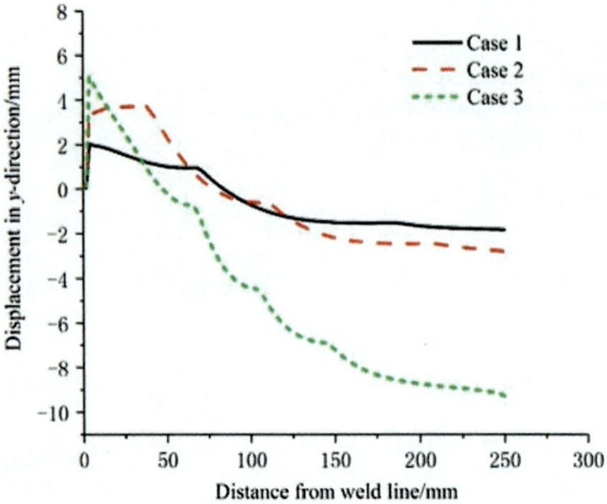

Fig. 2.9 Residual deformation along C–C cross section after release of the clamps for different MIG welding parameters

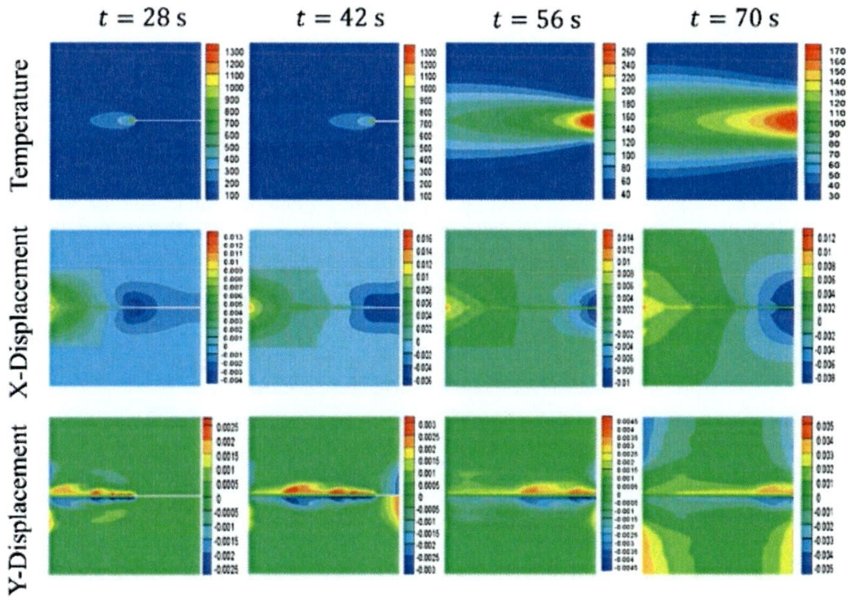

Fig. 2.10 Contours of temperature and displacement fields at time $t = 28$ s, 42 s, 56 s and 70 s, respectively

2.5 Conclusions

A nonlocal peridynamic method is proposed to simulate the MIG welding process and evaluate the heat transfer and the mechanical deformation of manufactured structures. For the thermo-mechanical problems, the formualtions of the heat transfer mechanism and mechanical distortion mechanism with thermal effect are presented by a peridynamic heat conduction model and a peridynamic microplastic model, respectively. To model the addition of weld metal to the workpiece for the MIG welding process, the birth–death material method in the peridynamic framework is developed by introducing a scalar field, which can simplify the calculation process and improve the computational efficiency. Moreover, an explicit algorithm is given out for dealing with the weak coupled formulations of the thermo-mechanical problems. In addition, a representative numerical example of two plates being joined together by the MIG welding is considered to certify the validity of the present method by comparing with the reference solution. The results indicate that the maximum temperature caused by MIG welding reached up to 233% higher than the melting point of the material. The residual tensile deformations are formed in the welded area while the residual compressive deformations are formed far from the wild line after the release of the clamps due to the plastic deformation.

References

1. Chen, Y., Chen, S., Li, L.: Influence of interfacial reaction layer morphologies on crack initiation and propagation in Ti/Al joint by laser welding-brazing. Mater. Des. **31**(1), 227–233 (2010)
2. Satoh, G., Yao, Y.L., Qiu, C.: Strength and microstructure of laser fusion-welded Ti-SS dissimilar material pair. Int. J. Adv. Manuf. Tech. **66**, 469–479 (2013)
3. Fang, Y.J., Jiang, X.S., Mo, D.F., Zhu, D.G., Luo, Z.P.: A review on dissimilar metals' welding methods and mechanisms with interlayer. Int. J. Adv. Manuf. Tech. **102**, 2845–2863 (2019)
4. Zhang, Z., Zhang, H.W.: Material behaviors and mechanical features in friction stir welding process. Int. J. Adv. Manuf. Tech. **35**, 86–100 (2007)
5. Zhu, Z., Lee, K.Y., Wang, X.: Ultrasonic welding of dissimilar metals, AA6061 and Ti6Al4V. Int. J. Adv. Manuf. Tech. **59**, 569–574 (2012)
6. Farajkhah, V., Liu, Y., Gannon, L.: Finite element study of 3D simulated welding effect in aluminium plates. Ships Offshore Structures **12**(2), 196–208 (2017)
7. Zhou, K., Yao, P.: Overview of recent advances of process analysis and quality control in resistance spot welding. Mech. Syst. Signal Pr. **124**, 170–198 (2019)
8. Masubuchi, K. (eds.): Analysis of Welded Structures. Oxford, Pergamon (1980)
9. Chao, Y.J., Qi, X.: Thermal and thermos-mechanical modeling of friction stir welding of aluminum alloy 6061–T6. J. Mater. Process Manuf. Sci. **7**(2), 215–233 (1998)
10. Riahi, M., Nazari, H.: Analysis of transient temperature and residual thermal stresses in friction stir welding of aluminum alloy 6061–T6 via numerical simulation. Int. J. Adv. Manuf. Tech. **55**(1), 143–152 (2010)
11. Kumaresan, D., Asraff, A. K., Muthukumar, R.: Numerical investigation on hear transfer and residual stress in a butt welded plate, Journal of Pressure Vessel Technology, Transactions of the ASME, 133(4), (2011).
12. Tang, W., Guo, X., McClure, J.C., Murr, L.E., Nunes, A.: Heat input and temperature distribution in friction stir welding. J. Mater. Process Manuf. Sci. **7**(2), 163–172 (1998)
13. Zhang, Z., Zhang, H.W.: A fully coupled thermos-mechanical model of friction stir welding. Int. J. Adv. Manuf. Tech. **37**, 279–293 (2008)
14. Gannon, L., Liu, Y., Pegg, N., Smith, M.: Effect of welding sequence on residual stress and distortion in flat-bar stiffened plates. Mar. Struct. **23**(3), 385–404 (2010)
15. Pardo, E., Weckman, D.C.: Prediction of weld pool and rein-forcement dimensions of GMA welds using a finite element model. Metall Trans. B. **20**, 937–947 (1989)
16. Silling, S.A.: Reformulation of elasticity theory for discontinuities and long-range forces. J. Mech. Phys. Solids **48**(1), 175–209 (2000)
17. Silling, S.A., Askari, E.: A meshfree method based on the peridynamic model of solid mechanics. Comput. Struct. **83**(17–18), 1526–1535 (2005)
18. Bobaru, F., Foster, J.T., Geubelle, P.H., Silling, S.A.: Handbook of Peridynamic Modeling. CRC Press, New York (2017)
19. Tupek, M.R., Radovitzky, R.: An extended constitutive correspondence formulation of peridynamics based on nonlinear bond-strain measures. J. Mech. Phys. Solids **65**, 82–92 (2014)
20. Han, F., Lubineau, G., Azdoud, Y., Askari, A.: A morphing approach to couple state-based peridynamics with classical continnum mechanics. Comput. Methods Appl. Mech. Engrg. **301**, 336–358 (2016)
21. Lai, X., Liu, L.S., Li, S.F., Zeleke, M., Liu, Q., Wang, Z.: A non-ordinary state-based peridynamics modeling of fractures in quasi-brittle materials. Int. J. Impact Eng. **111**, 130–146 (2018)
22. Ren, H.L., Zhuang, X.Y., Rabczuk, T.: Dual-horizon peridynamics: a stable solution to varying horizons. Comput. Methods Appl. Mech. Eng. **318**, 762–782 (2017)
23. Huang, D., Lu, G.D., Qiao, P.Z.: An improved peridynamic approach for quasi-static elastic deformation and brittle fracture analysis. Int. J. Mech. Sci. **94–94**, 111–122 (2015)
24. Oterkus, S., Madenci, E., Agwai, A.: Peridynamic thermal diffusion. J. Comput. Phys. **265**, 71–96 (2014)

25. Bobaru, F., Duangpanya, M.: A peridynamic formulation for transient heat conduction in bodies with evolving discontinuities. J. Comput. Phys. **231**(7), 2764–2785 (2012)
26. Ouchi, H., Katiyar, A., York, J., Foster, J.T., Sharma, M.M.: A fully coupled porous flow and geomechanics model for fluid driven cracks: a peridynamic approach. Comput. Mech. **55**, 561–576 (2015)
27. Zhang, H.W., Li, H., Ye, H.F., Zheng, Y.G.: A coupling peridynamic approach for the consolidation and dynamic analysis of saturated porous media. Comput. Mech. **64**, 1097–1113 (2019)
28. Li, H., Zhang, H.W., Zheng, Y.G., Zhang, L.: A peridynamic model for the nonlinear static analysis of truss and tensegrity structures. Comput. Mech. **57**, 843–858 (2016)
29. Chen, Z.G., Zhang, G.F., Bobaru, F.: The influence of passive film damage on pitting corrosion. J. Electrochem. Soc. **163**, C19–C24 (2016)

Chapter 3
Study on Improving the Performance of Nitinol Cardiovascular Stent by "Fiber Laser—Electropolishing"

Xiaoyu Sun, Xiuting Wei, Zhiyong Li, Deda Lou, Yongqi Wang, and Hanqing Liu

Abstract High performance cardiovascular stent plays an important role in solving cardiovascular disease. In order to improve the surface integrity and biocompatibility of cardiovascular stents, the collaborative manufacturing process of "fiber laser-electropolishing" was studied. Through experiments and comparative analysis, the surface integrity and biocompatibility of cardiovascular stent by collaborative process were better than that of single laser cutting. The surface of the cardiovascular stent was bright and smooth, without dross and heat damage. The surface roughness of the cardiovascular stent reached tens of nanometers, and a layer of titanium dioxide protective film was generated on the surface. So the process could improve the surface integrity and biocompatibility of cardiovascular stent. In addition, the process used alcohol-salt non-toxic electrolyte for electropolishing, which improved the environmental friendliness of the process. Finally, this process can produce high-quality cardiovascular stent, which is of great significance to the cardiovascular stent manufacturing industry and has bright application prospects.

Keywords Nitinol cardiovascular stents · "Fiber laser-electropolishing" collaborative manufacturing · Surface integrity · Process technology · Non-toxic electrolyte

3.1 Introduction

The incidence rate of cardiovascular diseases is high and the increasing trend is accelerating [1–3]. Most patients with cardiovascular disease need to be implanted with cardiovascular stents. Therefore, cardiovascular stents have great social needs. However, the manufacturing process of cardiovascular stent determines its surface

X. Sun · X. Wei (✉) · Z. Li · Y. Wang · H. Liu
School of Mechanical Engineering, Shandong University of Technology, Zibo 255049, China
e-mail: wxt@sdut.edu.cn

D. Lou
Weihai Visee Medical Device Co., Ltd., Weihai 264200, China

J. Xu and K. M. Pandey (eds.), *Mechanical Engineering and Materials*,
Mechanisms and Machine Science 100,
https://doi.org/10.1007/978-3-030-68303-0_3

integrity, which is very important for the performance and therapeutic efficacy of the stents. Shape memory alloy (SMA) is widely used in cardiovascular stents, and nitinol SMA is the first choice because of its super elasticity and good biocompatibility [4].

At present, laser cutting is widely used in the manufacturing of nitinol cardiovascular stent. In the process of laser cutting, there is no mechanical contact, and the deformation of stent is small. Moreover, the edge defects are few, and nitinol stent with complex shape can be manufactured [5]. However, the research of Li et al. [6] showed that the laser processing of metal was essentially a kind of hot processing technology, and it would also produce thermal damage on the processed surface. Hock et al. [7–12] studied the water guide laser cutting and dry laser cutting respectively. The results showed that the water guide laser cutting had narrow cutting seam, with no dross and no obvious heat affected zone (HAZ). However, the processing time and equipment price of water guide laser cutting are much higher than that of dry laser cutting. Meng et al. [13] carried out research on cutting technology of stainless steel cardiovascular stent by using millisecond optical fiber laser. Through experiments and analysis, the best cutting technology parameters were obtained, and high-quality cardiovascular stent was manufactured. Lv et al. [14] studied the heat affected zone of cutting nitinol alloy sheet by laser. The results showed that the argon assisted cutting could obtain better surface than the air or nitrogen assisted cutting. Raval et al. [15] used electrochemical polishing to treat cardiovascular stents, and obtained high-quality cardiovascular stents. But they used acid system which was difficult to control, and the electrolyte was harmful to the environment. Therefore, it is urgent to explore the process technology of improving the surface integrity of cardiovascular stent based on laser cutting, to eliminate or reduce thermal damage, to enhance the surface integrity, and to improve the performance and therapeutic efficacy of the stents.

In order to improve the surface integrity and biocompatibility of the cardiovascular stent, the collaborative manufacturing process of "fiber laser- electropolishing" is used in this paper. The laser processing system used microsecond optical fiber laser cutting, and the electropolishing system used alcohol-salt non-toxic electrolyte. Then through this process technology, we explore the effects of this technology on the surface morphology, heat affected zone (HAZ), recasting layer, dross, surface roughness and surface chemical composition of the stents.

3.2 Experimental Materials, Equipment and Methods

3.2.1 Materials

In this study, the cardiovascular stent was made of Nitinol (56.14 at. % Ni - 43.86 at. % Ti) thin-walled tubes (hereinafter referred to as Nitinol tubes) produced by Jiangsu Peier Tech company, with a diameter of 2.6 mm, a wall thickness of 0.2 mm, and a tolerance of ± 0.02 mm. The original nitinol tube for fiber laser cutting is shown in

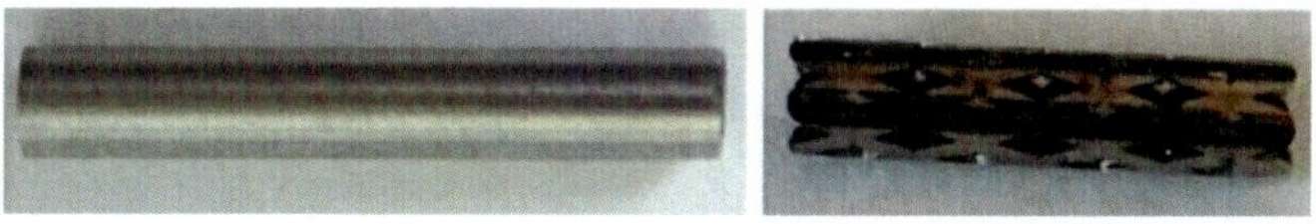

(a) Nitinol tube (b) Stent after laser cutting

Fig. 3.1 Stents

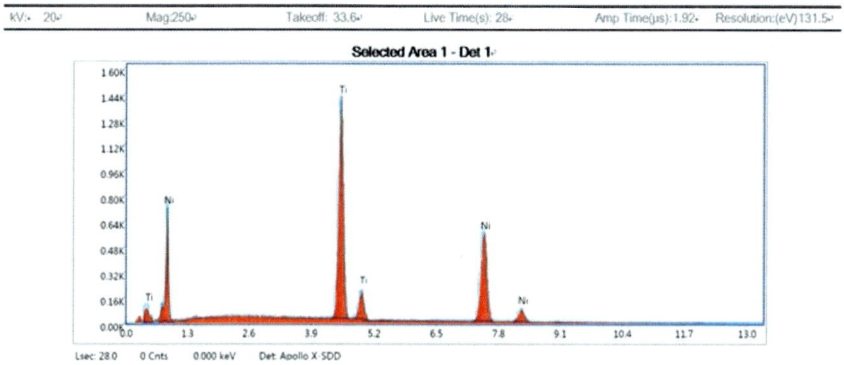

eZAF Smart Quant Results

Element	Weight %	Atomic %	Net Int.	Error %	Kratio	Z	A	F
NiL	61.07	56.14	386.00	6.83	0.3231	1.0048	0.5266	1.0000
TiK	38.93	43.86	1278.08	2.60	0.3876	0.9978	0.9547	1.0449

Fig. 3.2 The composition and content of nitinol tube

Fig. 3.1a, and its composition and content are shown in Fig. 3.2. The cardiovascular stent (hereinafter referred to as stent) after fiber laser cutting is shown in Fig. 3.1b, and it is used as the raw materials for electropolishing.

3.2.2 Equipment and Methods

The wet laser cutting system in this experiment is shown in Fig. 3.3a, which is mainly composed of computer controller, Theta TLS-HT1100 fiber laser, focusing lens and motor with clamping device. The laser parameters in the experiment are shown in Table 3.1. The pulse width was 14 μs, the laser power was 135 w, the pulse frequency was 9500 Hz, and cutting speed was 5 mm/s. The auxiliary gas in laser cutting was argon, and its pressure was 0.7 MPa. The water source and the tail end of nitinol tube were connected by soft water pipe, and the water pressure was 0.06 MPa. The electropolishing system is shown in Fig. 3.3b, which is mainly composed of DC power supply, electromagnetic stirrer, electropolishing cell, temperature sensor and fixture. The electropolishing experiment was carried out at room temperature (28

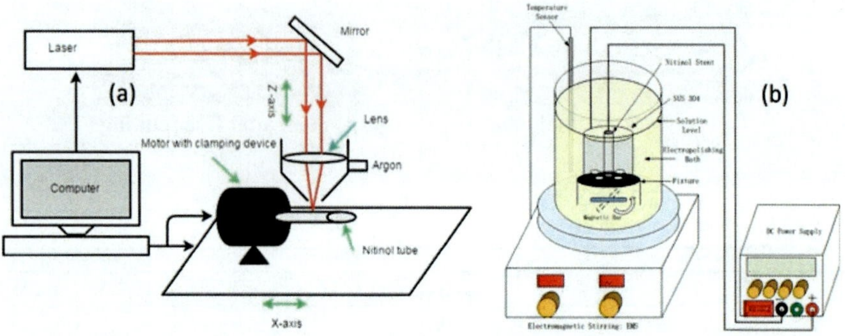

Fig. 3.3 Fiber laser wet cutting system (**a**) and electropolishing system (**b**)

Table 3.1 Laser parameters used in the experiment

Pulse width (μs)	Laser power (W)	Frequency (Hz)	Speed (mm/s)	Argon pressure (MPa)	Water pressure (MPa)
14	135	9500	5	0.7	0.06

Table 3.2 Electrolyte composition and polishing parameters

Distill water (vol.%)	NaCl (mol)	Ethylene glycol (L)	polar distance (mm)	Votage (V)	Time (min)
1	1	1	20		15

± 1 °C) using a non-toxic alcohol-salt system, and the electrolyte composition and polishing parameters are shown in Table 3.2.

3.2.3 Specimens Preparation and Inspection Methods

Before laser cutting, the whole nitinol tube was clamped on the laser cutting machine. During the cutting, the nitinol tube was cut into standard specimens (stents) according to preset parameters. After the cutting, the marked stents were put into argon for storage and inspection. Before electropolishing, the stents by laser cutting were put into ethanol for ultrasonic cleaning for 6 min, and then they were electropolished. After electropolishing, these stents after polished were put into ethanol for ultrasonic cleaning in 90 s, and then for subsequent inspection.

The MicroXAM-100 white light interferometer was used to observe the surface roughness of the stents. The specimens were made into metallographic specimens according to the requirements, and the FEI Quanta 250FEG field emission scanning electron microscope was used to observe the heat affected zone (HAZ), recast layer, dross, surface morphology and energy spectrum analysis of the stent. The Thermofly 250xi X-ray photoelectron spectroscopy (XPS) was used to analyze the chemical composition and content of the stent after laser cutting.

3.3 Results and Discussions

3.3.1 Surface Image and Morphology

The image of the stent after laser cutting is shown in Fig. 3.1b. There was burn at the kerf of the stent surface, and the burn was more serious at the corner of the cutting route because the energy density at the turning point was large. This also shows that microsecond laser processing is still hot processing, and the heat plays a major role in the damage of the stent. As can be seen from Fig. 3.5a, the SEM photo of the stent surface after laser cutting also showed that the surface of the stent was uneven with grooves and the colors were not uniform. The surface image of the stent after electropolishing is as shown in Fig. 3.4. The surface of the stent was bright and flat without burn marks. From the SEM photo of the stent surface after electropolishing in Fig. 3.5b, it could be seen that the surface of the stent by Fiber Laser-Electropolishing was smoother and the color was more uniform.

Fig. 3.4 Stent after electropolishing

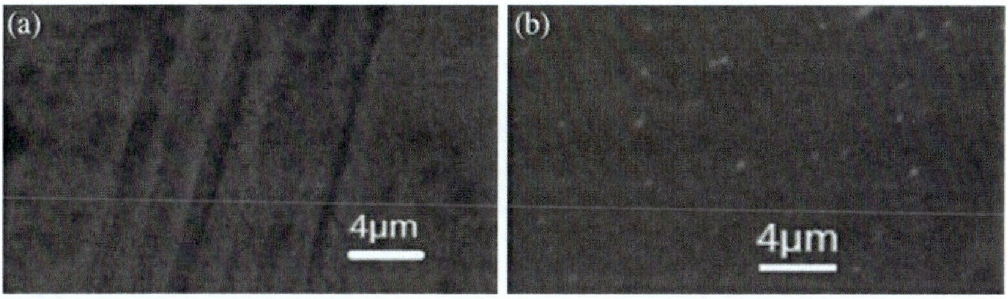

Fig. 3.5 SEM photo of stent surface before (**a**) and after (**b**) electropolishing

3.3.2 *Recast Layer and HAZ*

After the stent is cut by fiber laser, its structure usually includes three layers: recast layer, heat affected zone (HAZ) and matrix. And its SEM photo is shown in Fig. 3.6a. During the laser cutting, the molten material would be cooled again on the cut surface, forming a recast layer on the cut surface. A part of the surplus heat of laser cutting was transferred to the matrix, which changed the structure of the laser processing area and forming the HAZ. Figure 3.6b shows the SEM photo of the stent surface after electropolishing. It could be seen that the stent surface and cutting surface after electropolishing had no recast layer and HAZ. So the surface quality has been greatly improved.

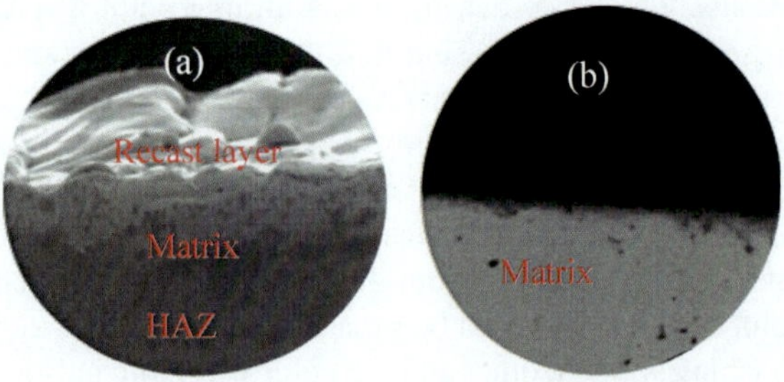

Fig. 3.6 Recast layer and HAZ after laster cutting (**a**) and surface after electropolishing (**b**)

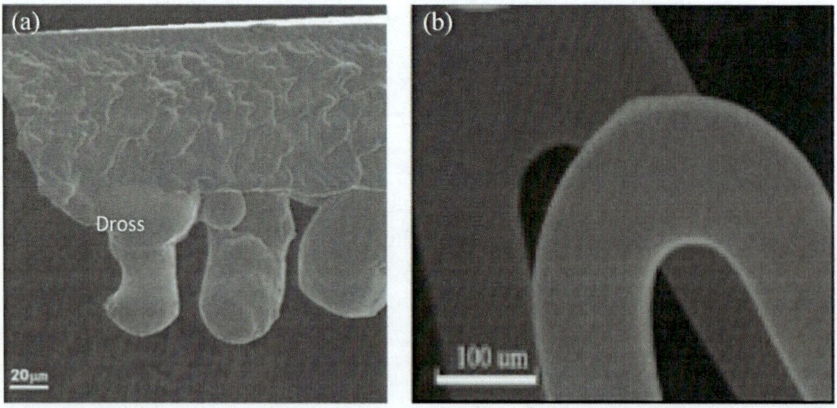

Fig. 3.7 Dross on the inner wall after laser cutting (**a**) and part of stent after polishing (**b**) by SEM

3.3.3 Dross

As shown in Fig. 3.7a, a large amount of dross appeared on the inner surface of the stent after laser cutting, which was caused by the melting of materials in the laser processing area and the accumulation of molten materials on the inner surface under the action of gravity and argon pressure. The dross seriously affected the performance of the stent. However, there was no dross on the inner surface of the stent after electropolishing as shown in Fig. 3.7b, and the quality of the inner surface was greatly improved.

3.3.4 Surface Roughness

Surface roughness plays an important role in the fatigue life and biocompatibility of cardiovascular stents, and it is also an important index to measure the surface quality. After laser cutting, the mean value of the surface roughness measured by the white light interferometer was about Ra 5.2543 μm. After laser cutting, the surface of the stent was relatively rough, which could also be seen from Figs. 3.5a, 3.6a and 3.7a. The surface roughness of the stent had been greatly improved with Ra 53.4 nm, as shown in in Fig. 3.8. This also showed that the electropolishing could improve the surface roughness of the stent very well.

3.3.5 Surface Chemical Composition

The chemical composition and content of the stents and nitinol tube are shown in Table 3.3. It could be seen from the table that the surface chemical composition of the stent after laser cutting had basically not changed, while the surface chemical composition of the stent after electropolishing had changed a lot, and the titanium dioxide (TiO_2) was generated on the surface of the stent, as shown in Fig. 3.9. In the process of electropolishing, part of titanium ions generated by anodic dissolution were combined with chloride ions in the electrolyte to form titanium tetrachloride, which reacted with water to form more stable titanium dioxide in turn, as shown in Eqs. (3.1) and (3.2).

$$Ti^{4+} + 4Cl^{-}TiCl_4(l) \tag{3.1}$$

$$TiCl_4(l) + 2H_2OTiO_2(s) + 4H^{+} + 4Cl^{-} \tag{3.2}$$

In addition, there was a large number of glycol in the electrolyte. Titanium tetrachloride reacted with glycol to form titanium compounds in the process of electropolishing. With the progress of electropolishing, titanium compounds interacted with

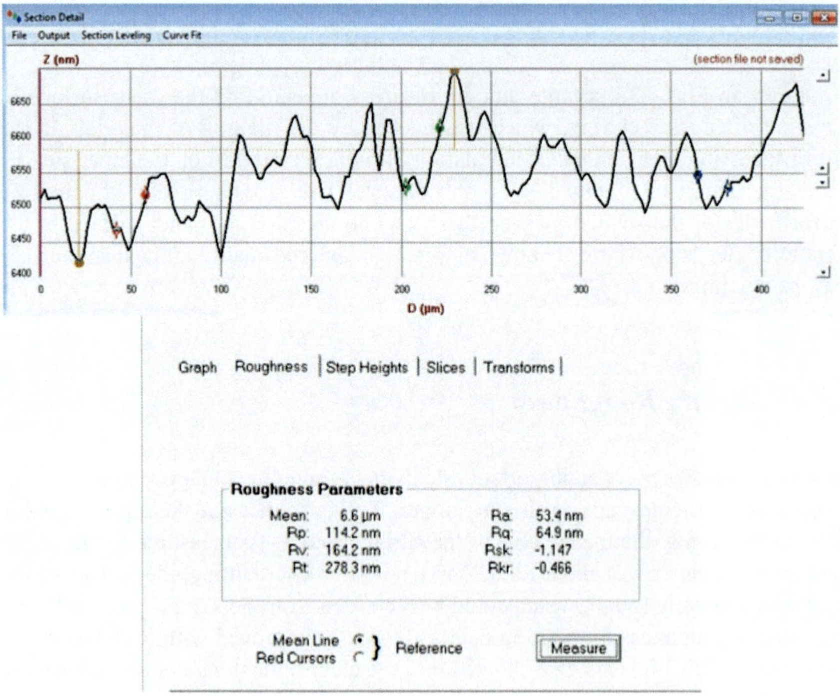

Fig. 3.8 Roughness measurement after electropolishing

Table 3.3 Surface chemical composition and content

Material element	Ti	Ni	Cl 2_p	C 1_s	O 1_s	TiO_2
Nitinol tube	43.86	56.14	0	0	0	0
Stent after laser cutting	47.29	52.71	0	0	0	0
Stent after electropolishing	0.7	38.70	0.97	13.36	9	37.27

water to form stable titanium dioxide, as shown in Eqs. (3.3) and (3.4). Titanium dioxide layer plays an important role in the biocompatibility of the stents [16], so the biocompatibility of the stents after electropolishing has been greatly improved.

$$TiCl_4(l) + \mathrm{HO} - \mathrm{CH_2CH_2} - \mathrm{OHTi(O} - \mathrm{CH_2CH_2} - \mathrm{OH)_4(l)} + 4\mathrm{H}^+ + 4\mathrm{Cl}^-, \tag{3.3}$$

$$\mathrm{Ti(O} - \mathrm{CH_2CH_2} - \mathrm{OH)_4(l)} + \mathrm{2H_2OTiO_2(s)} + \mathrm{4HO} - \mathrm{CH_2CH_2} - \mathrm{OH}. \tag{3.4}$$

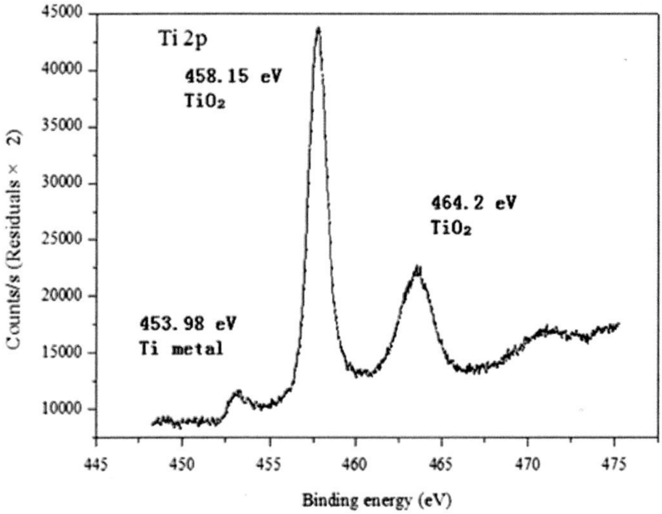

Fig. 3.9 XPS of the stent surface after electropolishing

3.4 Conclusions

In this paper, "fiber laser—electropolishing" collaborative manufacturing process was proposed to study the influences of the process on the surface integrity (mainly including surface morphology, HAZ, recast layer, dross, surface roughness) and biocompatibility of cardiovascular stent. The main conclusions are as follows:

(1) The "fiber laser—electropolishing" collaborative manufacturing process had changed the surface integrity of the stent very well. There was no heat affected zone, no recast layer and no dross on the surface of the stent. The surface morphology and the surface roughness had been improved very well, and the surface roughness value had reached Ra 53.4 nm.
(2) TiO_2 layer was generated on the surface of the stents by the collaborative manufacturing process, which greatly improved the biocompatibility of the stents.
(3) The process adopted alcohol-salt non-toxic electrolyte system, which was harmless to the environment.

Acknowledgements This study was supported by the NSFC (Grant No.51775321), and the Key Technology Research and Development Program of Shandong (public welfare) (Grant No. 2017GGX30116).

References

1. Garg, S., Serruys, W.P.: Coronary stents current status. J. Am. Coll. Cardiol. **56**, 1–42 (2010)
2. Issued by the Ministry of Health of the People's Republic of China. China Health Statistics Yearbook (2004–2011)
3. Prepared by the state health and Health Commission of the People's Republic of China. China Health Statistics Yearbook (2019). China Union Medical University Press.
4. Duerig, T., Pelton, A., Stöckel, D.: An overview of nitinol medical applications. Mater. Sci. Eng. A **273–275**, 149–160 (1999)
5. Wu, M.H.: Fabrication of nitinol materials and components. In: Proceeding of the International Conference on Shape Memory and Superelastic Technologies, pp. 285–292. Trans Tech Publications. Kunming, China (2001).
6. Li, C., Nikumb, S., Wong, F.: An optimal process of femtosecond laser cutting of NiTi shape memory alloy for fabrication of miniature devices. Opt. Lasers Eng. **44**(10), 1078–1087 (2006)
7. Li, C., Johnson, D., Kovacevic, R.: Modeling of waterjet guided laser grooving of silicon. Int. J. Mach. Tools Manuf **43**(9), 925–936 (2003)
8. Mai, T., Richerzhagen. B., Snowdon, P.C., Wood, D., Maropoulos, P.G.: The laser MicroJet (LMJ): A multi-solution technology for high quality micro-machining. Proc. SPIE **6459**(1), 64590P-64590P-7 (2007)
9. Manley, J., Housh, R., Wagner, F., Richerzhagen, B.: Water-guided lasers create clean cuts. Laser Focus World **40**(5), S15–S18 (2004)
10. Richerzhagen, B., Housh, R., Wagner, F., Manley, J.: Water jet guided laser cutting: a powerful hybrid technology for fine cutting and grooving. In: Progress in Theoretical Chemistry and Physics (2004)
11. Richerzhagen, B., Kutsuna, M., Okada, H., Ikeda, T.: Waterjet-guided laser processing. Proc. SPIE **4830**(1), 91–94 (2003)
12. Hock, K., Adelmann, B., Hellmann, R.: Comparative study of remote fiber laser and water-jet guided laser cutting of thin metal sheets. Phys. Procedia **39**, 225–231 (2012)
13. Meng, H.Y., Liao, J.H., Guan, B.G., et al.: Fiber Laser Cutting Technology on Coronary Artery Stent. Chin. J. Lasers **34**(5), 733–736 (2007)
14. Lv, S., Wang, Y.: An investigation of pulsed laser cutting of titanium alloy sheet. Opt. Lasers Eng. **44**(10), 1067–1077 (2006)
15. Raval, A., Choubey, A., Engineer, C., Kothwala, D.: Development and assessment of 316LVM cardiovascular stents. Mater. Sci. Eng. A **386**(1–2), 331–343 (2004)
16. Wever, D.J., Veldhuizen, A.G., de Vries, J., Busscher, H.J., Uges, D.R.A., van Horn, J.R.: Electrochemical and surface characterization of a nickel-titanium alloy. Biomaterials **19**(7–9), 761–769 (1998)

Chapter 4
Study on a Digital Inspection Method for Aircraft Tubing Assembly

GuangJun Li, JingChao Yang, Dan Huang, and Wei Han

Abstract Based on the difference analysis between measuring datum and assembling datum of aircraft tubing, reference points of these two kinds of datum were unified. The key technical issues were investigated from aspects of assembling datum analysis, centerline data extraction, measuring datum fitting and assembly accuracy analysis. The digital inspection software of tubing for assembly was developed based on CATIA V5. The shape deviation of tubing was divided into the axial deviation, the radial deviation and the angle deviation to realize the accurate description of the shape deviation data. And then, these three deviations were matched with the butt clearance, misalignment and angle deviation in actual assembly of tubular components. Finally, error analysis report of tubing assembly and accurate description of tubing shape deviation data can be generated, so as to solve the problem of misjudgment in traditional inspection methods and provide supports for product quality improvement and data statistical analysis.

Keywords Aircraft tubing assembly · Digital inspection · Shape deviation · Error analysis

4.1 Introduction

Tubing is one of the most significant parts applied in the pipeline systems of aircraft. It is spread all over a whole aircraft with the features of large amount and compact structure, known as the "bleeding" components of aircraft to guarantee conveying fluids from one location to another [1, 2]. On one hand, they bring the gas/oil to the engine, the landing gear and the cockpit, etc., satisfying the requirements of

G. Li (✉) · D. Huang · W. Han
Chengdu Aircraft Industry (Group) Corporation Ltd., Chengdu 610092, China
e-mail: lgjaviation@126.com

J. Yang
State Key Laboratory of Solidification Processing, School of Materials Science, Northwestern Polytechnical University, Xi'an 710072, China
e-mail: Yang-shuaii@126.com

J. Xu and K. M. Pandey (eds.), *Mechanical Engineering and Materials*,
Mechanisms and Machine Science 100,
https://doi.org/10.1007/978-3-030-68303-0_4

hydraulic, fuel manifold, air conditioning, portable water galley, auxiliary power unit, oxygen line, manometric line, and thrust reverser door actuator systems. On the other hand, they are assembled on the aircraft's overall frame as a structure to increase the overall stiffness and strength of the aircraft. Not only does the exterior of these tubular components withstand high and low temperature and high frequency vibration, but the interior also withstands the pulsating impact of the fluid medium, so that their working conditions are very harsh [3]. Thus, the manufacturing and assembling of aviation tubular components are highly concerned.

At present, tubular bent components are fabricated by numerical control bending [1, 4]. Their shapes are inspected by means of laser-measurement [5, 6], and special software is used to compare the difference between the measured shape and the theoretical shape [7–9], so as to realize the digital inspection of aviation tubing. The above technologies only inspect the tubing as a part, and they adopt the best fitting in the comparison process. Thus, there is no clear detection datum, so that the assembling positioning datum in the aircraft is not considered.

Due to the difference between the inspection datum just for the tubing in the status of parts and assembly positioning datum of the tubing in the assembling status, the deviation of tubing shape and that of assembling positioning can not correspond individually. There are often misjudgement situations of "positive judgment for the unqualified (PJN)", i.e., parts that can not meet the assembly requirements are judged to be qualified and "misjudgement for the qualified (MJQ)", i.e., parts that can meet the assembly requirements are judged to be unqualified. Therefore, it is necessary to integrate the inspection datum of parts and the assembling positioning datum and to divide the deviations of tubing shape, and finally solving the problem of misjudgement of tubing accuracy.

4.2 Tubing Inspection Method

The tubing in the aircraft is generally made of metal, for instance, titanium alloy [10], aluminum alloy [11] and stainless steel [12], etc. Its shape is composed of line sections and bent sections alternately and its cross section is the standard circle. Accordingly, the tubing shape can be directly determined through the centerline, and the inspection datum and assembling positioning datum can be unified. As shown in Fig. 4.1, the central point of the fixed joint and the locating point of the clamp are used as the reference points to calculate the shape deviation of the tubing. The correspondence between the deviation of tubing shape, viz., axial deviation, radial deviation and angle deviation and the deviation of tubing assembling position, viz., the butt clearance, misalignment and angle deviation can be obtained. The occurrence of PJN and MJQ can be avoided, and the requirements of manufacturing and assembling of the tubing are satisfied. In order to realize this digital inspection, the key technical issues, viz., assembling datum analysis, centerline data extraction, measuring datum fitting and assembly accuracy analysis, need to be investigated.

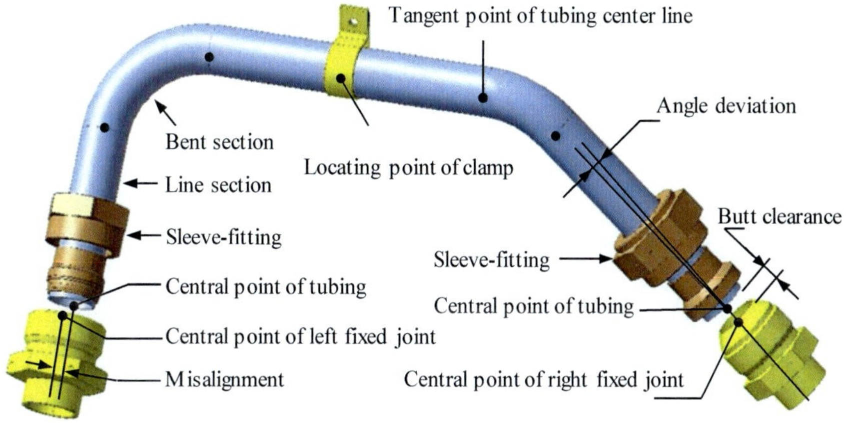

Fig. 4.1 Digital inspection plan for tubing assembly

4.2.1 Analysis of Tubing Assembly Datum

Generally, three-step method is used in aircraft tubing assembly. The first step is to align the left and right tubing end with the fixed joint in aircraft and make them concentric; the second step is to fix the middle part of tubing on the locating point of the clamp; the third step is to tighten the nuts of the left and right tubing end. During the assembly process, it is necessary to check that the angle deviation, butt clearance and misalignment between the central points of the tubing end and those of the fixed joints, avoiding exceeding the specified values. Therefore, tubing assembling positioning datum is determined by two central points of fixed joints and one locating point of a clamp.

4.2.2 Data Extraction on the Centreline of Tubing

Due to the geometric and structural features of the tubing, the deviation of tubing centreline can directly represent the deviation of tubing shape. Additionally, the central coordinates of the circle can be measured by the existing commercialized fork laser scanning equipment. By continuously scanning the contour of lines of tubing, the coordinates of central points of two tubing ends and those of several consecutive measured points on the lines can be directly obtained; the centreline's equation in the line section and the coordinates of the tangent points to the bent section are obtained by the calculation method, viz., finding the mean values of the multiple points for the centreline. Therefore, the two ends coordinates in the centreline and the coordinates of the tangent points to the bent section can be used to calculate and express the actual shape of the tested tubing completely, as shown in Fig. 4.2.

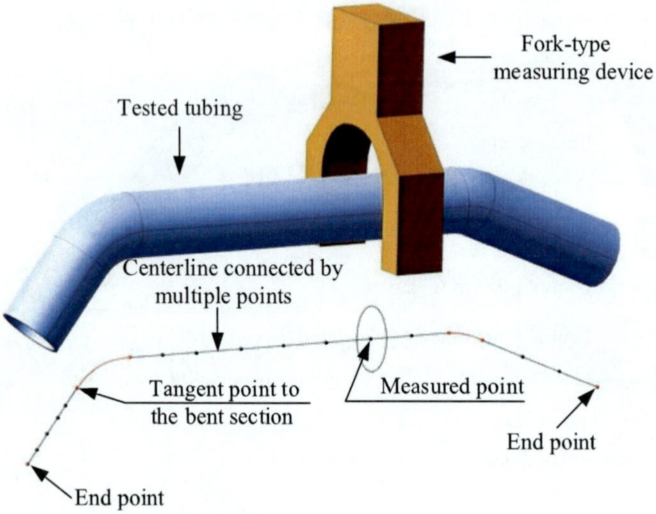

Fig. 4.2 Data extraction on the centreline of tubing

4.2.3 Inspection Datum Fitting

The theoretical position of the central point of the aircraft fixed joint coincides with that of the central point of the tubing end. Accordingly, the theoretical endpoints of tubing and the locating point of the clamp can be directly used to fit the inspection datum, so as to realize the unification of the inspection datum and the assembling positioning datum of tubing, as shown in Fig. 4.3. Four steps are as follows:

(1) Determination of the datum points: the theoretical endpoints of tubing and the locating point of the clamp. According to the actual assembly situation, the

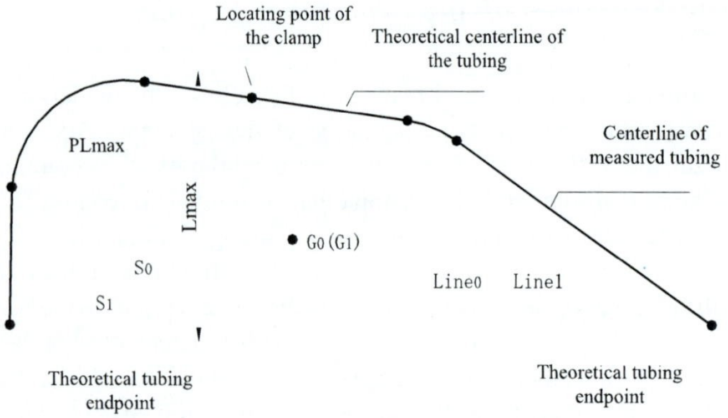

Fig. 4.3 Fitting the inspection datum of tubing

theoretical location coordinates of the clamp positioning point on the tubing centreline can be determined. If the locating point of the clamp cannot be determined, the tangent point to the bent section with largest distance from the two tubing endpoints, PLmax, can be used to replace that, so that obtaining the theoretical triangle with the largest area and minimize the inspection error.

(2) Determination of the theoretical barycenter of the triangle: the theoretical triangle S_0 is composed of two theoretical tubing endpoints and the locating point of the clamp, and its barycenter is determined as G_0.

(3) Determination of the measured barycenter of the triangle: the measured triangle S_1 is composed of two measured tubing endpoints and the locating point of the clamp, and its barycenter is determine as G_1.

(4) Comparison and matching of the datum: rotating and moving the measured triangle S_1, and making S_0 coplanar with S_1, G_0 coincide with G_1 and Line0 connecting theoretical tubing endpoints parallel to Line1 connecting measured tubing endpoints, so as to realize the locating of the measured and the theoretical tubing based on the three assembly points.

4.2.4 Inspection Datum Fitting

Because the actual shape of the measured tubing can be completely expressed by the coordinates of two endpoints and bent tangent points, only the above tubing points need to be taken as deviation calculation points. And the shape deviation of tubing was divided into the axial deviation, the radial deviation and the angle deviation, as shown in Fig. 4.1.

Axial deviation: projecting the calculation deviation points of the measured tubing on the centreline of the theoretical tubing, the distance between the projection point and theoretical point is the axial deviation. If the deviation calculation point is the end point of tube, the axial deviation corresponds to the butt clearance of tubing assembly.

Radial deviation: projecting the deviation calculation point of the measured tubing on the cross section of theoretical tubing, the distance between the projection point and the theoretical point is the radial deviation. If the deviation calculation point is the tubing endpoint, the radial deviation corresponds to the misalignment of tubing assembly.

Angle deviation: the angle value between the centreline of the deviation calculation point of the measured tubing and the centreline of theoretical tubing is considered as the angle deviation. If the deviation calculation point is the tubing endpoint, the angle deviation corresponds to the angle deviation of tubing assembly.

In order to improve the inspection accuracy, the digitized equipment is used to measure the tube shape and the accuracy shall not be less than one third of the deviation of the tubing shape, at least 0.1 mm. In the process of numerical calculation, the calculation accuracy shall not be less than one thousandth of the deviation of tubing shape, at least 0.0001 mm.

4.3 Application of Tubing Inspection Method for Assembly

The ROMER Absolute Arm portable coordinate measuring machine of Hexagon Metrology is used for tube inspection [13], as shown in Fig. 4.4. The measurement accuracy of ROMER Absolute Arm is determined on the basis of B89.4.22 and VDI / VDE 2617–9 standards. The accuracy of different models is slightly different. Taking the model of RA-7530 as an example, the accuracy of contact probe is 0.054 mm, and that of fork probe is 0.09 mm. In the process of use, if the error is too large due to improper operation, the arm will prompt the error to operate again to ensure the reliability of the measurement results. The specific measurement steps of the bent tube are as follows:

(1) Create a measurement project, use laser fork probe to measure the tube, and obtain the IP point of the tube;
(2) Replace the contact probe to measure all accessories to obtain the plane, cylinder, circle and other elements of each accessory;
(3) If the pipe is a straight pipe section, the external cylinder characteristics of the tube end are measured with a contact probe.

By establishing the software interface of the laser vector measuring machine, the digital inspection software for tubing assembly of is developed based on CATIA V5. This software can read the theoretical data of CATIA model and the measured data of the tubing and automatically generate the analysis report of tube error by analyzing the error based on the assembly datum. The digital inspection software for tubing assembly has been applied in the development of tubular parts of civil aircraft, and it

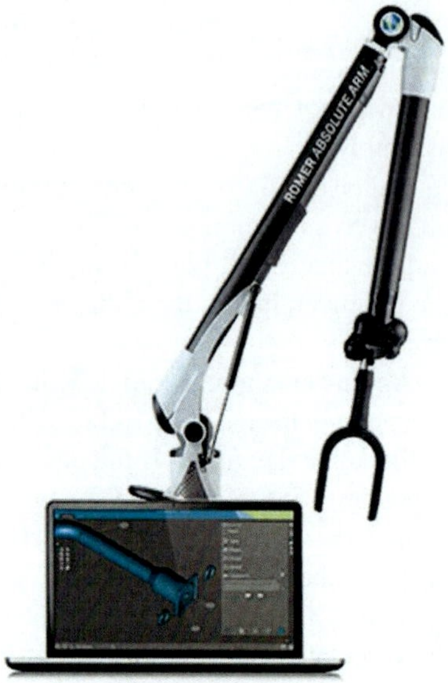

Fig. 4.4 ROMER Absolute Arm portable coordinate measuring machine with tube inspection system [13]

has completed a number of digital inspection work for aluminum alloy and stainless steel tubes of 6–32 mm diameter.

Since the multi-point average centerline method is used in the calculation of the tubing lines' coordinates, there may be large errors in the inspection of small angle bent tube. Due to this technical limitation, it is recommended to use it in the inspection of tubular parts whose bending angle is more than 2°.

4.4 Conclusions

The greatest advantage of the digital inspection technology for aircraft tubing assembly is that it considers the actual necessary of assembly, unifies the inspection datum and the tubing assembly datum, and decomposes the shape deviation into axial deviation, radial deviation and angle deviation. And the accurate description for tubing shape deviation, which corresponds to the butt clearance, the misalignment and the angle deviation in tubing assembly, is realized. This solves the problems of "PJN" and "MJQ" in the traditional inspection method, so that reducing the production cost and improving the inspection efficiency, finally providing supports for product quality improvement and data statistical analysis.

It should be noted that the digital inspection technology for aircraft tubing assembly is the supplement and improvement to the existing digital inspection technology, which can not completely replace the existing inspection methods, and has great application requirements of tubing assembly in the complex assembly process. If the fixed joint position of the assembled tubing is not accurate, the tubing inspected by this method may still not meet the assembly requirements. Anyhow, this inspection method can expand the application scope of ordinary coordinate measurement, provide another accurate measurement method for the shape inspection of aviation tubing, and promote the CAD / CAM digital production of aviation tubing.

Acknowledgements The authors gratefully acknowledge the financial support from the special fund project for civil aircraft (MJ-2016-G-64) and the National Natural Science Foundation of China (51775441 and 51835011).

References

1. Yang, H., Li, H., Zhang, Z., Zhan, M., Liu, J., Li, G.: Advances and trends on tube bending forming technologies. Chin. J. Aeronaut. **25**, 2–12 (2012)
2. Li, H., Yang, J.C., Chen, G.Y., Liu, X., Zhang, Z., Li, G.J., Liu, W.H.: Towards intelligent design optimization: Progress and challenge of design optimization theories and technologies for plastic forming. Chinese J. Aeronaut. in press (2020)
3. Yang, J.C., Li, H., Huang, D., Li, G.J., Yuan, S.: Forming of thin-walled AA6061-T4 tubular joint by elastomeric bulging: experiment and computation. Int. J. Adv. Manuf. Technol. **107**, 25–38 (2020)

4. Zhang, Z., Wu, J., Liang, B., Wang, M., Yang, J., Muzamil, M.: A new strategy for acquiring the forming parameters of a complex spatial tube product in free bending technology. J. Mater. Process. Technol. **282**, 116662 (2020)
5. Estler, W.T., Edmundson, K.L., Peggs, G.N., Parker, D.H.: Large-scale metrology—an update. CIRP Ann. Manuf. Technol. **51**(2), 587–609 (2002)
6. Peggs, G.N., Maropoulos, P.G., Hughes, E.B., Forbes, A.B., Robson, S., Ziebart, M., Muralikrishnan, B.: Recent developments in large-scale dimensional metrology. Proc. Inst. Mech. Eng. Part B J. Eng. Manuf. **223**(6), 571–595 (2009)
7. Taekwang, H., Jun, M., Jørgen, B., Torgeir, W., Geir, R., Jyhwen, W.: In-line springback measurement for tube bending using a laser system. Procedia Manuf. **47**, 766–773 (2020)
8. Zhang, D., Li, Y., Liu, J., Xie, G., Su, E.: A novel 3D optical method for measuring and evaluating springback in sheet metal forming process. Measurement **92**, 303–317 (2016)
9. Katona, S., Lušić, M., Koch, M., Wartzack, S.: Integrating optical 3D measurement techniques in pipe bending: a model-based approach minimizing waste by deriving real functional design behavior. Procedia CIRP **50**, 808–812 (2016)
10. Singh, P., Pungotra, H., Kalsi, N.S.: On the characteristics of titanium alloys for the aircraft applications. Mater Today: Proc **4**(8), 8971–8982 (2017)
11. Devezas, T.: Trends in aviation: rebound effect and the struggle composites x aluminum. Technol. Forecast. Soc. Chang. **160**, 120241 (2020)
12. Li, H., Yang, H., Lu, R.D., Fu, M.W.: Coupled modeling of anisotropy variation and damage evolution for high strength steel tubular materials. Int. J. Mech. Sci. **105**, 41–57 (2016)
13. Jiang, Y.M., Zhang, J., Zhao, H.Y., Ou, J.X., Liu, S.L.: The digital metrology for aircraft welded tube based on the articulated arm coordinate measuring machine. Mach. Des. Manuf. Eng. (1), 71–74 (2020)

Chapter 5
Two-Step Exothermic Chemical Reaction for Heat Transfer Investigation in a Convective Cylinder

R. S. Lebelo and O. D. Makinde

Abstract In this article, the investigation of transient heat in a stockpile whose materials react spontaneously with the oxygen trapped within it is carried out. A reactive material in this context refers to any substances that contain carbon or hydrocarbons and readily react with oxygen in contact. The process whereby a carbon containing material reacts automatically with oxygen is called exothermic chemical reaction. The study is modelled in a domain that is cylindrical and loses heat by convection to the surrounding environment. The combustion process for the heat transfer study is very complicated and nonlinear in nature, and therefore, the partial differential equation governing the problem cannot be solved exactly, but numerically. For this purpose, the Finite Difference Method (FDM) was employed to solve the problem. The heat transfer investigation is performed by observing effects of selected embedded thermo-physical parameters like the reaction rate on the temperature of the system. The thermo-physical parameters may either increase or reduce the temperature of the system during combustion of reactive materials. The results obtained indicate that the rate of reaction and some parameters increase the temperature profiles during combustion, whereas the heat loss parameter lowers the levels of the temperature. The results are presented graphically and discussed appropriately.

Keywords Two-step exothermic chemical reaction · Heat transfer investigation · Convective cylinder · Finite difference method

5.1 Introduction

Heat transfer investigation of in stockpiles of combustible materials attracted the attention of many researchers in the past years. This is because of applications of heat

R. S. Lebelo (✉)
Vaal University of Technology, Vanderbijlpark 1911, South Africa
e-mail: sollyl@vut.ac.za

O. D. Makinde
Stellenbosch University, Saldanha 7395, South Africa

J. Xu and K. M. Pandey (eds.), *Mechanical Engineering and Materials*,
Mechanisms and Machine Science 100,
https://doi.org/10.1007/978-3-030-68303-0_5

transfer in engineering and industrial products manufacturing [1, 2]. Examples of heat transfer applications include, among others, heavy oil recovery, storage of cellulose materials, solids combustion, biomass and coal pyrolysis, waste incineration, and coal gasification [3]. Furthermore, ignorance of heat transfer and thermal stability applications during the manufacturing of industrial products using reactive materials may lead to bending or deformation and decolorization those products [4].

Combustible materials are those that react spontaneously with oxygen because of exothermic chemical reaction [5]. This study considers a two-step exothermic chemical reaction like the one occurring in methane combustion. Heat transfer is considered in a stockpile whose materials are reactive. An example can be a stockpile of wood or hay, and the study is modelled in a cylindrical pipe, such as the one carried out in [6]. The investigation of heat transfer in stockpiles was conducted experimentally in [7, 8], where physical properties like the material's coarseness, surface area thereof, its humidity and the surrounding temperature were studied. In this study, a theoretical approach is conducted where the thermo-physical parameters including the reaction rate and activation energy, which are embedded on the governing differential equation, are analyzed. The analysis is carried out by varying selected parameters to see how they affect temperature profiles during combustion.

Transient heat analysis for exothermic chemical reaction in two-step was studied in [9], where the combustion process was modelled in a rectangular slab and coupled convective and radiative boundary conditions were applied. The analysis considered a temperature dependent thermal conductivity, and the reactant diffusion was also included, but in this case, a constant thermal conductivity is observed. In [10] the analysis was conducted in a spherical domain of reactive materials, where heat loss by convection was considered. The mathematical modelling of a two-step exothermic chemical reaction was given by Makinde et al. [11], and their investigation was conducted with or without reactant consumption, with no extensive consideration of the reactant consumption. The objective of this study is to investigate the transfer of heat in a combusting stockpile of reactive materials which loses heat to the immediate environment by convection in an exothermic chemical reaction undergoing two steps. From the literature above, this investigation was not intensely carried out.

5.2 Mathematical Modelling

An example of a two-step exothermic chemical reaction taking place within a stockpile of combustible materials modelled in a cylindrical domain is given as Eqs. (5.1) and (5.2):

$$2CH_4 + 0.22N_2 + 3(O_2 + 3.76N_2) \rightarrow 2CO + 4H_2O + 11.50N_2 \quad \text{(Step 1)} \tag{5.1}$$

$$2CO + O_2 + 3.76N_2 \rightarrow 2CO_2 + 3.76N_2 \quad \text{(Step 2)} \tag{5.2}$$

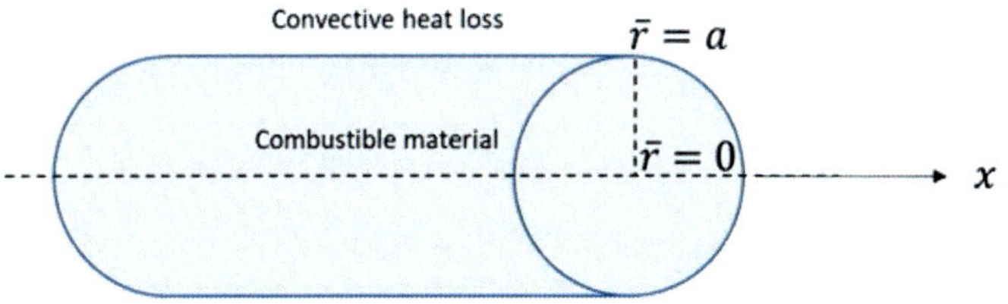

Fig. 5.1 The geometry of the problem

Above equations represent a two-step combustion of the methane gas in air, where the second step helps to eliminate the toxic gas carbon monoxide. Heat loss to the environment is assumed to be convective and follows Newton's law of cooling with the expression $-\frac{h}{k}[T - T_w]$, where h is the heat transfer coefficient, k is the thermal conductivity of the material, T is the absolute temperature of the cylinder, and T_w is the ambient temperature. The geometry of the problem is illustrated in Fig. 5.1 below.

Equation (5.3) below represents the transient heat energy equation for chemical reaction [10, 12–15],

$$
\begin{aligned}
pc_p \frac{\partial T}{\partial \bar{t}} = \frac{k}{\bar{r}} \frac{\partial}{\partial \bar{r}}\left(\bar{r} \frac{\partial T}{\partial r}\right) + Q_1 A_1 C_1 \left(\frac{KT}{vl}\right)^m e^{-E_1/RT} \\
+ Q_2 A_2 C_2 \left(\frac{KT}{vl}\right)^m e^{-E_2/RT} - \frac{h}{k}(T - T_w)
\end{aligned} \tag{5.3}
$$

The respective initial and boundary conditions are given in Eq. (5.4):

$$
T\left(\bar{r}, 0\right) = T_0; \frac{\partial T}{\partial \bar{r}}\left(0, \bar{t}\right) = 0; T\left(a, \bar{t}\right) = T_w \tag{5.4}
$$

In this case, T_0 is the cylinder initial temperature, Q_1 is the first step heat of reaction,A_1 is the first step rate constant and C_1 the first step reactant's concentration. It follows that Q_2,A_2 and C_2, are respectively, the second step's heat of reaction, rate constant, and reactant's concentration. E_1 and E_2 are the activation energies for the first and second steps, respectively. m is for the type of kinetics, where $m = -2$ is for light induced kinetics, the Arrhenius kinetics is represented by $m = 0$ and $m = 0.5$ presents bimolecular type.

The following dimensionless variables were introduced as Eq. (5.5):

$$
\theta = \frac{E_1(T - T_b)}{RT_b^2}, \theta_0 = \frac{E_1(T_0 - T_b)}{RT_b^2}, r = \frac{\bar{r}}{a}, \varepsilon = \frac{RT_b}{E_1}, \varepsilon = \frac{E_2}{E_1}, t = \frac{k\bar{t}}{pc_p a^2}, \varphi = \frac{a^2 h}{k},
$$

$$
\omega = \frac{Q_2 A_2 E_2}{Q_1 A_1 E_1} e^{(E_1 - E_2)/RT}, \alpha = \left(\frac{KT_b}{vl}\right)^m \frac{Q_1 A_1 E_1 a^2 C_1}{kRT_b^2} e^{-E_1/RT}. \tag{5.5}
$$

Equations (5.3) and (5.4) now have the following forms (Eq. (5.6)):

$$\frac{\partial\theta}{\partial t} = \frac{\partial^2\theta}{\partial r^2} + \frac{1}{r}\frac{\partial\theta}{\partial r} + \alpha(1+\varepsilon\theta)^m e^{[\theta/(1+\varepsilon\theta)]} + \alpha\omega(1+\varepsilon\theta)^m e^{[\epsilon\theta/(1+\varepsilon\theta)]} - \varphi\theta. \tag{5.6}$$

The initial and boundary conditions are shown in Eq. (5.7):

$$\theta(r,0) = \theta_0;\ \frac{\partial\theta}{\partial r}(0,t) = 0;\ \theta(1,t) = 0. \tag{5.7}$$

Here θ and θ_0 are the respective dimensionless temperature and initial temperature. The dimensionless parameters α, ε, r and ϵ are Frank-Kamenetskii (reaction rate), activation energy, radial distance, and activation energy ratio, respectively. Lastly, the dimensionless parameters for two-step reaction and heat loss are represented by ω and φ.

5.2.1 Numerical Approach

The numerical solution for the governing equation using the semi-implicit FDM followed the following route.

$$\begin{aligned}\frac{\theta^{N+1}-\theta^N}{\Delta t} &= \frac{\partial^2}{\partial r^2}\theta^{N+\tau} + \frac{1}{r}\frac{\partial}{\partial r}\theta^N + \alpha\left\{(1+\varepsilon\theta)^m e^{[\theta/(1+\varepsilon\theta)]}\right\}^N \\ &\quad + \alpha\omega\left\{e^{[\varepsilon\theta/(1+\varepsilon\theta)]}\right\}^N - \varphi\theta^N \end{aligned} \tag{5.8}$$

The τ is an arbitrary number such that $0 \le \tau \le 1$ and $\tau = 1$ for the convenience of using larger time steps. Rearranging Eq. (5.8), multiplying through by Δt and taking $\gamma = \frac{\Delta t}{\Delta r^2}$ yielded the expression for $\theta^{N+\tau}$ as follows:

$$\begin{aligned}-\tau\gamma\theta_{j+1}^{N+1} + (1+2\tau\gamma)\theta_j^{N+1} - \tau\gamma\theta_{j-1}^{N+1} &= -\gamma(1-\tau)\theta_{j+1}^N + [1-2\gamma(1-\tau)-\varphi]\theta_j^N \\ &\quad - \gamma(1-\tau)\theta_{j-1}^N + \frac{1}{2r}\gamma\left(\theta_{j+1}^N - \theta_{j-1}^N\right) \\ &\quad + \alpha\Delta t\left\{(1+\varepsilon\theta)^m e^{[\theta/(1+\varepsilon\theta)]} + \omega e^{[\theta/(1+\varepsilon\theta)]}\right\}^N \end{aligned} \tag{5.9}$$

A tri-diagonal matrix system was derived from Eq. (5.9) and Maple software was used to solve the system.

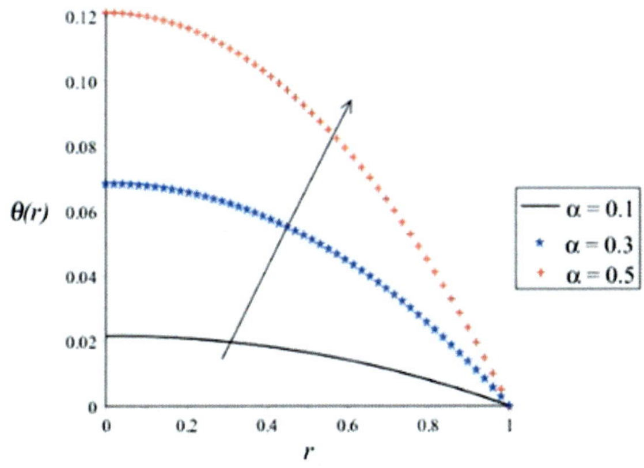

Fig. 5.2 α effects

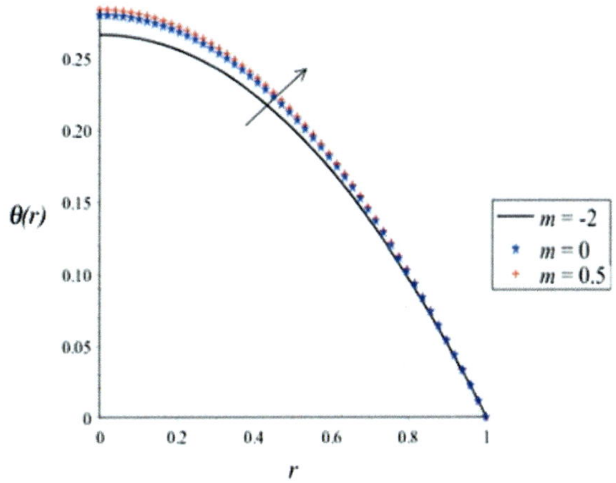

Fig. 5.3 m effects

5.3 Results and Discussion

Effects of selected kinetic parameters on the temperature behavior are presented in this section. Arbitrary parameter values chosen were:

$$m = 0.5, \omega = 0.1, \epsilon = 0.1, \alpha = 0.1, \varphi = 1, \varepsilon = 0.1, t = 1.$$

Fig. 5.4 ε effects

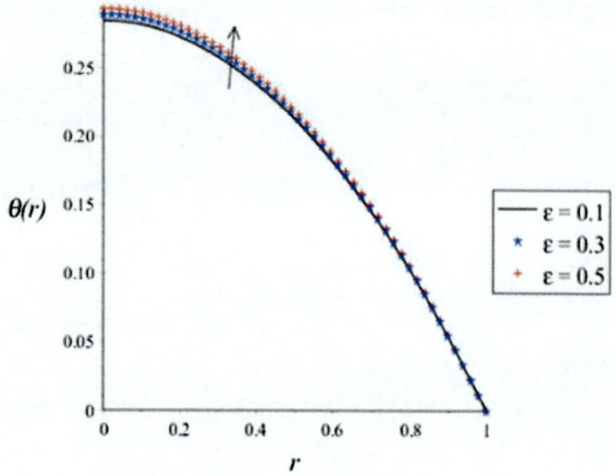

Fig. 5.5 ω effects

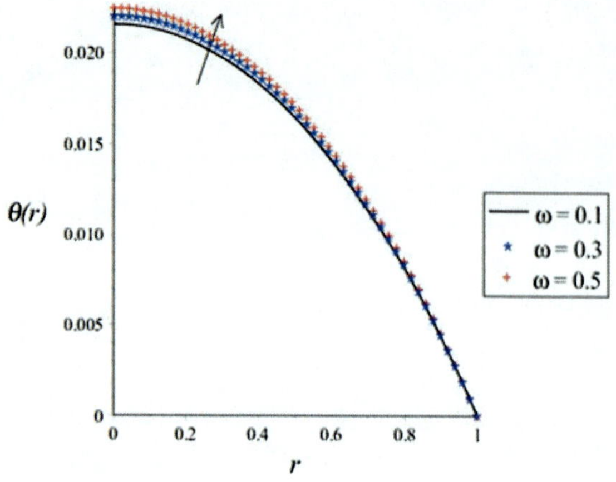

Fig. 5.6 φ effects

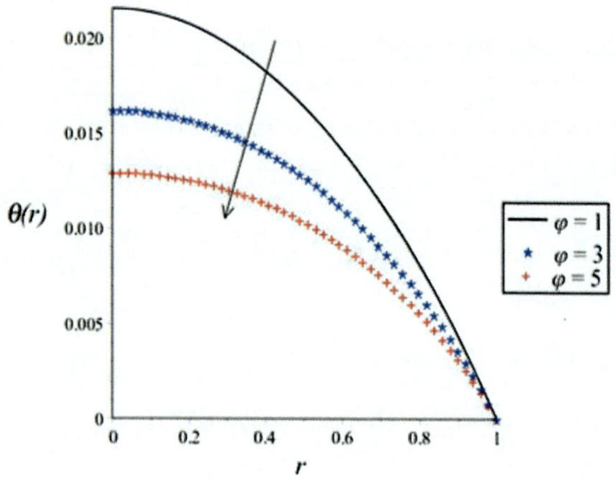

5.3.1 *Effects of Thermo-physical Parameters on Temperature (Unsteady State)*

Figures 5.2, 5.3, 5.4, 5.5 and 5.6 show the effects of α, m, ε, ω and φ on the temperature behavior during combustion. From Figs. 5.2, 5.3, 5.4 and 5.5, it can be observed that an increase in α (reaction rate parameter), m (chemical kinetics type), ε (activation energy) and ω (two-step exothermic chemical reaction) shows that the profiles of temperature increase correspondingly. Therefore, a combusting stockpile will increase its heat release rate under the effects of these parameters and fire may ultimately ignite if no control measures are done. In other words, these parameters enhance spontaneous reaction of oxygen trapped within the stockpile with its carbon or hydrocarbon substances to sustain the combustion process, which results with heat release. A high temperature profile increase in Fig. 5.2 shows that α's effect on the exothermic chemical reaction is so intense to fast-track the combustion process for emission of more heat within the stockpile. Figure 5.3 indicates that the chemical reaction that is exothermic is very quick in the kinetic type of bimolecular and that it slows down in the light induced one. In Fig. 5.6 a different scene is illustrated which shows that an increase in φ (heat loss parameter) reduces the levels of temperature. The parameter in the latter figure enhances the loss of more heat from the cylinder's surface to reduce the accumulation of heat inside the stockpile, which is the reason for the decline in the levels of temperature.

5.3.2 *Steady State Heat Transfer Analysis*

The temperature independent state is represented by the following ordinary differential equation (Eq. (5.10)):

$$\frac{\partial^2\theta}{\partial r^2} + \frac{1}{r}\frac{\partial\theta}{\partial r} + \alpha(1+\varepsilon\theta)^m e^{[\theta/(1+\varepsilon\theta)]} + \alpha\omega(1+\varepsilon\theta)^m e^{[\epsilon\theta/(1+\varepsilon\theta)]} - \varphi\theta = 0, \tag{5.10}$$

with the boundary conditions as shown in Eq. (5.11)

$$\frac{\partial\theta}{\partial r}(0) = 0; \theta(1) = 0. \tag{5.11}$$

The numerical solution in this case is obtained by using the Runge–Kutta Fehlberg (RKF45) method coupled with Shooting technique. Equation (5.10) is reduced to 1st order ordinary differential equation by letting $\theta = s_1, \theta' = s_2$. We express Eqs. (5.10)–(5.11) as Eqs. (5.12), (5.13) and (5.14):

$$s_1' = s_2 \tag{5.12}$$

$$s_2' = -\frac{1}{r}\frac{\partial s_1}{\partial r} - \alpha(1+\varepsilon s_1)^m e^{[s_1/(1+\varepsilon s_1)]} - \alpha\omega(1+\varepsilon s_1)^m e^{[\varepsilon s_1/(1+\varepsilon s_1)]} + \varphi s_1, \tag{5.13}$$

$$s_2(0) = 0, s_1(1) = 0. \tag{5.14}$$

The solutions are illustrated in Figs. 5.7 and 5.8 showing the steady state situation as $t \to \infty$. It is noted that high values of the temperature are experienced at the center of the cylinder and that the values decrease towards the surface thereof, due to convective heat loss. Figure 5.7 demonstrates temperature distribution in a two-dimensional domain, whereas Fig. 5.8 shows a three-dimensional one.

Fig. 5.7 Steady state 2-D $t \to \infty$

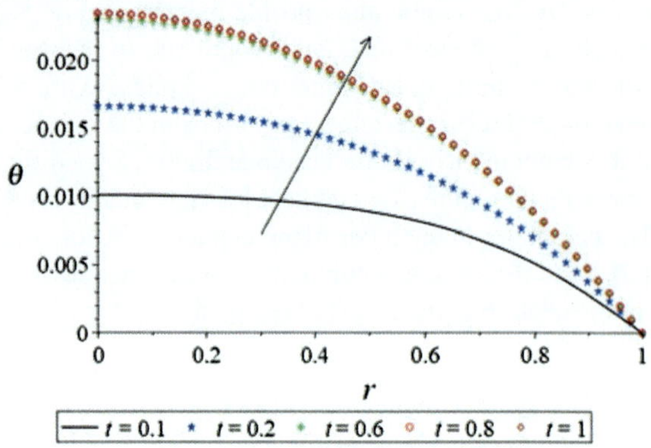

Fig. 5.8 Steady state 3-D $t \to \infty$

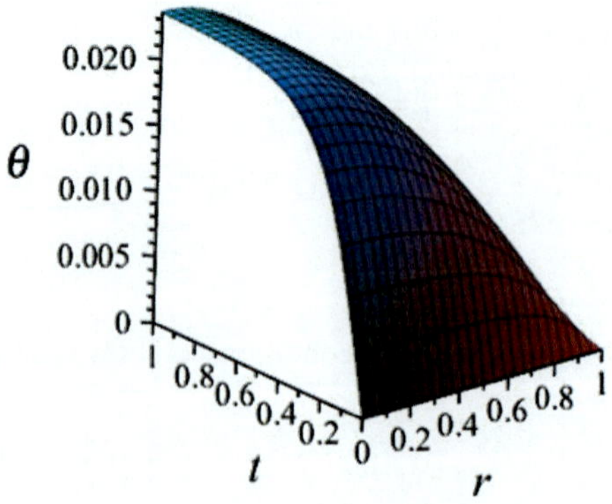

5.4 Conclusion

The work in this article investigated the heat in transit of a chemical reaction undergoing a two-step mechanism occurring in a stockpile whose materials react spontaneously with the trapped oxygen. The study was modelled in a cylindrical domain. A theoretical analysis using mathematical application was conducted to understand the heat transfer process by observation of effects selected thermo-physical parameters have on the temperature during combustion. The study considered both the temperature dependent and independent scenarios to demonstrate the heat transfer process clearly. It was observed that parameters which increase the temperature during combustion are α (reaction rate parameter), m (chemical kinetics type), ε (activation energy) and ω (two-step exothermic chemical reaction). The raise in temperature of the system is not good for environmental purposes, because global warming and depletion of the ozone layer are the results. Global warming has contributed so much to climate change. The parameter that lowers the system's temperature was identified as φ (heat loss parameter). In general, the results obtained agree with the literature in [2, 3, 6, 8]. A future study can consider material's thermal conductivity that is temperature dependent.

References

1. Hamza, B.H., Massawe, E.S., Makinde, O.D.: Analysis of transient heating due to exothermic reaction in a stockpile of combustible material. Int. J. Phys. Sci. **6**(18), 4337–4341 (2011)
2. Lebelo, R.S., Okosun, K.O., Mtunzi, F., Muthuvalu, M.S., Mahlobo, R.K.: Thermal stability and reactant consumption analysis in a reactive cylinder of variable thermal conductivity. Int. Rev. Mech. Eng. **11**, 787–796 (2017)
3. Lebelo, R.S.: Thermal stability investigation in a reactive sphere of combustible material. Adv. Math. Phys. **1**, 1–6 (2016)
4. Bebernes, J., Eberly, D.: Mathematical Problems from Combustion Theory. Springer, New York (1989)
5. Frank-Kamenetskii, D.A.: Diffusion and Heat Transfer in Chemical Kinetics. Plenum Press, New York (1969)
6. Lebelo, R.S., Makinde, O.D., Chinyoka, T.: Thermal decomposition analysis in a sphere of combustible materials. Adv. Mech. Eng. **9**(2), 1–14 (2017)
7. Lohrer, C., Krause, U., Steinbach, J.: Self-ignition of combustible bulk materials under various ambient conditions. Process Saf. Environ. Prot. **83**(2), 145–150 (2005)
8. Hensel, W., Krause, U., Loffler, U.: Self-ignition of materials (including dust). In: Hartwig, M., Steen, H. (eds.) Handbook of explosion prevention and protection. Wiley-VCH 2.7, Weinheim, pp. 227–270 (2004)
9. Makinde, O.D., Tshehla, M.S.: Analysis of thermal stability in a convecting and radiating two-step reactive slab. Adv. Mech. Eng. **1**, 1–9 (2013)
10. Lebelo, R.S., Moloi, K.C., Okosun, K.O., Mukamuri, M., Adesanya, S.O., Muthuvalu, M.S.: Two-step low-temperature oxidation for thermal stability analysis of a combustible sphere. Alexandria Eng. J. **57**(4), 2829–2835 (2018)
11. Makinde, O.S., Adegbie, K.S., Fasoranbaku, O.A.: Mathematical modelling of two-step exothermic reactions with and without reactants consumption. FUTA J. Res. Sci. **1**, 44–53 (2013)

12. Gajjela, N., Matta, A., Kaladhar, K.: The effects of Soret and Dufour, chemical reaction, Hall and ion currents on magnetized micropolar flow through co-rotating cylinders. AIP Adv. **7**(11), 1–16 (2017)
13. Nagaraju, G., Garvandha, M.: Magnetohydrodynamic viscous fluid flow and heat transfer in a circular pipe under an externally applied constant suction. Heliyon **5**(2), 1–25 (2019)
14. Ramana Murthy, J.V., Nagaraju, G., Sai, K.S.: Numerical solution for MHD flow of Micro polar fluid between two concentric rotating cylinders with porous lining. Int. J. Nonlinear Sci. **13**(2), 183–193 (2012)
15. Govardha, K., Nagaraju, G., Kaladhar, K., Balasiddulu, M.: MHD and radiation effects on mixed convection unsteady flow of micropolar fluid over a stretching sheet. Procedia Comput. Sci. **57**, 65–76 (2015)

Chapter 6
Tool Model Building and Research on Cutting Simulation Experiment of Ti6Al4V

Qimeng Liu, Jinkai Xu, Xu Wang, and Huadong Yu

Abstract In this paper, the theoretical model of cutting tool is established. Based on the structural dimension of the end face of the tool and the axis of the tool as the rotating axis, the mathematical modeling of the spiral cutting edge line is carried out. The cutting bottom edge of the tool is composed of the intersection lines of the chip containing groove on the bottom surface, the cutting surface in front of the bottom surface and the cutting surface behind the bottom surface, which makes the size structure of the simulation tool model consistent with that of the actual tool model. High speed milling simulation is carried out on Ti6Al4V material by ABAQUS software. The simulation is based on Johnson cook material constitutive model and damage model. At the same time, the reliability of simulation results of cutting force and cutting temperature is verified by experiments. The results show that the error of cutting force and cutting temperature is less than 15%, especially the error of cutting temperature is less than 10%.

Keywords Ti6Al4V · Tool model · Milling simulation · Cutting force · Cutting temperature

6.1 Introduction

Ti6Al4V is often called superalloy, and widely used at the forefront of science and technology because of its excellent mechanical and chemical properties. Ti6Al4V is also a kind of engineering material with excellent performances such as high specific strength, high thermal strength, good fracture resistance and good corrosion resistance, and has been rapidly popularized and applied in aviation, aerospace, electronics, shipping and many other industries [1–3]. However, Ti6Al4V has the characteristics of high friction coefficient, high cutting temperature, low thermal conductivity and high chemical activity, leading to poor processing performance and

Q. Liu · J. Xu · X. Wang · H. Yu (✉)
Ministry of Education Key Laboratory for Cross-Scale Micro and Nano Manufacturing, Changchun University of Science and Technology, Changchun 130022, China
e-mail: 172024452@qq.com

J. Xu and K. M. Pandey (eds.), *Mechanical Engineering and Materials*, Mechanisms and Machine Science 100,
https://doi.org/10.1007/978-3-030-68303-0_6

serious tool wear. The main factors affecting tool life are cutting force and cutting temperature because cutting force during the micro milling process of Ti6Al4V will lead to higher cutting temperature, especially in high-speed cutting process. Failing to set the cutting parameters properly may cause the cutting force and cutting temperature to increase sharply, resulting in reduced tool life. ABAQUS cutting simulation is one of the effective means to predict and simulate the cutting characteristics of Ti6Al4V before machining [4, 5].

In recent years, many scientists have done a lot of research on the establishment of tool model and the influence of tool wear on machining characteristics. Professor Zhao Jun [2] established an accurate model of milling cutter by determining the cutting edge line and spiral groove end section, which provided a theoretical basis for the design of milling cutter. According to its invariance, Professor Sun Xiuyuan [3] uses the natural axis coordinate system to model the attitude of the grinding wheel, and obtains a general normal vector equation of the grinding wheel end face and the central coordinate equation of the grinding wheel end face. Arisoy YM based on the three-dimensional finite element simulation, the induced microstructure of Ti-6Al-4 V alloy is predicted, and the simulation results are obtained [4]. Wang [5] studied the changes of grain size and microhardness in the process of chip formation and surface formation in Ti-6Al-4 V high speed cutting, and made in-depth theoretical research. Rahman [6, 7] believes that small CNC machine tools have high accuracy conditions and can improve the accuracy of micro cutting. When mamedov, a is micro cutting, the tool stability is high. This machining method can effectively improve the machining accuracy and surface quality. They have established a three-dimensional analytical cutting force model for micro end milling operation. This model has guiding significance for processing and production [8–10]. Wu et al. [11] researched the burr formation mechanism in micro cutting. Jing et al. [12] researched the modelling and experimental analysis of the effects of run out, minimum chip thickness and elastic recovery on the cutting force in micro-end-milling. They have practical guiding significance for the research of cutting mechanism and chip forming in micro cutting, both in simulation and practical application.

In order to study the cutting simulation accuracy of Ti6Al4V, based on other research scholars, this paper makes a tool model to make the simulation tool structure consistent with the actual tool structure, and uses the Abaqus simulation software, numerical control processing equipment and measuring equipment to carry out simulation and experimental research on high speed micro-cutting of Ti6Al4V. According to the simulation and experimental data, cutting force and cutting temperature were compared and analyzed to further verify the validity and reliability of the simulation, and examine the influence of cutting parameters on cutting force and cutting temperature in the simulation.

6.2 Milling Finite Element Model

6.2.1 Establishment of Cutting Simulation Model

In this study, the edge burr formation of Ti6Al4V under high-speed micro-milling is studied using Abaqus. Firstly, the geometric models of the workpiece and tool were established in CATIA.

Tool model establishment: the micro-cutting tool of the integral end milling cutter was selected for the simulation. The key dimensions of the milling cutter are shown in Fig. 6.1.

The helix edge of the milling cutter can be defined as a helix which is composed of the helix motion with the axis of the center axis of the cutter as the rotating axis at the point of outline [3].

As shown in Fig. 6.2, a point on the spiral edge line can be divided into three vector points v_r, v_a and v_t (as defined in Eq. (6.1)) according to polar coordinates, in which v_t is tangential vector of the spiral line, v_a is axial vector of the cutter, and v_r is radial vector of the cutter. If the angular velocity of the spiral point is ω, the three decomposed vector points then are v_r, v_a and v_t.

$$\begin{aligned} v_t &= \rho(z)\omega \\ v_r &= \rho'(z)z'\omega \\ v_a &= z'\omega \end{aligned} \tag{6.1}$$

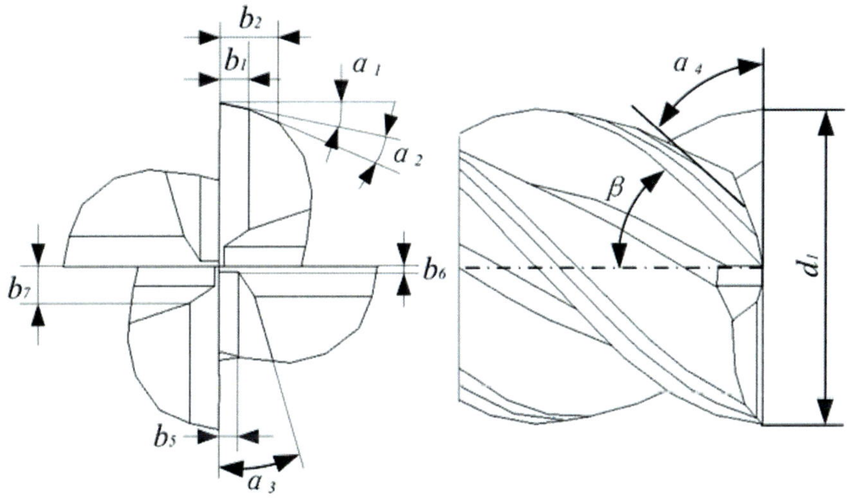

Fig. 6.1 Key dimensions of milling cutters

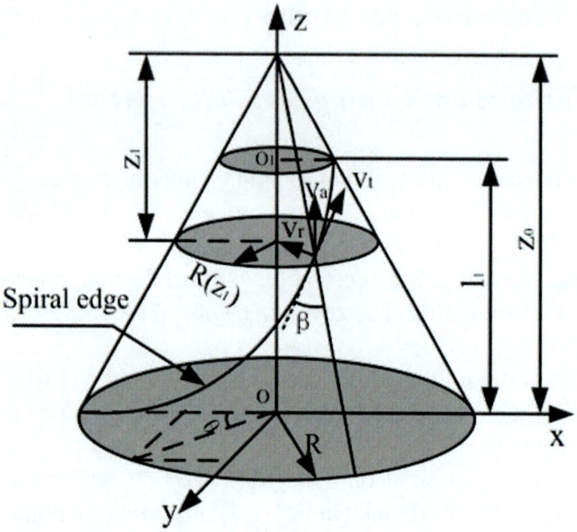

Fig. 6.2 Mathematical Model of Helical Edge Line of the Cutting Tool

The geometric equation of the cutting edge line of the cutter is divided into two parts because both bottom and side cutting edges of the cutter are involved in removing material during the micro-milling process. Assuming that the cutting edge line on the cutting tool is l, then $l = l_1 + l_2$, l_1 is the geometric equation of the cutting edge line on the side of the cutting tool and l_2 is the geometric equation of the cutting edge line on the bottom of the cutting tool. Compared with the side cutting edge, the bottom cutting edge has little effect on machining quality. The cutting coordinate system is shown in Fig. 6.3. Equation (6.2) is the equation of the edge line.

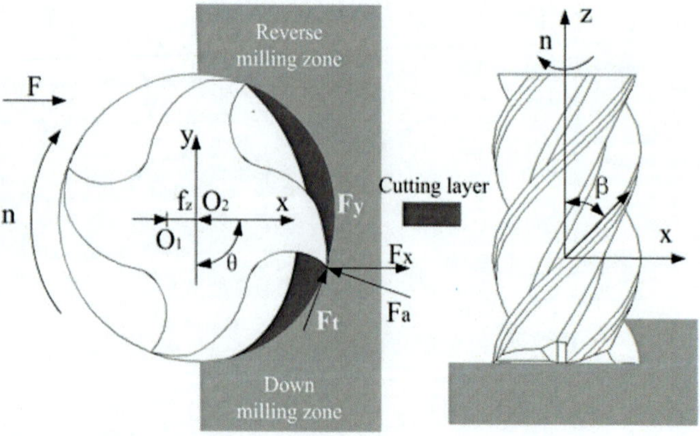

Fig. 6.3 Milling Cartesian coordinate system

$$
I_1 \begin{cases} x = \frac{d}{2}\cos\theta/2 \\ y = \frac{d}{2}\sin\theta \\ z = \frac{d}{2}\theta/\tan\beta \end{cases} \tag{6.2}
$$

In Fig. 6.3 and Eq. (6.2), *d/2* is tool radius, θ is contact angle, β is tool helix angle, *n* is spindle speed, F is feed speed, and f_Z is the feed per tooth.

Bottom edge of the cutting tool is composed of intersection lines among the bottom chip flute, the front and back faces of the bottom. The space pentahedron composed of the front face and the bottom chip groove of the cutting edge of the tool bottom is called the bottom chip groove of the milling cutter. The geometric size of the chip flute structure directly relates to the spatial geometry of the cutting edge of the tool. It not only affects the chip removal performance of cutting tools, but also affects the stiffness of the tool bottom edge. The chip flute at the bottom of the milling cutter is grinded by ultra-precision grinding with multi-process. Figure 6.4a is the path of conical grinding wheel grinding chip-holding groove. The path of grinding chip-holding groove surface is divided into: Path 1: The large end face of the conical grinding wheel is used to grind the front face f_1 of the cutter chip groove; Path 2: Big end circle of conical grinding wheel is used to grind tool chip-bearing groove surface f_2; Path 3: Grinding wheel cone is used to grind tool bottom f_3. When milling material, the tool bottom edge rake face f_1 is the main cutting surface in end milling, while the chip groove face f_2 and the bottom face f_3 are the auxiliary cutting surface. Therefore, the bottom of Gash can be simplified to a pentahedron, as illustrated in Fig. 6.4b [6].

As shown in Fig. 6.5 below, the space vertices of the geometry of chip groove on the tool bottom are $p_1, p_2, p_3, p_4, p_5, p_6$, which are in the *x–y-z* coordinate system. The front angle γ' of the bottom cutting edge is the angle between the plane xOy and the $p_1p_2p_3$ in the bottom chip groove. Combining with Fig. 6.1, the angle α_3 of the expansion angle of the transverse die of the bottom chip groove is the angle between the plane $p_1p_2p_3$ and $p_4p_5p_6$, the width of the expansion of the transverse die of the

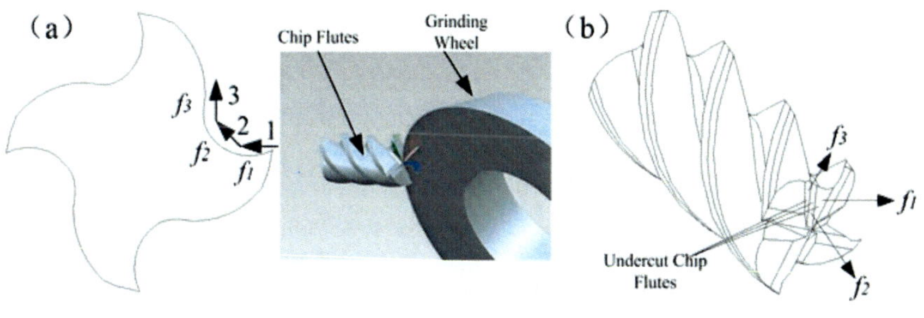

Fig. 6.4 Pentahedron model of grinding wheel envelope: **a** The path of conical grinding wheel grinding chip-holding groove; **b** the bottom of Gash

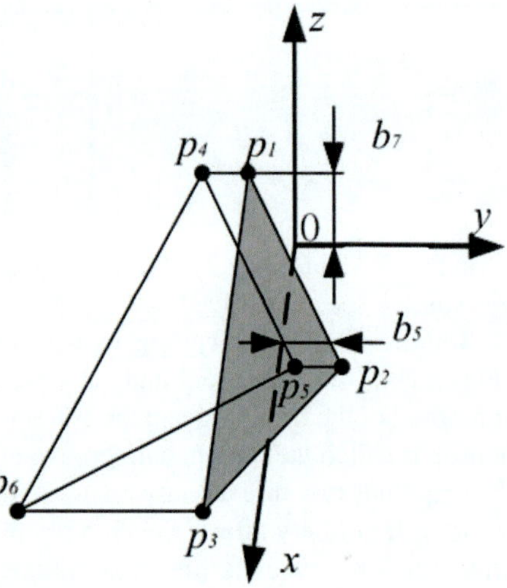

Fig. 6.5 Space Pentahedron Model of Base Chip Groove and Tool Diagram

bottom chip groove is b_5, the eccentricity of the bottom chip groove b_6 is the bottom edge, and the bottom chip groove b_7 is the long tooth passing. The angle α4 in the chip groove at the bottom of the tool is the angle between the straight line p_1p_2 and p_2p_3 [6–10].

Equation (6.3) is the coordinate values of each point in Fig. 6.5.

$$\begin{aligned}
&(x_{p1}, y_{p1}, z_{p1}) = (-b_7, -b_6, 0)\\
&(x_{p2}, y_{p2}, z_{p2}) = [d_1/2, (d_1/2 + b_7)\tan\gamma'/\cos(a_2) - b_6,\\
&(d_1/2 + b_7)\tan(\alpha_4)]\\
&(x_{p3}, y_{p3}, z_{p3}) = (d_1/2, -b_6, 0)\\
&(x_{p4}, y_{p4}, z_{p4}) = (-b_7, -b_5 - b_6, 0)\\
&(x_{p5}, y_{p5}, z_{p5}) = [d_1/2, (d_1/2 + b_7)\tan\gamma'/\cos(a_2) - b_6 - b_5,\\
&(d_1/2 + b_7)\tan(\alpha_4)]\\
&(x_{p6}, y_{p6}, z_{p6}) = [d_1/2, -b_5 - (d_1/2 + b_7)\tan(\alpha_3) - b_6, 0]
\end{aligned} \tag{6.3}$$

According to Eq. (6.3), six coordinate points of the bottom chip groove were calculated, and the space pentahedron of the bottom chip groove was drawn in CATIA three-dimensional surface drawing module. The bottom chip groove model shown in Fig. 6.4b can then be calculated by surface cutting combined with the tool cylinder.

Finally, the milling cutter with the following parameters was established: the diameter = 6 mm, the front angle = 30 degrees, the rear angle = 14 degrees, the helix angle = 45 degrees and the radius of the arc of the tool tip = 5 μm. Figure 6.6 depicts the comparison between simulation and experimental cutters.

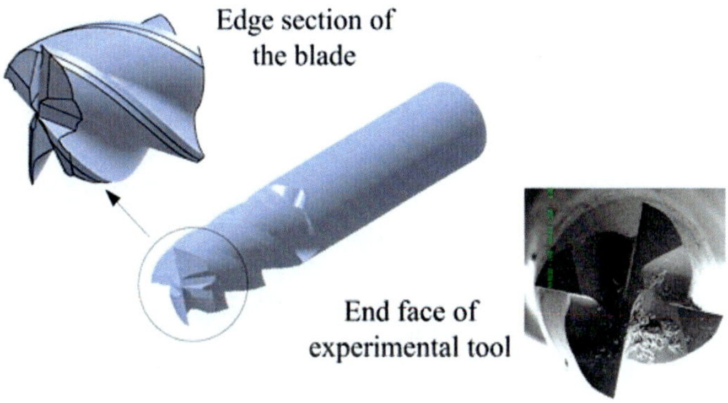

Fig. 6.6 End-to-face comparison of simulation and experimental cutters

6.2.2 *Meshing*

Workpiece model building: workpiece can be built by Abaqus simulation software module. The workpiece is 12 mm in length, 6 mm in width, and 3 mm in height.

Tool simplification needs to be conducted in the simulation by importing the cutting part model of the tool into Abaqus only. As the deformation of the tool is very small, the influence of deformation can be ignored. Therefore, the tool is set as a rigid body. Figure 6.7 is three-dimensional geometry model of cutting simulation in Abaqus.

Fig. 6.7 Geometric model of 3D cutting simulation

Fig. 6.8 Meshing model of the workpiece and tool

The workpiece model mesh is set to 20,000, hexahedron element shape, the tool model mesh size is 0.015 and the unit shape is set to tetrahedron in grid control attributes. At the same time, the gridding of milling layer area and cutter is refined. The mesh size of the refined part is 0.005. The mesh model of the workpiece and cutter is shown in Fig. 6.8.

6.2.3 *Attribute Definition of Cutting Simulation Model*

Tool Material Properties: the density is $1.47e^{-8}$ kg/m^3, the conductivity is 75.3 W/m/C, the specific heat is 220000000 J/kg/C, the elastic modulus is 583000 MPa, the Poisson's ratio is 0.33, the inelastic share is 0.8, and the thermal expansion is $5.4e^{-6}$/°C.

Material Attributes of the Workpiece.

Finite Element Model of Temperature Field. As an adiabatic process, because the heat generated during cutting cannot be timely dissipated, the heat conduction equation can be expressed as Eq. (6.4):

$$\lambda\left(\frac{\partial^2 T}{\partial x^2}+\frac{\partial^2 T}{\partial y^2}+\frac{\partial^2 T}{\partial z^2}\right)+\dot{q}=\rho C\dot{T} \tag{6.4}$$

Table 6.1 Material physical property parameters

Λ(W m^{-1} K^{-1})	Thermal conductivity	6.8
T (°C)	Temperature	100
E (GPa)	Elastic modulus	108
P (Kg m^{-3})	Density	4428.8
C (J kg^{-1} K^{-1})	Specific heat capacity	520

where, λ is the material thermal conductivity, T is the temperature, $\dot{q}$ represents that the heat source intensity is time-dependent, ρ is the material density, C is the specific heat capacity, and $\dot{T}$ shows that the temperature is time-dependent. x, y, and z are vectors of directions in the cartesian coordinate system. The default value of inelastic heat fraction (0.8) is accepted. Physical property parameters of materials are shown in Table 6.1 [10–12].

Johnson–Cook material constitutive model. Material constitutive model defines mechanical properties of materials. The correctness of constitutive model directly affects the reliability of simulation results. The constitutive model must reflect the relationship among temperature, strain rate, stress and strain of the model itself in the simulation of metal cutting. In the current simulation constitutive model, Johnson–Cook material constitutive model covers the relationship among temperature, strain rate, stress and strain. The Johnson–Cook constitutive model is shown in the following Eq. (6.5):

$$\sigma = \left(A + B\varepsilon^{n}\right)\left(1 + C\ln\frac{\dot{\varepsilon}}{\dot{\varepsilon}_0}\right)\left[1 - \left(\frac{T - T_r}{T_m - T_r}\right)^{m}\right] \quad (6.5)$$

where, $(A + B\varepsilon^{n})$ refers to the relationship between strain and stress in a material, $\left(1 + C\ln\frac{\dot{\varepsilon}}{\dot{\varepsilon}_0}\right)$ is the relationship between stress and strain rate, $\left[1 - \left(\frac{T-T_r}{T_m-T_r}\right)^{m}\right]$ represents the relation between stress and temperature, σ is the stress, ε is the strain; $\dot{\varepsilon}$ is the strain rate, $\dot{\varepsilon}_0$ is the reference strain rate, T is the temperature, T_r is the room temperature, T_m is the melting point of the material, A is the yield strength under quasi-static conditions, B and n are the parameters related to strain hardening, m is the strain softening parameter, C is the strain strengthening parameter. These values are shown in Table 6.2 [12].

Table 6.2 Constitutive model parameters of the Johnson–Cook materials

A	Initial yield stress (MPa)	792
B	Hardening modulus (MPa)	510
C	Strain strengthening coefficient	0.033
n	Strain hardening coefficient	0.47
m	Strain softening coefficient	1.02

Table 6.3 Johnson–Cook damage model parameters

d_1	−0.09
d_2	0.25
d_3	−0.5
d_4	−0.014
d_5	3.87

Johnson–Cook Damage Model. In this paper, Johnson–Cook failure model is used to define failure strain and separate chip from workpiece, as shown in Eq. (6.6).

$$\overline{\varepsilon_f}^{pl} = \left[d_1 + d_2 \exp\left(d_3 \frac{P}{Q}\right)\right]\left[1 + d_4 \ln\left(\frac{\dot{\varepsilon}^{pl}}{\dot{\varepsilon}_0}\right)\right]\left(1 + d_5 \hat{\theta}\right) \tag{6.6}$$

where, P is the pressure; Q is the Mises stress; d_1–$d5$ is the failure normal (room temperature), the parameters are shown in Table 6.3; $\dot{\varepsilon}_0$ is the reference strain rate; $\dot{\varepsilon}_0^{pl}$ is the plastic strain rate; and $\hat{\theta}$ is determined by current temperature, melting point and room temperature conditions.

Friction Contact Model. As shown in Fig. 6.9, due to the extrusion and friction between the rake face and the workpiece, a second deformation zone is formed. The characteristic of this zone is that the chip bottom layer gradually slips near the rake face, and the material separation gradually slows down in unit time. The temperature generated in the contact zone between the tool and the chip increases sharply due to friction, which directly affects the service life of the tool. The coupling change between physical dynamic changes in this process directly affects the plastic deformation of metal materials [10].

In cutting simulation, because the contact relationship between workpiece, cutting tool and chip is complex, the relationship between them should be simplified. The contact area between tool and workpiece is mainly divided into viscous friction zone and sliding friction zone. The common Coulomb model of the two friction zones is expressed by the following Eq. (6.7).

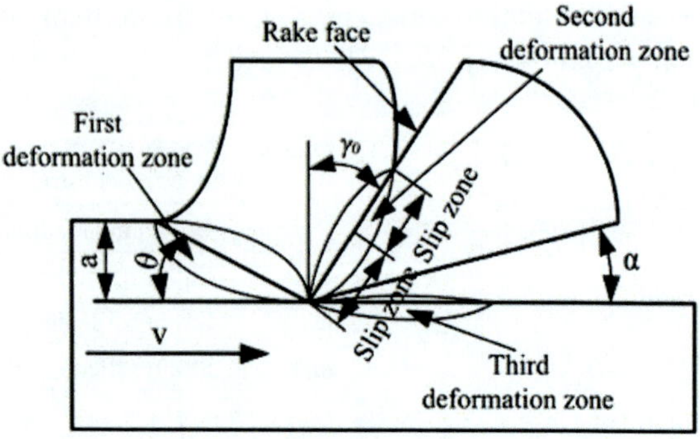

Fig. 6.9 Schematic diagram of friction during metal cutting

$$\begin{aligned} \tau &= \tau_c, \text{ viscous friction zone when } \mu\sigma_n \geq \tau_c \text{ , It is viscous friction zone} \\ \tau &= \sigma_n, \text{ when } \mu\sigma_n < \tau_c, \text{ It is the sliding friction zone} \end{aligned} \tag{6.7}$$

where τ is the frictional stress, τ_c is the ultimate shear flow stress, μ is the friction coefficient and σ_n is the normal stress.

6.2.4 Modeling of Cutting Force in Cutting Process

As shown in Fig. 6.10 cutting model, under the condition of cutting, due to the small scale of cutting, the material area removed by the tool rake face will be very small, the cutting thickness is almost equivalent to the radius of the tool edge, and the transverse surface of the cutting edge of the tool is circular arc shape. Therefore, the radius of the tool edge circle can not be ignored, and the influence of the edge circle size must be considered [16].

In the cutting parameters, the cutting depth is almost the same as the cutting edge radius. Compared with the tool rake angle, the cutting edge geometry has a great influence on the cutting. With the different cutting edge geometry, the actual cutting rake angle of the contact area among the tool, workpiece and chip formation will also be different, showing a large negative rake angle, which can be defined as Eq. (6.8):

$$\gamma_e = -\arcsin\left(\frac{2r_n - a_c}{2r_n}\right) \tag{6.8}$$

where: r_n is the radius of round edge tool; γ_e is the actual average cutting rake angle of contact area among tool, workpiece and chip formation; ac is the cutting depth.

From the above Eq. (6.8), it can be deduced that the smaller the cutting depth, the more obvious the effect of actual negative rake angle in the whole cutting process, and the greater the impact.

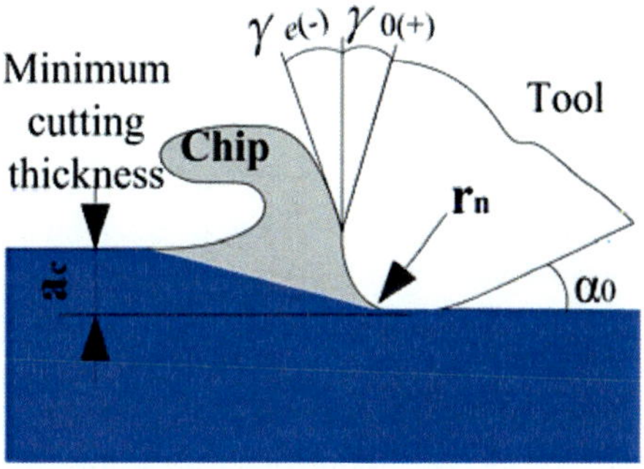

Fig. 6.10 Cutting model

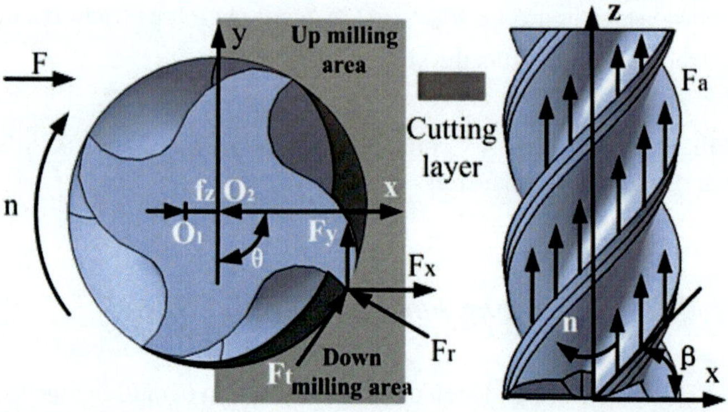

Fig. 6.11 Force distribution of milling cutter

Figure 6.11 shows the distribution of force in three directions on the cutting edge of milling cutter, where F_a is the axial cutting force, F_t is the tangential cutting force, and F_r is the radial cutting force.

Assuming that one of the four edge end mills of micro monolithic cemented carbide is taken as the reference point and its cutting angle is set as θ, then the formula of the cutting angle of the other three cutting edges is as follows: (1) $\theta_i(0) = \theta + i2\pi/z$, i = 0, 1, 2, ...(z−1), where Z is the number of tool edges. Since the number of cutting edges Z selected in this paper is 4, the cutting angles of the four cutting edges are in turn: (2) $\theta_0 = \theta, \theta_1 = \theta + \pi/2, \theta_2 = \theta + \pi, \theta_3 = \theta + 3\pi/2$.

As shown in Fig. 6.11, if the helix angle of the spiral end mill is β, the delay angle $\psi = 2a_p \tan \beta/D$, where a_p is the axial cutting depth and D is the diameter of the milling cutter. Therefore, the cut in angle of cutting edge i in the axial cutting depth direction of milling cutter is shown in Eq. (6.9).

$$\theta_i(a_p) = \theta + 2\pi i/z - \phi = \theta + i\pi - 2a_p \tan\beta/D \tag{6.9}$$

Because the milling is intermittent and the cutting thickness is a progressive process (as shown in Fig. 6.11), the transient chip thickness is as Eq. (6.10):

$$h_i[\theta_i(a_p)] = f_z \sin\theta_i(a_p) \tag{6.10}$$

where: f_z is the feed rate per tooth.

The micro elements are divided along the axis direction of the milling cutter. The cutting forces acting on the micro unit with the height of dz from the end face of the cutter are respectively set as radial cutting force dF_r, axial cutting force dF_a and tangential cutting force dF_t. From Eq. (6.10), it can be analyzed that the cutting force on one side of the up milling area first decreases and then increases, while the cutting force on the side of the forward milling area increases first and then decreases, while the cutting force on the tool end face is very small, so it can be ignored in calculation.

Therefore, the expression of cutting force should be as Eq. (6.11):

$$
\begin{aligned}
dF_{ri}(\theta, z) &= [k_{rc} h_i(\theta_i(z))]dz \\
dF_{ti}(\theta, z) &= [k_{tc} h_i(\theta_i(z))]dz \\
dF_{ai}(\theta, z) &= [k_{ac} h_i(\theta_i(z))]dz
\end{aligned}
\tag{6.11}
$$

Among them, k_{rc} is the radial shear force coefficient of the tool; k_{tc} is the tangential shear force coefficient of the tool; k_{ac} is the axial shear force coefficient of the tool, and the shear force coefficient of each direction is defined in Eq. (6.12):

$$
\begin{aligned}
k_{rc} &= \frac{\tau_s}{\sin\theta_n \cos j} \times \frac{\sin(\beta_n - a_n)}{\sqrt{\cos^2(\theta_n + \beta_n - a_n) + \tan^2\eta \sin^2\beta_n}} \\
k_{tc} &= \frac{\tau_s}{\sin\theta_n} \times \frac{\cos(\beta_n - a_n) + \tan j \tan\eta \sin\beta_n}{\sqrt{\cos^2(\theta_n + \beta_n - a_n) + \tan^2\eta \sin^2\beta_n}} \\
k_{ac} &= \frac{\tau_s}{\sin\theta_n} \times \frac{\cos(\beta_n - a_n)\tan j - \tan\eta \sin\beta_n}{\sqrt{\cos^2(\theta_n + \beta_n - a_n) + \tan^2\eta \sin^2\beta_n}} \\
h_i(\theta, z) &= f_z \sin(\theta_i(z))
\end{aligned}
\tag{6.12}
$$

where: fz is the feed rate of each tooth; a_n is the normal forward angle; β_n is the friction angle; $hi(\theta,z)$ is the chip thickness; η is the chip flow angle, θ_n is the normal shear angle; j is the inclination angle; τ_s is the shear yield strength.

The total cutting force in each direction is as Eq. (6.13):

$$
\begin{aligned}
F_x(\theta) &= \sum_{i=0}^{4} F_{xi} \\
F_y(\theta) &= \sum_{i=0}^{4} F_{yi} \\
F_z(\theta) &= \sum_{i=0}^{4} F_{zi}
\end{aligned}
\tag{6.13}
$$

Fig. 6.12 Experimental platform

6.3 Cutting Simulation and Experimental Analysis of Ti6Al4V

6.3.1 Experimental Condition

A force platform and a temperature measuring instrument are installed on the work platform of the machine tool. Besides, and a scanning electron microscope is employed to present the morphological structures of chip [11]. The experimental platform is shown in Fig. 6.12 and Table 6.4.

6.3.2 Comparison of Chip between Cutting Simulation and Experiment of Ti6Al4V

In the forming cutting process, the workpiece material at the cutting edge of the milling cutter is divided into two parts: one part flows above the front cutter face; the other part flows below the back cutter face to form chips. Cutting force and cutting temperature are important output variables in the simulation process. The simulation results have important theoretical analysis influence on tool wear and workpiece machining quality. Figure 6.13 shows the comparison of simulation and experimental

Table 6.4 Experimental platform

Tool type	Diameter: 6 mm, 4 edge, a front angle of 30°, a back angle of 14°and a helix angle of 45, surface coating: TiSiN carbide milling cutter	Equipment	Performance parameter
Spindle speed (r/min)	4000, 8000, 12,000, 16,000	Machining center	Spindle:20000 rpm, Stroke: 900 × 700 × 400 mm
		Dynamometer	Fx, Fy, Fz:-250 ~ 250 N
Feed rate (mm/min)	200, 400, 600, 800	Temperature measuring instrument	50–3000 °C
Cutting depth (μm)	20, 40, 60, 80	Scanning electron microscope	Maximum magnification: 50,000 times

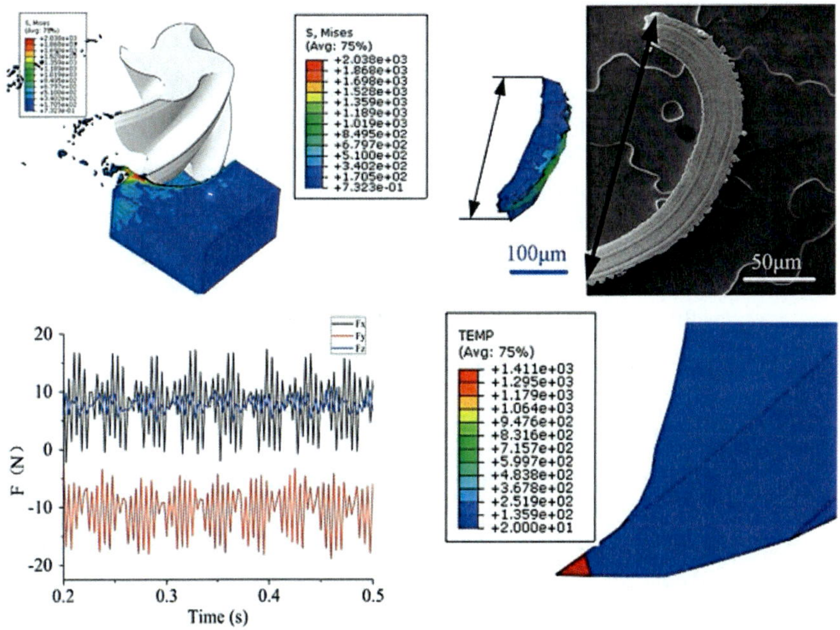

Fig. 6.13 Comparison of simulation and experimental chip forming data and drawing of cutting force and temperature

chip forming data and the drawing of cutting force and temperature when $n = 16{,}000$ r/min, $f = 400$ mm/min, $a_p = 40$ μm. The simulation and experimental parameters are the data extraction of chip separation, cutting force and temperature of simulation and experimental cutting Ti6Al4V after the end of the experiment. The results show that the simulation results of the established tool model are very close to the chip shape and length values of the experimental results. The results of simulation using the tool and cutting model attribute parameters are credible [6–12].

6.3.3 Analysis of Cutting Force Comparison between Simulation and Experiment

10 groups of Single Factors Are Selected for Calculation of Simulation and Experimental Cutting Forces. Finally, the Results Are Compared, as Shown in Table 6.5. The Error is Calculated as Follows. Equation (6.14)

$$Error = \frac{\left|x_{Sim} - x_{Exp}\right|}{x_{Exp}} \times 100\% \tag{6.14}$$

Figure 6.14 shows the relative error percentage between simulated and experimental cutting forces. Table 6.5 shows that the simulation errors are small except for the larger errors of the points (test numbers 2, 5 and 6).

In addition, 10 groups of simulation and experimental data can be obtained. Based on the accurate modeling of milling cutter model, the error of cutting force in three directions of simulation data and experimental data is less than 15%. It shows that the consistency of milling cutter model and actual cutter has a direct impact on the simulation results, and the data is more reliable [12].

6.3.4 Analysis of Cutting Temperature Comparison between Simulation and Experiment

Figure 6.15 shows the simplified schematic diagram of milling temperature model. According to the model structure of milling cutter and cutting principle, the milling cutter can be divided into numerous units along the spiral edge when analyzing the cutting temperature produced in milling process. The cutting process of each unit can be simplified as oblique cutting. Figure 6.15a and b show the temperature generated in the cutting area during milling. Results showed that temperature rise mainly occurred in three areas during milling, as shown in Fig. 6.15c including cutting area of rake face where temperature rose resulting from friction behavior between chip and rake face; cutting area of flank face where temperature increased due to friction between flank face and the workpiece; and plastic deformation area of the shear surface where

Table 6.5 Main machining force comparison between experimental, simulated and statistical values

NO	Cutting parameters			Simulation (N)			Experiments (N)			Error (%)		
	N r/min	F mm/min	a_p mm	Fx	Fy	Fz	Fx	Fy	Fz	Fx	Fy	Fz
1	4000	400	0.04	25	−25	21	24	−23	19	4.2	8.7	10.5
2	8000	400	0.04	17	−22	11.5	18	−20	10	5.6	10	15
3	12,000	400	0.04	22	−16	9	22.8	−17	8	3.5	5.9	12.5
4	16,000	400	0.04	16	−17	10	18.5	−15.8	10.5	13.5	7.6	4.8
5	16,000	200	0.04	11	−12.5	9	12.3	−11.6	10.5	10.6	13.6	14.3
6	16,000	600	0.04	12	−10	11	13.5	−9	11.8	11.1	11.1	14.4
7	16,000	800	0.04	13.5	−10.5	11.5	15	−12	12.5	10	12.5	8
8	16,000	400	0.02	2.9	−1.8	2	3	−2	2.3	3.3	10	13
9	16,000	400	0.06	24	−20.5	18	26.2	−22	16	8.4	6.8	12.5
10	16,000	400	0.08	25	−20	20	24	−23	18	4.2	13	11.1

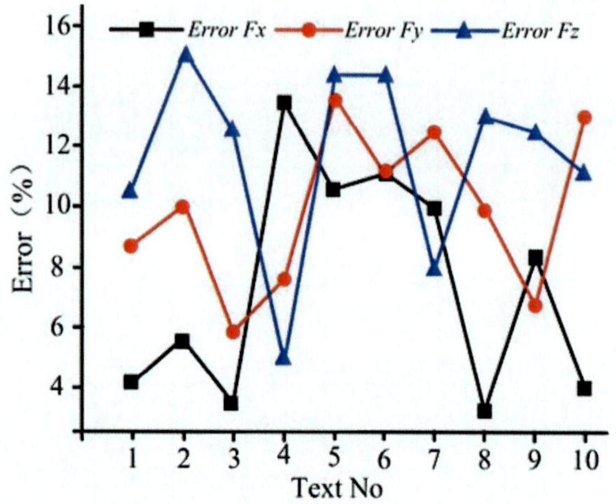

Fig. 6.14 Relative error comparison between simulations and experiments

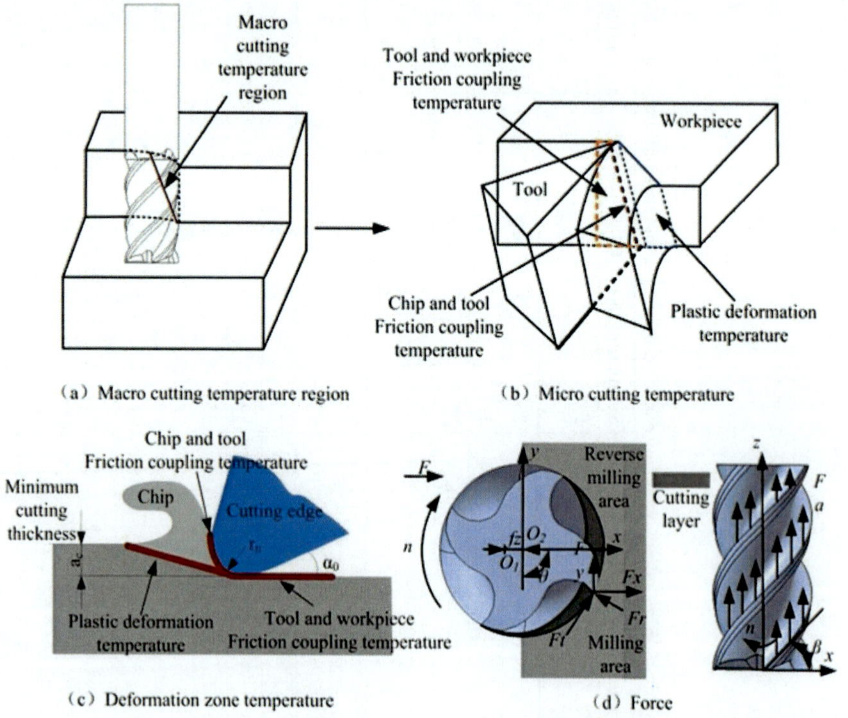

Fig. 6.15 Milling temperature model: **a** Macro cutting temperation; **b** Micro cutting temperature; **c** Deformation zone temperature; **d** Force

Table 6.6 Temperature comparison between experimental, simulated and statistical values

Std Order	Cutting parameters			Simulation	Experiments	Error (%)
	n (r/min)	F (mm/min)	a_p (mm)	Temperature (°C)	Temperature (°C)	
1	4000	400	0.04	90.26	100	9.7
2	8000	400	0.04	286.1	280	2.1
3	12,000	400	0.04	582.8	601	3
4	16,000	400	0.04	1411	1499	5.9
5	16,000	200	0.04	342.4	351	0.3
6	16,000	600	0.04	1038	1022	1.6
7	16,000	800	0.04	1683	1657	1.6
8	16,000	400	0.02	116.3	126	7.7
9	16,000	400	0.06	1156	1170	1.2
10	16,000	400	0.08	2299	2267	1.4

temperature increase occurred because of shear slip of the metal plastic deformation [3].

In the milling process, temperature was found to change with cutting thickness because the instantaneous cutting thickness of the tool spiral edge varied. As the milling cutter rotated during machining, the corresponding cutting thickness being machined by spiral edge was different. Meanwhile, the temperature streamline in the temperature rise area was non-uniform, and the cutting temperature along the helix changed continuously. Therefore, the temperature conduction of the spiral edge in the process of machining can be simplified as a length change.

10 groups of Single Factors Are Selected for Calculation of Simulation and Experimental Cutting Temperature. Finally, the Results Are Compared, as Shown in Table 6.6. The Error is Calculated as Eq. (6.15).

$$Error = \frac{\left|x_{Sim} - x_{Exp}\right|}{x_{Exp}} \times 100\% \tag{6.15}$$

Figure 6.16 shows the relative error percentage between simulated and experimental cutting temperature. Figure 6.16 shows that the simulation errors are small except for the larger errors of the points (test numbers 1 and 8).

In addition, 10 groups of simulation and experimental data can be obtained. Based on the accurate modeling of milling cutter model, the error of cutting temperature in three directions of simulation data and experimental data is less than 10%. It shows that the consistency of milling cutter model and actual cutter has a direct impact on the simulation results, and the data is more reliable.

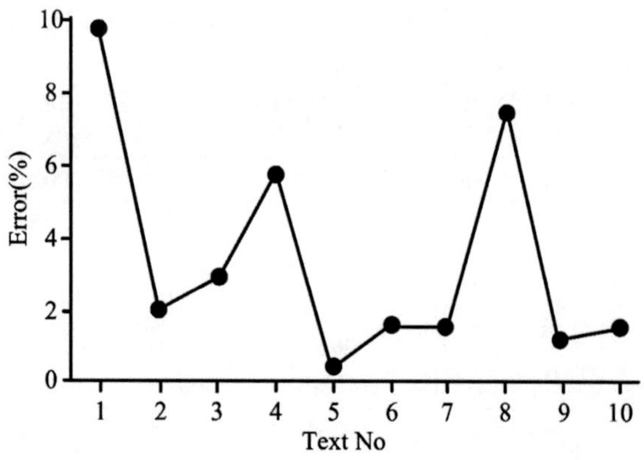

Fig. 6.16 Relative error comparison between simulations and experiments

6.4 Conclusions

The research on tool model building method and simulation model can provide basic technical reference for titanium alloy machining. In this study, Ti6Al4V was taken as the object of research to study the characteristics in cutting Ti6Al4V by combining Abaqus with processing experiments. The reliability of the cutting simulation results was investigated by carrying out experiments. The conclusions are as follows:

(1) Under the condition that the dimension of the tool model is consistent with that of the experimental tool, the close degree between the simulation data and the experimental data is obviously improved. This kind of simulation milling cutter model can be established flexibly by using the tool model building method in this paper.
(2) Within the range of parameters and material constitutive properties selected in the experiment, the error of cutting force data of cutting simulation and machining experiment is less than 15%. The experiment proves that the data obtained by Abaqus simulation is reliable.
(3) Within the range of parameters and material constitutive properties selected in the experiment, the error of cutting temperature data of cutting simulation and machining experiment is less than 10%. The experiment proves that the data obtained by Abaqus simulation is reliable.

Acknowledgements The authors sincerely thank Professor Hua-dong Yu of Changchun University of Science and Technology for his critical discussion and reading during manuscript preparation. This work was supported by "Supported by national basic scientific research program" (JCKY2017208B006).

References

1. Cai, M., Gong, Y., Yu, N., Cao, Q.: Experimental study on surface quality of single crystal DD98 in micro-milling processes. China Mech. Eng. **28**(11), 1261–1265 (2017)
2. Pei, Z.Q., Zhao, J.: Parameterized Design and Optimization of Solid Carbide End Mills. Shan Dong University, Shan Dong (2012) (in Chinese)
3. Sun, X.Y., Zhang, G.J., Zheng, J.M., Zhang, B., Zhang, T.L.: Modeling of spiral groove grinding for integral carbide end milling cutter. Tool Technol. **53**(10), 61–64 (2019). (in Chinese)
4. Arisoy, Y.M., Özel, T.: Prediction of machining induced microstructure in Ti-6Al-4V alloy using 3-D FE-based simulations: effects of tool micro-geometry, coating and cutting conditions. J. Mater. Process Technol. **220**, 1–26 (2015)
5. Wang, Q.Q., Liu, Z.Q., Wang, B., Song, Q.H., Wan, Y.: Evolutions of grain size and micro-hardness during chip formation and machined surface generation for Ti-6Al-4V in high-speed machining. Int. J. Adv. Manuf. Technol. **82**, 1725–1736 (2016)
6. Rahman, M.A., Rahman, M., Kumar, A.S., Lim, H.S.: CNC Microturning: an application to miniaturization. Int. J. Mach. Tools Manuf. **45**(6), 631–639 (2005)
7. Rahman, M.A., Rahman, M., Kumar, A.S.: Influence of cutting edge radius on small scale material removal at ultra-precise level. Procedia CIRP **77**, 658–661 (2018)
8. Rahman, M.A., Rahman, M., Kumar, A.S.: Material perspective on the evolution of micro- and nano-scale cutting of metal alloys. J. Micromanuf. **1**(2), 97–114 (2018)
9. Zaman, M.T., Kumar, A.S., Rahman, M., Sreeram, S.: A three-dimensional analytical cutting force model for micro end milling operation. Int. J. Mach. Tools Manuf **46**(3–4), 353–366 (2006)
10. Mamedov, A., Layegh, K.S.E., Lazoglu, I.: Instantaneous tool deflection model for micro milling. Int. J. Adv. Manuf. Technol. **79**, 769–777 (2015)
11. Wu, X., Li, L., He, N.: Investigation on the burr formation mechanism in micro cutting. Precision Eng. **47**, 191–196 (2017)
12. Jing, X.B., Lv, R.Y., Chen, Y., Tian, Y.L., Li, H.Z.: Modelling and experimental analysis of the effects of run out, minimum chip thickness and elastic recovery on the cutting force in micro-end-milling. Int. J. Mech. Sci. **176**, 105540 (2020)

Chapter 7
Prediction of Bearing Remaining Useful Life Based on LSTM Network

Xin Wang, Jiazheng Guo, Jian Wang, Changying Liu, and Chuang Du

Abstract Bearing is one of the important components in the mechanical system, and the reliability of the bearing is significant. The paper aimed to simulate remaining useful life (RUL) prediction process in the engineering project. In the process of predicting the RUL of the bearing, the noisy in the data leads to the over-fitting phenomenon. In order to solve the problem, it is necessary to train a Variational Auto-Encoder (VAE) for denoising short time series before the prediction starts. In the paper, the VAE was selected to denoise the bearing performance degradation index sequence. The denoised data curve was better than the original, which was smooth. The denoised data were used to solve the bearing RUL prediction problem based on LSTM network, and the over-fitting phenomenon did not occur. Compared with the prediction results of directly using Support Vector machine Regression (SVR) and Artificial Neural Network (ANN), the deviation of the prediction based on Long Short-term Memory (LSTM) results from the observation values were smaller, and the method can be applied in the engineering project.

Keywords Variational Auto-Encoder · LSTM network · Bearing remaining · Useful life prediction

7.1 Introduction

Bearing RUL prediction is one of the important components in Prognostic and Health Management (PHM) system. In recent decades, varieties of methods have been proposed to study bearing RUL prediction. The earliest appeared was to predict the RUL of bearings based on statistical analysis. On the one hand, the formula for

X. Wang · J. Guo · C. Du
Key Laboratory of CNC Equipment Reliability, Ministry of Education, School of Mechanical and Aerospace Engineering (SMAE), Jilin University, Changchun, People's Republic of China

J. Wang · C. Liu (✉)
School of Instrument Science and Electrical Engineering, Jilin University, Changchun, People's Republic of China
e-mail: liuchangy@jlu.edu.cn

J. Xu and K. M. Pandey (eds.), *Mechanical Engineering and Materials*,
Mechanisms and Machine Science 100,
https://doi.org/10.1007/978-3-030-68303-0_7

calculating the remaining life of bearings was proposed based on the L-P theory. After continuous revision, ISO proposed a standardized formula for calculating the life of bearings [1]. Using the formula to predict bearing life results only represented the average bearing life, and it cannot reflect the extreme bearing life. On the other hand, Weibull proposed the Weibull model for bearing life prediction, which was still used in the field of life prediction. The prediction of bearing life using the Weibull model required a large amount of testing data as support, and the prediction results will be accurate [2–4]. In the past thirty years, the finite element analysis method has been applied to the prediction of the RUL of the bearing. By the finite element analysis, the stress and strain datasets for outer race, inner race and ball are obtained and used to calculate the fatigue life [5]. In recent years, with the rise of artificial intelligence, the combination of experiments and artificial intelligence algorithms to solve bearing life problems has been a popular solution. A novel prediction method based on MSPSO and MK-LSSVM was proposed for the life prediction of rolling element bearing by Zhang. The analysis results of actual rolling bearing vibration data verify the effectiveness of the method in rolling bearing life prediction [6]. The DPNN was used to predict bearing RUL. The experimental results show that DPNN can predict effectively the RUL of bearing with high prediction accuracy and strong robustness [7]. An online method of RUL prediction was proposed by the Kolmogorov–Smirnov test theory and LSTM network [8]. In the above methods, when the performance degradation index changes abruptly, the accuracy of prediction results will decrease.

In the actual prediction process, the problems of unpredictability and decreased prediction accuracy caused by sudden changes in performance degradation indicators, the paper proposed a denoising method of bearing performance index sequence based on VAE, and the smooth bearing performance index sequence facilitated the LSTM network to predict the RUL of the bearing. The RUL prediction results based on the LSTM network were compared with the results using SVR and ANN, which verified that the accuracy of the RUL prediction results based on the LSTM network was higher than others. The methods in the paper not only improved the accuracy of bearing RUL prediction, but also provided reference for solving other parts RUL prediction problems and the reliability problem of mechanical system.

7.2 Theory

7.2.1 LSTM Network

LSTM network is a recurrent neural network for time series prediction. The structure diagram of the LSTM unit was shown in Fig. 7.1. The unit includes input gate, output gate and forget gate [9–11].

The new signal Xt and the previous unit output ht-1 were used as input to the unit. After the forgetting gate operation, the information of the previous unit was

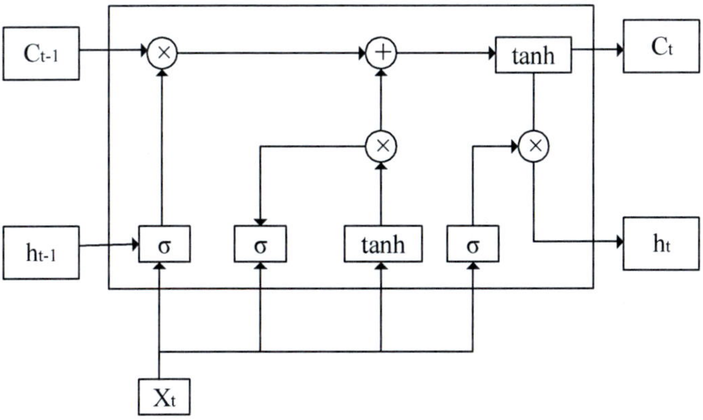

Fig. 7.1 LSTM network unit construction

transferred to the unit, as shown in Eq. (7.1).

$$f_t = \sigma(W_f \cdot [h_{t-1}, x_t] + b_f) \tag{7.1}$$

The unit information was updated after input gate operation, as shown in Eqs. (7.2), (7.3) and (7.4).

$$i_t = \sigma(W_i \cdot [h_{t-1}, x_t] + b_i) \tag{7.2}$$

$$\tilde{C}_t = \tan h(W_c \cdot [h_{t-1}, x_t] + b_c) \tag{7.3}$$

$$C_t = f_t * C_{t-1} + i_t * \tilde{C}_t \tag{7.4}$$

The output information was determined after the output gate operation, as shown in Eqs. (7.5) and (7.6).

$$o_t = \sigma(W_o \cdot [h_{t-1}, x_t] + b_o) \tag{7.5}$$

$$h_t = o_t \cdot \tanh(C_t) \tag{7.6}$$

where: f_t, W_f, and b_f are the parameter of forgetting gate,
i_t, W_c, W_i, b_i, b_c are the parameter of input gate,
o_t, W_o, b_o are the parameter of output gate,
h_t, h_{t-1}, X_t, C_t, C_{t-1} are the input and output of LSTM network.

7.2.2 *VAE*

VAE is a deep generation model, which is mainly used for data compression and generation in the field of image processing. Its structure is shown in Fig. 7.2. The variational autoencoder is composed of an encoder and a decoder. The principle is that the data vector x is input into the encoder and compressed into a low-dimensional hidden vector z. The expression of the hidden vector z was shown in Eq. (7.7). The low-dimensional hidden vector z is mapped to the high-dimensional space through the decoder to form a vector similar to the data vector x [12–14].

$$z = z_{mean} + z_{sigma} \cdot N(0, 1) \tag{7.7}$$

The encoder and the decoder can choose ordinary neural network or deep learning neural network. The paper chose LSTM network as the encoder and the decoder, because the bearing performance degradation index sequence is time series.

7.3 Experiment Data

The experiment data came from the bearing life test data of Xi'an Jiaotong University. A total of 3 different operating conditions were set in the accelerated degradation experiments, and 5 bearings were tested under each operating condition. The operating conditions include:

Conditions 1: 2100 rpm (35 Hz) and 12 kN.
Conditions 2: 2250 rpm (37.5 Hz) and 11 kN.
Conditions 3: 2400 rpm (40 Hz) and 10 kN.

During the experiment, the vibration acceleration in the horizontal direction and the vibration acceleration in the vertical direction of the bearing were collected. The RMS of bearing vibration signals did not change significantly before the failure period, so the paper used the K-Means algorithm to find the starting sampling point of the bearing failure period. The K-Means algorithm was an unsupervised learning algorithm. Compared with other clustering algorithms, its advantage is that it does

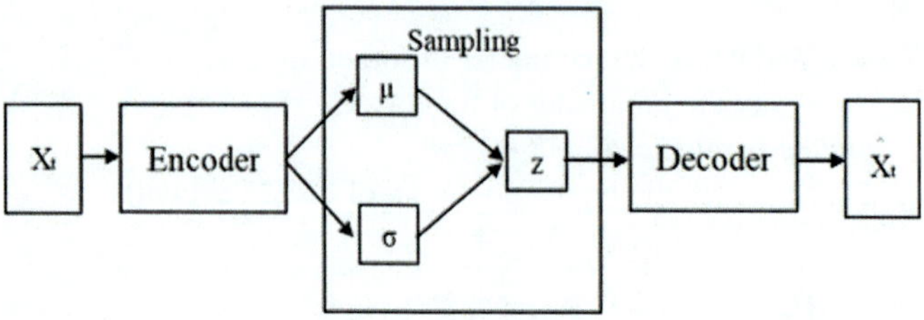

Fig. 7.2 VAE structure

Table 7.1 Sampling points at which bearings began to fail and the corresponding RUL

Conditions	No	Sample point	RUL (min)
Conditions 1	Bearing1_1	94	29
	Bearing1_2	129	32
	Bearing1_3	128	25
	Bearing1_5	40	12
Conditions 2	Bearing2_1	467	24
	Bearing2_3	500	21
	Bearing2_4	32	10
Conditions 3	Bearing3_1	2519	19
	Bearing3_3	345	8
	Bearing3_5	95	19

Table 7.2 The training set and the testing set

Training set	Testing set
Bearing1_1, Bearing1_2, Bearing1_3, Bearing2_1, Bearing2_3, Bearing2_4, Bearing3_3, Bearing3_5	Bearing1_5, Bearing3_1

not need to mark the training data variables. The K-Means algorithm divided all sampling points into 4 categories. The first sampling point of the third category (the expiration period) was used as the initial sampling point of each bearing failure period. The results were shown in Table 7.1. In the paper, the starting sampling point of each bearing failure period was taken as the starting point of prediction.

There were 209 sampling points in the dataset. The dataset was divided into training set and testing set, including 176 sampling points in the training set and 33 sampling points in the testing set. The details were shown in Table 7.2.

7.4 Methods

7.4.1 The RUL Prediction Using LSTM Network Directly

The bearing RUL prediction method based on LSTM network was that the bearing performance degradation index was input to the LSTM network to complete the prediction of bearing RUL. The specific process was as follows:

a. The training set and the testing set were constructed and normalized. In the experiment data, the training set and the testing set were normalized.
b. The network was trained. The network-related parameters were set, the 'tanh' function was chosen as the activation function, the 'mse' function was chosen as

the loss function, and the 'adam' algorithm was chosen as the optimizer. There were 20 hidden nodes in the network, the input tensor size was (176, 10, 2), and the number of training steps was 400.

c. Bearing RUL prediction. The testing set inputs the trained network to complete bearing life prediction.

7.4.2 *The RUL Prediction Combined LSTM Network with VAE*

In order to simulate the actual prediction process, the vibration signals of bearing were divided into several sections, the length of which was 10. Other denoising methods, for example, EMD, EEMD, CEEMD, cannot handle this kind of short time data, so the VAE model was used to denoise in the paper. The training set data before the denoised in Table 7.2 and the data that was denoised by VMD formed a new training set, and it was used to train the VAE to obtain a denoising model.

The setting of the VAE model was as follows: the encoder input size was (176, 10, 2), the encoder hidden node was 5, the encoder output size was (176, 3, 2), the encoder loss function was the cross-entropy function, and the optimizer was 'rmsprop' algorithm, and the training steps were 800. The decoder settings were the same as the encoder.

The data used for bearing RUL prediction was processed by the VAE, and it was input into the bearing RUL prediction model to complete the bearing RUL prediction.

7.5 Result

7.5.1 *The Result of Bearing RUL Prediction based on LSTM*

According to the above process, the bearing lives of Bearing3_1 and Bearing1_5 were predicted by LSTM network. The network training effect was shown in Fig. 7.3.

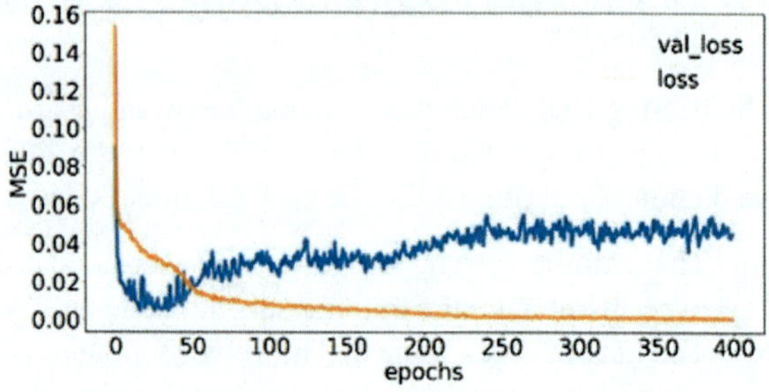

Fig. 7.3 The loss of training set and testing set during the training process

The training set error continued to decrease and then stabilized, but the loss of testing set during training process began to decreases, and then rises. This phenomenon indicated the overfitting of the model. The reason for overfitting was that the training set was noisy. Therefore, the training set and the testing set need to be denoised.

7.5.2 The Result of Bearing RUL Prediction Combined LSTM Network and VAE

Aiming at the phenomenon of overfitting, the paper used the VAE to denoise training data. The denoising results of the testing set were shown in Fig. 7.4. The denoising curve was smoother than the original data curve, and the denoised curve based on VAE was closer to original data than the denoised curve based on VMD, which indicated that the VAE model had a better denoising effect than VMD model. The training set after denoising was applied for the prediction of bearing RUL.

The prediction results of the bearing RUL based on LSTM network were shown in Fig. 7.5. The predicted results were roughly distributed on both sides of the real results, which indicated that the predicted results were valid.

In the RUL prediction process, the results of the training set loss and testing set loss were shown in Fig. 7.6. The results showed that the model had no underfitting and overfitting.

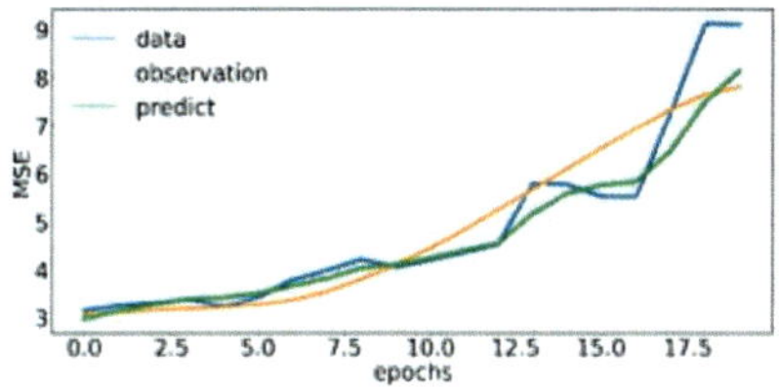

(a) Bearing3_1 denoising result of vertical vibration signal RMS values

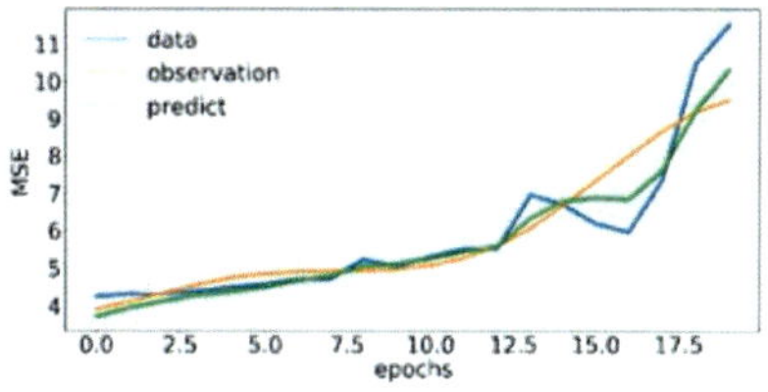

(b) Bearing3_1 denoising result of horizontal vibration signal RMS values

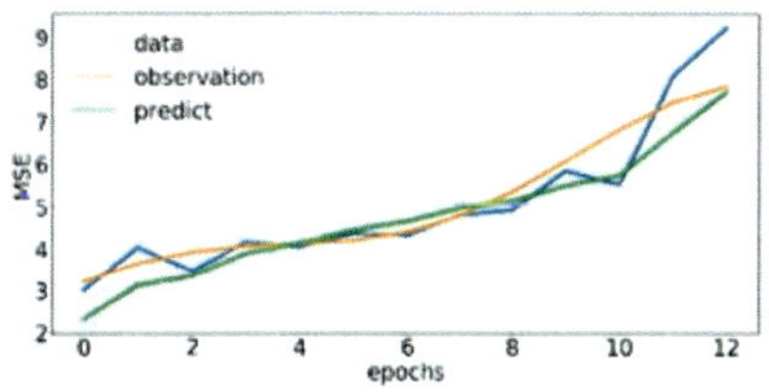

(c) Bearing1_5 denoising result of vertical vibration signal RMS values

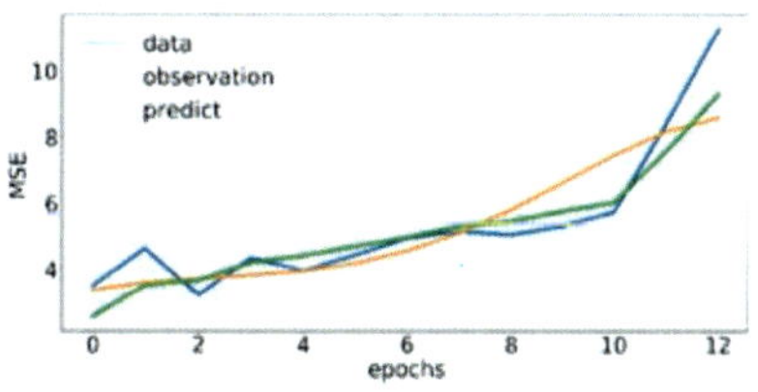

(d) Bearing1_5 denoising result of horizontal vibration signal RMS values

Fig. 7.4 The testing set data denoising results

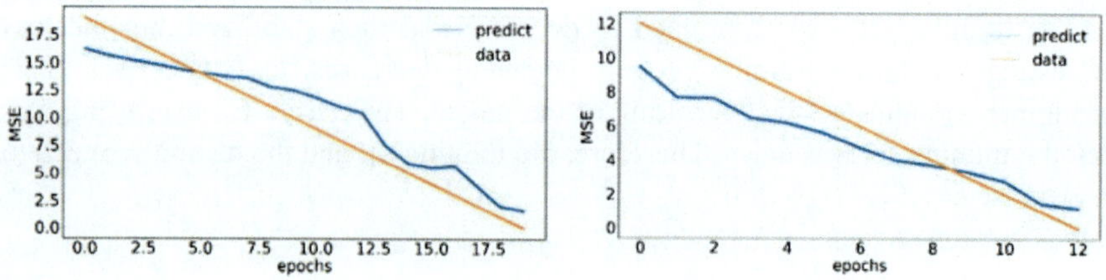

(a) Bearing3_1 RUL prediction results based on LSTM

(b) Bearing1_5 RUL prediction results on LSTM

Fig. 7.5 The prediction results of bearing RUL based on LSTM

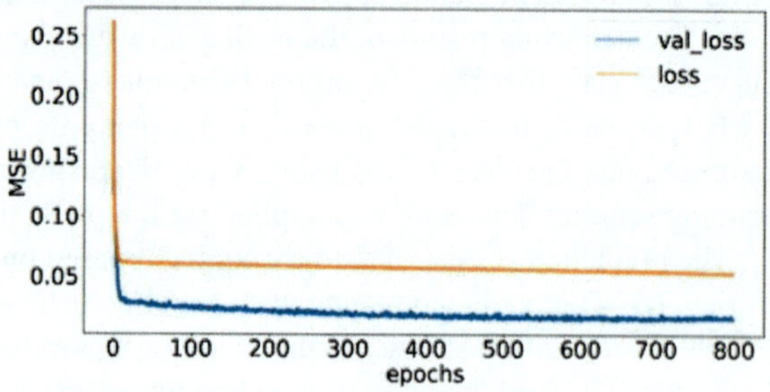

Fig. 7.6 The loss of training set and testing set during the prediction process based on LSTM

7.5.3 Compared with the Result Based SVR and ANN

In order to prove the accuracy of the result based on LSTM, the SVR and ANN were chose to predicted RUL, and all results were analyzed [4, 15]. The results based on SVR model and ANN were shown in Fig. 7.7. The prediction results were distributed around the observation values, and the model had no over-fitting phenomenon, which indicated that the models can also solve the RUL prediction problem.

The mean square error (MSE) reflected the abnormal values in the prediction results. The mean absolute error (MAE) reflected the error between the predicted values and the observation values. The MSE and MAE between the predicted values of the three methods and the observation values were shown in Table 7.3, which showed that bearing RUL prediction based on LSTM was superior to other prediction methods.

7.6 Conclusions

The paper predicted the bearing RUL based on the LSTM network. In order to simulate the actual prediction process, the vibration signals of bearing were divided into several sections. When the prediction model was trained, the model had overfitting. Therefore, the paper used the VAE to pre-process the bearing data, it had a better

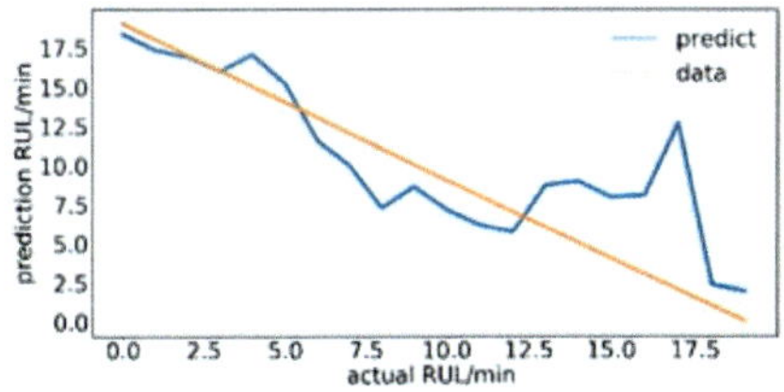

(a) Bearing 3_1 RUL prediction results based on SVR

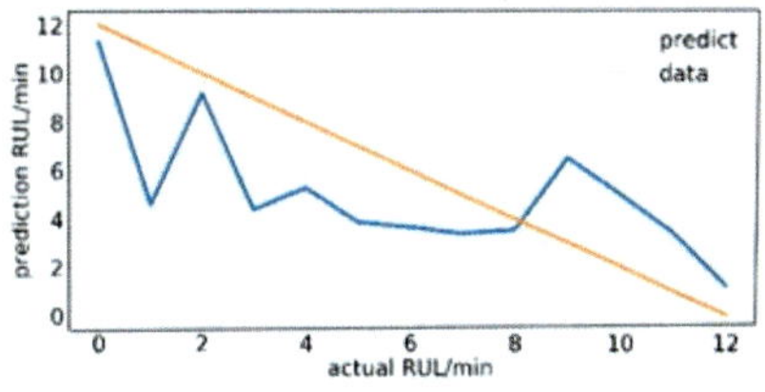

(b) Bearing 1_5 RUL prediction results based on SVR

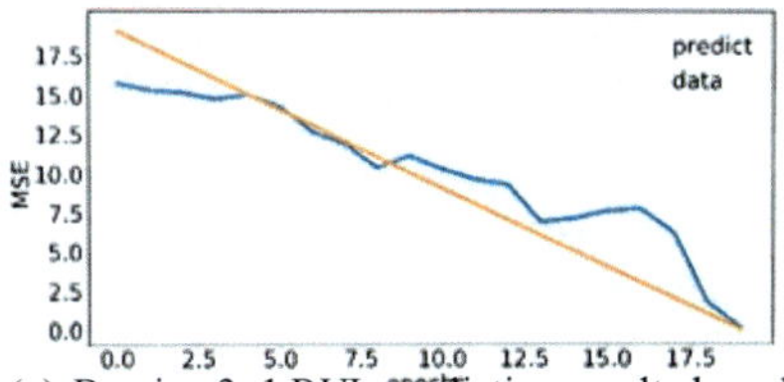

(c) Bearing 3_1 RUL prediction results based on ANN

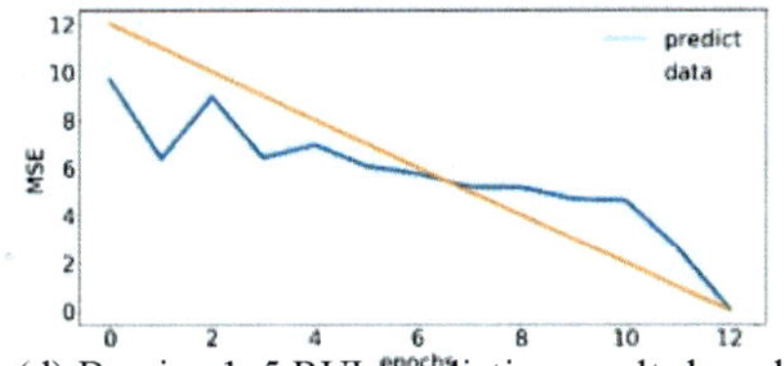

(d) Bearing 1_5 RUL prediction results based on ANN

Fig. 7.7 Bearing RUL prediction results based on the SVR model and ANN

Table 7.3 The MSE and MAE between the predicted values and the observation values

Method	MSE		MAE	
	Bearing3_1	Bearing1_5	Bearing3_1	Bearing1_5
LSTM	3.54	3.07	1.52	1.48
SVC	10.98	9.06	2.39	2.54
ANN	4.61	3.88	2.07	1.52

denoising effect than VMD model. The bearing RUL prediction was completed by LSTM network. SVR and ANN were used to predict the RUL of the bearing, and then comparing the prediction RUL based on the three methods with the actual RUL, the prediction results based on LSTM had a smaller the MSE between prediction values and observation values. The method solved bearing RUL offline prediction, in the future, it can be combined with real-time monitoring system to solve the problem of bearing life online prediction.

References

1. Talllan, T.E.: Weibull Distribution of Rolling Contact Fatigue Life and Deviations. ASLE Trans. **5**, 183–196 (1962)
2. Zhang, F.S., Liu, T.T., Liu, J.T., Cui, F.K.: Research on bearing life prediction based on three parameters Weibull distribution. Adv. Mater. Res. **136**, 162–166 (2010)
3. Yin, F.L., Wang, Y.S., Zhang. C.H,, Zhang, X.P.: An improved parameter estimation method for three-parameter weibull distribution in the life analysis of rolling bearing. Adv. Mater. Res.

569, 442–446 (2012)
4. Ben Ali, J., Chebel-Morello, B., Saidi, L., Malinowski, S., Fnaiech, F.: Accurate bearing remaining useful life prediction based on Weibull distribution and artificial neural network. Mech. Syst. Signal Process. **56–57**, 150–172 (2015)
5. Deng, S., Hua, L., Han, X.H., Huang. S.: Finite element analysis of fatigue life for deep groove ball bearing. Proc. Inst. Mech. Eng. Part L: J. Mater.: Design Appl. **227**(1), 70–81 (2013)
6. Zhang, Y., Tang, B.P., Xiong, P.: Rolling element bearing life prediction based on multi-scale mutation particle swarm optimized multi-kernel least square support vector machine. Chinese J. Sci. Instrum. **37**, 2489–2496 (2016)
7. Hong, S., Yin, J.W.: Remaining Useful Life Prediction of Bearing Based on Deep Perceptron Neural Networks. In: Proceedings of the 2018 2nd International Conference on Big Data and Internet of Things (BDIOT 2018), pp. 175–179. Association for Computing Machinery, New York, NY, USA (2018)
8. Ge, Y., Guo, L.Z., Dou, Y.: Remaining useful life prediction of machinery based on K-S distance and LSTM neural network. Int. J. Performability Eng. **15**, 895–901 (2019)
9. Jiang, J.R., Lee, J.E., Zeng, Y.M.: Time series multiple channel convolutional neural network with attention-based long short- term memory for predicting bearing remaining useful life. Sensors **20**(1), 166 (2020)
10. Wang, Q., Bu, S.Q., He, Z.Y.: Achieving Predictive and Proactive Maintenance for High-Speed Railway Power Equipment with LSTM-RNN. IEEE Trans. Ind. Inf. **16**(10), 6509–6517 (2020)
11. Essien, A., Giannetti, C.: A Deep Learning Model for Smart Manufacturing Using Convolutional LSTM Neural Network Autoencoders. IEEE Trans. Ind. Inf. **16**(9), 6069–6078 (2020)
12. Walker J., Doersch C., Gupta A., Hebert M.: An Uncertain Future: Forecasting from Static Images Using Variational Autoencoders. In: Leibe B., Matas J., Sebe N., Welling M. (eds) Computer Vision – ECCV 2016. ECCV 2016. Lecture Notes in Computer Science, vol 9911. Springer, Cham (2016)
13. Ohno, H.: Auto-encoder-based generative models for data augmentation on regression problems. Soft Comput. **24**, 7999–8009 (2020)
14. Xie, R.M., Jan, N.M., Hao, K.R., Chen, L., Huang, B.: Supervised variational autoencoders for soft sensor modeling with missing data. IEEE Trans. Industr. Inf. **16**(4), 2820–2828 (2020)
15. Wei, J.W., Dong, G.Z., Chen. Z.H.: Remaining useful life prediction and state of health diagnosis for lithium-ion batteries using particle filter and support vector regression. IEEE Trans. Ind. Electron. **65**(7), 5634–5643 (2018)

Chapter 8
Uncertainty Quantification of Spray Nozzle Fatigue Analysis Under Transient Load

Zhenyu Liu, Lijuan Li, Liping Zhang, Hui Li, Xuejiao Shao, Lu Jiang, and Ying Zhang

Abstract In this paper, the randomness concept of input parameters is introduced into the design of nuclear equipment, and the uncertainty analysis of input parameters is carried out for fatigue analysis. Taking the pressurizer spray nozzle as the research object, considering the random fluctuation of temperature and pressure of the transient load, the reliability and distribution of the fatigue usage are obtained by performing fatigue analysis according to the fatigue analysis method of nuclear power equipment with Monte Carlo sampling method. Taking section No.3 as an example, the thermal analysis and stress analysis models are established. Through the parameter sensitivity analysis, the relative sensitivity of the fatigue usage to each input parameter is obtained. From the results, it is clear which parameters have a significant impact on the results. The results can further optimize the structural design of the nuclear power equipment and improve its fatigue resistance, which has great application value to the life extension of nuclear power equipment.

Keywords Uncertainty quantification · Nuclear equipment · Parameter sensitivity · Fatigue reliability · Reactor structural mechanics

8.1 Introduction

The main structures of the nuclear power equipment are very complicated, which are subjected to the fluctuating load from various transient conditions during the service life thus generating alternating stress. Damage occurs at one or some high stress areas and then accumulates, resulting in performance degradation, crack initiation, expansion until complete fracture. As a result, fatigue damage occurs, causing serious accidents. Due to severe safety requirements, the effects of these loads must be fully considered in the reactor design, because fatigue damage can occur even when the stress is much less than the yield strength of the material [1]. However,

Z. Liu (✉) · L. Li · L. Zhang · H. Li · X. Shao · L. Jiang · Y. Zhang
Science and Technology On Reactor System Design Technology Laboratory, Nuclear Power Institute of China, No. 328 Changshun Avenue, Chengdu 610200, Sichuan, P. R. China
e-mail: zhenyuliuchn@foxmail.com

J. Xu and K. M. Pandey (eds.), *Mechanical Engineering and Materials*,
Mechanisms and Machine Science 100,
https://doi.org/10.1007/978-3-030-68303-0_8

for the fatigue analysis of structures, the deterministic method is used in the current nuclear equipment design [2, 3]. It is considered that the material properties, structural dimensions and bearing loads are all specific values, regardless of the random fluctuations of the input, thus not taking into account the calculation error caused by the uncertainty of the design parameters.

In order to reflect the fatigue performance of the analyzed object more realistically, it is necessary to consider the uncertainty of the input parameters [4]. The uncertainty quantification study of input parameters of fatigue analysis can not only analyze the failure probability, which can guide the structural design according to the allowable failure probability value, but also fully consider the sensitivity of each parameter to further optimize the structural design [5–7].

In this paper, the nuclear power equipment pressurizer spray nozzle is taken as the research object, and the uncertainty analysis of the input parameters of fatigue analysis is carried out. The randomness is introduced into the transient fluctuation, and the sensitivity of the transient parameters is analyzed, thus proposing advice for structural optimization. The research results show that the uncertainty analysis of the parameters can obtain the influence degree of each parameter on the analysis results, which can greatly guide the structure design and optimization.

8.2 Method

8.2.1 Nuclear Power Equipment Fatigue Analysis Method

Since the structure of nuclear power equipment is mostly designed by the finite life design method (safe life design method), the amplitude of the alternating stress is generally higher than the yield limit of the material. Therefore, in the fatigue analysis of nuclear power equipment, the maximum stress may not be considered only, but the total fatigue damage should be estimated according to a certain cumulative damage theory. The alternating load amplitude of the nuclear power equipment structure is sometimes non-constant. If the amplitude of the alternating stress is always calculated according to the maximum amplitude, it is too conservative. An approximate engineering treatment method is to use the linear fatigue cumulative damage theory. The fatigue damage of the material is independent under various stresses and the total damage can be linearly accumulated. The most representative method of these is the PalmgrenMiner theory, and Eq. (8.1) defines Miner damage rule:

$$U = \sum \frac{n_i}{N_i} \tag{8.1}$$

Among them, n_i is the cycle under the alternating stress; N_i is the allowable cycle under the alternating stress, which is determined by the *S–N* curve, U is the fatigue usage. Figure 8.1 is a schematic diagram of linear accumulation of fatigue damage.

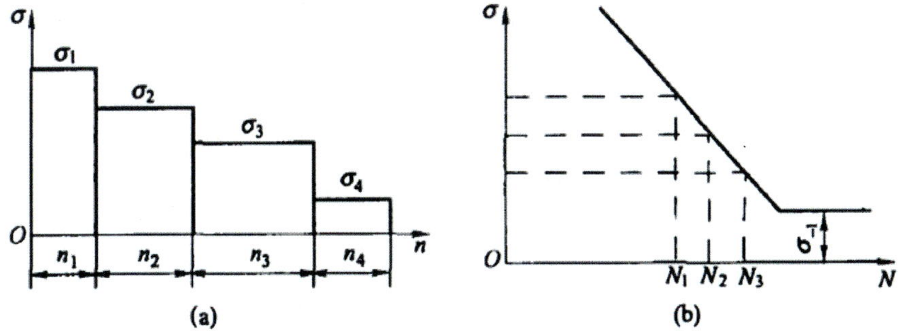

Fig. 8.1 The diagram of linear accumulation of fatigue damage

In the figure, Fig. 8.1a shows the alternating stress and Fig. 8.1b shows the *S–N* curve.

According to the requirements of fatigue analysis in ASME [8], the transient calculation of the model is first carried out to obtain the stress time history of each position on the structure. Then, the peak point of the stress is selected as the event point in the fatigue analysis. The stress at these event points is used as the input value of the event storage stress. Combining with the corresponding cycle number and the material fatigue curve, the fatigue analysis coefficient can be calculated by ANSYS fatigue analysis module. The calculation method of the fatigue usage is as follows.

Each point is subjected to a plurality of cyclic stresses, and the cumulative fatigue usage is calculated by the following steps (1) to (5):

(1) Select the extreme point of each stress parameter as the event point and the load point for each transient stress, and record each transient as *1,2,3,…,n*, and the corresponding cycle number is specified as $n_1,n_2,n_3,\ldots,n_n$;
(2) Determine the magnitude $S_{a1},S_{a2},S_{a3},\ldots,S_{an}$ of the alternating stress intensity required for the design calculation for each stress cycle;
(3) For the magnitude of the alternating stress intensity in step (2), when each type of cyclic stress is considered separately, the number of permitted cycles is determined according to the design fatigue curve $N_1,N_2,N_3,\ldots,N_n$;
(4) Calculate the fatigue usage using damage theory $U = \sum \frac{n_i}{N_i}$;
(5) The fatigue usage is less than 1.

8.2.2 *MCMC Sampling Technology*

Markov Chain-Monte Carlo (MCMC) [9–11] simulation is a method of determining the probability of failure *Pf* by mass sampling to each random variable, which is also called random sampling method or statistical test method. It is the most intuitive, accurate, and most information-intensive method, which is the most effective method for high-order nonlinear problems of structural reliability calculation.

The theoretical basis comes from the Law of Large Numbers of probability theory:

Let x_1, x_2, ..., x_n be n independent random variables, if they are from the same parent, which have the same distribution and the same finite mean and variance, denoted by μ and σ^2, respectively. Then, for any $\varepsilon > 0$, we have Eq. (8.2) below.

$$\lim_{n\to\infty} P\left(\left|\frac{1}{n}\sum_{i=1}^{n} x_i - \mu\right| \geq \varepsilon\right) = 0 \tag{8.2}$$

Bernoulli Trials: If the probability of occurrence of random event A is $P(\mathrm{A})$, in n independent experiments, the frequency of event A occurs is m, and the frequency is $W(\mathrm{A}) = m/n$. Then for any $\varepsilon > 0$, we have Eq. (8.3) below.

$$\lim_{n\to\infty} P\left(\left|\frac{m}{n} - P(A)\right| < \varepsilon\right) = 1 \tag{8.3}$$

The core content of the Monte Carlo method is to obtain the ratio of the number of structural failures to the total number of samples by a large number of samples, which is the failure probability of the structure. As the number of simulations increases, the Monte Carlo method will gradually become more accurate. Therefore, in structural reliability calculations, Monte Carlo calculations are considered to be quasi-accurate calculations, while the accuracy of the other approximation method is often verified by the Monte Carlo method.

8.2.3 Calculation Model

The pressurizer spray nozzle and the adjacent structure mainly include the upper head, the surfacing layer, the thermowell, the safety end, the spray head connecting pipe and the forging, etc. Considering that the analyzed object has the axisymmetric geometric characteristics, the calculation is performed using a two-dimensional axisymmetric model in ANSYS. Two models are built in total. The first is the thermal analysis model, which is established by the two-dimensional 8-node thermal element (PLANE77). The second is the thermal stress analysis model, which is established by the 8-node harmonic element (PLANE83), which is used to calculate the secondary stress due to thermal gradient and different material expansion coefficients. Both models contain non-pressure boundary portions such as nozzle tips and heat pipes, while also considering the effect of the weld overlay on heat transfer.

In order to eliminate the influence of the boundary effect, the nozzle and the lower part of the cylinder should be extended. Equation (8.4) defines the extension length.

$$L = 2.5\sqrt{R_m t} \tag{8.4}$$

where R_m is the average radius of the nozzle or the head, and t is the corresponding minimum wall thickness. After calculation, the extension length of the nozzle and the head should be at least 0.0582 m and 0.732 m, respectively. Then the actual extension length is 0.08 m and 0.803 m respectively.

The types of loads considered include self-weight, design external loads, LLSE, thermal expansion loads, and normal and upset condition transients. In thermal analysis, the transient temperature is applied to the inner surface, and other surfaces that are not in contact with water or steam are subjected to adiabatic boundary conditions. In the stress analysis, internal pressure is applied to the inner surface of the structure, and the external mechanical load and hydrostatic end force generated by the internal pressure are applied to the top of the extension of the nozzle. And a circumferential and vertical displacement constraint is applied on the bottom of the lower cylinder of the model.

8.3 Result and Discussion

In this paper, section No. 3 is taken as an example to carry out calculations. From the point of view of the pressure, the wall thickness of section No. 3 should be as thick as possible, while from the perspective of thermal stress, in order to avoid excessive thermal gradient of section No. 3, which will greatly affect its fatigue performance, its thickness should be reduced if necessary. Therefore, in order to achieve the balance between the two schemes in the structural design, it is necessary to carry out quantitative study of the transient load uncertainty, including the pressure and temperature, thus calculating the sensitivity of parameters to the fatigue strength using design transient and allowable cycle numbers. Finally, suggestions from the point of view of mechanics can be proposed for structural design and optimization according to the calculation results.

When the fatigue analysis is carried out to the pressurizer spray nozzle in normal and upset conditions, a total of 40 transients are involved in the fatigue calculation, and the fatigue usage of the inner point of section No. 3 is calculated to be 0.37851. After sorting out, transient combinations that contribute most to the final results are obtained, which can be seen in the Table 8.1.

Therefore, the sum of the fatigue usage of transients 4, 7, 10, 12 and 16 is 0.28417, which accounts for 75.08% of the final result. In this paper, randomness is introduced

Table 8.1 Several transient combinations that contribute most to fatigue usage

No	Transient 1	Time Point 1 (s)	Transient 2	Time Point 2 (s)	Cycle number	Fatigue usage
1	10	78.8	10	230.8	54,000	0.19528
2	16	1816.5	16	3609.5	21,000	0.03769
3	7	27.5	7	800.0	2500	0.03046
4	4	1200.0	12	39.0	2000	0.02074

into transient data fluctuations to study the sensitivity of fatigue strength to input parameters, including temperature and pressure.

Transient fluctuations are considered that temperature and pressure data points obey a normal distribution within a certain range. In order to process transient data efficiently, each transient is processed into dimensionless variables. In other words, the value of each data point is divided by the first data point of each transient. Finally, 10 temperature–time curves and 5 pressure–time curves are obtained after the process.

Taking the first data point as an example, the uncertainty quantization assumed in this paper of the parameters is shown in Table 8.2.

The above parameters are sampled using Monte Carlo method. The Latin square sampling method is chosen to avoid sample point repetition, and the sampling frequency is 3000 times. Based on OPTIMUS platform, fatigue analysis process is established for nuclear power equipment, as can be seen in Fig. 8.2, with which a data pool of input random variables is generated. Taking the first stochastic input variable as an example, the probability density function and the cumulative distribution function graph are shown in Fig. 8.3 (Taking the first data point as an example).

Table 8.2 Uncertainty quantization assumed in this paper

Transient number	Temperature 1 (°C)		Pressure (MPa)		Temperature 2 (°C)	
	Average	Standard deviation	Average	Standard deviation	Average	Standard deviation
4	213.89	10.69	2.07	0.10	213.89	10.69
7	213.89	10.69	2.07	0.10	213.89	10.69
10	345.00	17.25	15.51	0.78	345.00	17.25
12	345.00	17.25	15.51	0.78	345.00	17.25
16	345.00	17.25	15.51	0.78	345.00	17.25

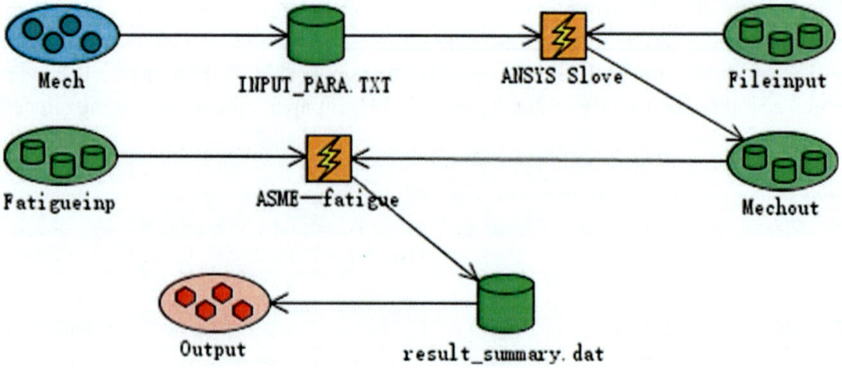

Fig. 8.2 Fatigue analysis process for nuclear power equipment based on OPTIMUS platform

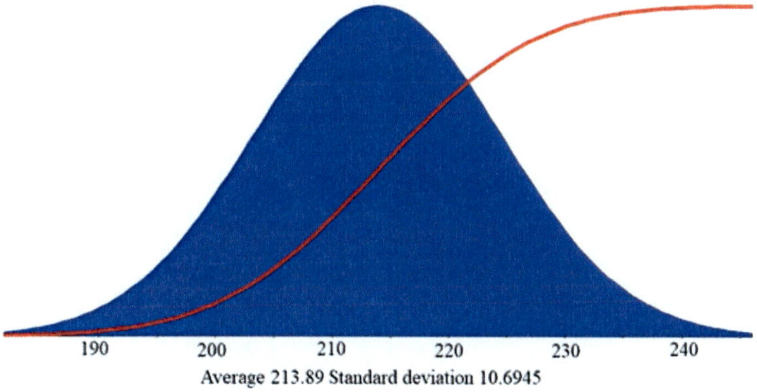

Fig. 8.3 Probability density function and cumulative distribution function of the first data point

Taking the obtained sample data as an input, the fatigue analysis of the nuclear power equipment is performed, and the fatigue usage of each sample is obtained.

At the same time, the parameter sensitivity analysis is carried out, and the correlation coefficient (Pearson) between the transient input parameters and the fatigue usage is obtained as well, as shown in Fig. 8.4 and Table 8.3.

Therefore, from the results, after five transient calculations, the fatigue usage of section No. 3 is more sensitive to changes in temperature 2 data, and less sensitive to transient pressure changes. Some inspiration can be obtained that in the structural design, from the perspective of design life, the section No. 3 should be appropriately thinned to reduce the thermal stress gradient during operation and reduce the fatigue usage under the condition of satisfying the pressure requirements.

Fig. 8.4 The result of correlation coefficient calculation

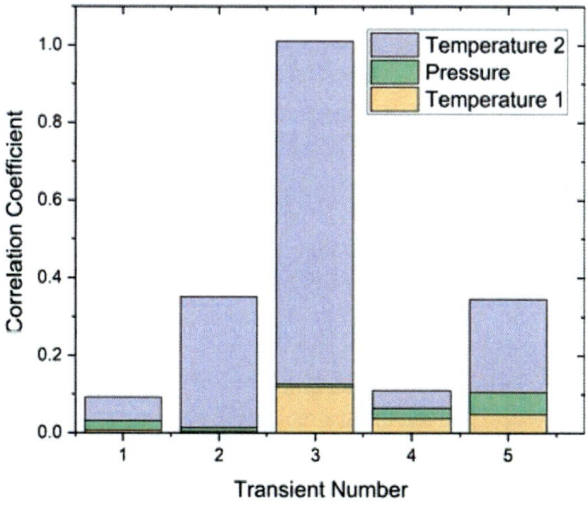

Table 8.3 The summary of correlation coefficient calculation

Transient Number		P3U_in	P3Sn1_in	P3Sn2_in
4	Temperature 1	0.00686	0.06615	−0.09945
	Pressure	0.02431	0.02709	0.05680
	Temperature 2	0.06089	0.00581	0.12589
7	Temperature 1	−0.00473	0.10689	−0.31247
	Pressure	−0.00885	-0.04816	−0.03382
	Temperature 2	0.33761	0.85050	0.44026
10	Temperature 1	−0.11815	-0.05421	0.22461
	Pressure	0.00722	0.18774	−0.02298
	Temperature 2	0.88492	0.23845	−0.13772
12	Temperature 1	−0.03692	-0.06088	0.01226
	Pressure	0.02690	0.12920	−0.06825
	Temperature 2	0.04568	0.16483	−0.14625
16	Temperature 1	−0.04783	0.02073	0.26300
	Pressure	0.05657	0.12006	0.01626
	Temperature 2	0.24098	0.11669	−0.10351

8.4 Conclusions

In this paper, the uncertainty is introduced into the transient load, and the uncertainty quantification study of spray nozzle fatigue analysis is carried out. The results are as follows:

(1) Based on the Optimus platform, this paper introduces the concept of uncertainty quantification into the complete process of fatigue analysis and evaluation of nuclear power equipment, establishes the fatigue strength reliability calculation process, and realizes the transition from the deterministic theory to the fatigue probability evaluation of nuclear power equipment. The results of the mechanical analysis are closer to the physical reality.

(2) This paper takes the pressurizer spray nozzle as the research object. The calculation results show that for the transient and cycle times of this paper, considering the fatigue design and ensuring the pressure bearing requirements, the wall thickness of the section No. 3 should be appropriately thinned, which can not only improve the economics of the structure, but also extend the life of the equipment structure.

(3) In the design phase, the uncertainty quantification study of input parameters can be used to further optimize the structure by considering the randomness of the parameters more widely. The analysis method of this paper can be extended to other equipment in the field of nuclear power, providing quantitative analysis means and effective tools for the reliability of fatigue analysis.

References

1. Wang, D.H., Zhang, L.P., Fu, X.L.: Study of elastic-plastic fatigue analysis in nuclear equipment. Atomic Energy Sci. Technol. **51**(11), 2036–2041 (2017)
2. Zhang, L.P., Zheng, L.G., Lu, Y.C., Liu, W.J.: Analysis of fast fracture of core region of RPV. Nuclear Power Eng. **34**(05), 30–32 (2013)
3. Shao, X.J., Zheng, L.G., Zhang, L.P., Yang, Y.: Elasto-plastic analysis for seal rings of reactor pressure vessels. Appl. Math. Mech. **35**(S1), 23–27 (2014)
4. Soize, C.: Uncertainty Quantification: An Accelerated Course with Advanced Applications in Computational Engineering. Springer International Publishing AG (2017)
5. Gong, J.X., Zhong, W.Q., Zhao, G.F.: Developments and applications of reliability theories for engineering structures (1). J. Build. Struct. **04**, 2–9 (2002)
6. Gong, J.X., Zhong, W.Q., Zhao, G.F.: Developments and applications of reliability theories for engineering structures (2). J. Build. Struct. **05**, 2–10 (2002)
7. Gong, J.X., Zhong, W.Q., Zhao, G.F.: Developments and applications of reliability theories for engineering structures (3). J. Build. Struct. **06**, 2–9 (2002)
8. ASME. ASME boiler and pressure vessel code: Section III. The American Society of Mechanical Engineers, New York (2004)
9. Metropolis, N., Ulam, S.: The monte carlo method. J. Am. Stat. Assoc. **44**(247), 335–341 (1949)
10. Metropolis, N., Rosenbluth, A.W., Rosenbluth, M.N., Teller, A.H., Teller, E.: Equation of state calculations by fast computing machines. J. Chem. Phys. **21**(6), 1087–1092 (1953)
11. Gelfand, A.E., Smith, A.F.M.: Sampling-based approaches to calculating marginal densities. J. Am. Stat. Assoc. **85**(410), 398–409 (1990)

Chapter 9
Numerical Simulation on the Slab Heat Transfer Behavior with Growth of Oxide Scale in a Pusher Type Reheating Furnace

Mingjie Feng, Cheng Mao, Yi Guo, and Yuxi Feng

Abstract Pusher type continuous reheating furnace is an important heating equipment. Its main function is to improve the billet temperature to meet the requirements of rolling process. It is inevitable that a certain thickness of iron oxide will be formed on the surface of the billet during reheating in the furnace, which will affect the heat transfer characteristics of the billet. It is quite important to accurately predict the heat transfer behavior of steel slab with growth of oxide scale throughout heating process. Therefore, a 3-dimensional transient heat transfer mathematical model for a pusher type reheating furnace has been established including the kinetic model of steel oxidation sub-model. The heat behavior and the temperature distribution in the slab with growth of oxide scale on the surfaces during the reheating process have been studied by numerical simulation. The results indicate that the established model is valid and has some more applied value. In the preheating zone, the effect of oxide scale on the heat transfer was negligibly small. But, in the heating zone, its influence cannot be ignored. The temperature of the slab with oxide scale is lower than that of without oxide scale during the reheating process, particularly in the high temperature zone.

Keywords Reheating furnace · Oxide scale · Temperature distribution · Heat transfer behavior

9.1 Introduction

The reheating furnaces are widely used in the steel industries to raise the temperature of the steel slab and meet the demands of steel rolling process [1–3]. The quality and productivity of the steel product may depend on the attainment of uniform temperature distribution within the steel slab and the target temperature of the steel slab at the

M. Feng (✉) · C. Mao · Y. Guo
College of Metallurgy, Northeastern University, Shenyang 110819, China
e-mail: fengmj@epm.neu.edu.cn

Y. Feng
College of Robotics Science and Engineering, Northeastern University, Shenyang 110819, China

J. Xu and K. M. Pandey (eds.), *Mechanical Engineering and Materials*,
Mechanisms and Machine Science 100,
https://doi.org/10.1007/978-3-030-68303-0_9

furnace exit, therefore the transient heat behavior of the steel slab in the reheating furnace has attracted a great deal of interest [4–6]. In general case, the reheating furnace is filled with high temperature oxidizing gas. During reheating process, the steel slab surfaces are exposed to the high temperature oxidizing gas environment, therefore, a layer of iron oxide will be generated on the slab surfaces and become more and more noticeable over time. In particular, when the steel slab moves to the heating zone and the soaking zone, it will become more serious. The oxide scale has poor thermal conductivity, large specific heat and less surface emission rate compared to steel slab, so the formation and growth of oxide scale may affect the heat transfer characteristics of the steel slab. For these reasons, it is quite important to accurately predict the heat transfer behavior of steel slab with growth of oxide scale throughout heating process. Numerical simulation is an effective method for solving problems of heat transfer in complex geometries. Over the past decades, a number of studies have been conducted employing numerical simulation to analyze the heating behavior of the steel slabs in the reheating furnace [7–11], but these studies rarely involve the problem of slab oxidation. In 2003, Chen et al. [12] presented a kinetic model for high temperature oxidation of carbon steels in oxidizing gas. Then, Jang et al. [13] predicted the heat flux distribution within the furnace and the temperature distribution in the slab throughout a walking-beam type reheating furnace process by use of the kinetic model. Subsequently, Dubey et al. [14] also studied three dimensional transient heat conduction within steel slab considering the growth of oxide scale on the slab surface during reheating on base of this model. The results show that the kinetic model of steel oxidation by Chen et al. is a great help to predict the heat transfer behavior of slab with growth of oxide scale throughout heating process. But, their computational procedures are too complex because of the geometric model including the oxide scale thickness.

In this wok, on base of the kinetic model of steel oxidation by Chen et al., the transient heat behavior and the temperature distribution in the slab throughout the reheating process in a pusher type furnace by considering growing oxide layer have been studied. But, the geometric model is not including the oxide scale thickness, the oxide scale growth and the thermal resistance models has been considered using a sub-model. In addition, the heating exchange between the steel slabs and its surroundings, including the hot combustion gases, the furnace wall, the skids, and the gas convection heat transfer, etc., has also been considered in the simulation. Our research objective is to provide reference for the design and operation of the reheating furnaces.

9.2 Geometry and Mathematical Model

The pusher type reheating furnace modeled in this paper is shown in Fig. 9.1. This furnace has about 32 m in length and 5.6 m in width, and the highest/lowest furnace roof is about 3.5/1.2 m inside. There are three zones in the reheating furnace as shown in Fig. 9.1a: preheating, heating, and soaking zones. The water-cooled beam and partition wall is located between the heating zone and soaking zone in the furnace.

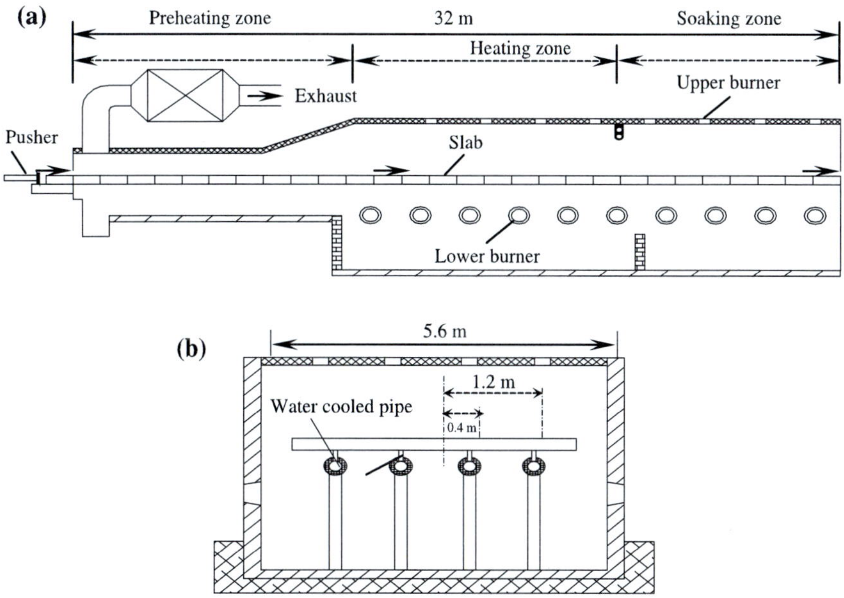

Fig. 9.1 Schematic diagram of the reheating furnace: **a** longitudinal and **b** transverse sections

The steel slabs, skids and support pipes are arranged in the furnace as shown in Fig. 9.1b. According to the rhythm of production, the steel slabs on the skids are regularly moved from entrance to exit by a pushing steel machine. The steel slab is a low alloyed steel of 16Mn. The length, width and thickness is 2.4 m, 1.0 m and 0.2 m, respectively. There are 32 slabs next to each other in the reheating furnace. The steel slab is assumed to be isothermal of 300 K when charged into the furnace and each slab residence time is about 220 min. The thermal properties of the slab are given in Table 9.1. The composition of gas and temperature distribution in the furnace based on experimental data is listed in Tables 9.2 and 9.3 respectively. The total pressure of the furnace gas chamber is assumed to be 1.01325×10^5 Pa.

Table 9.1 Thermal properties of steel slab and FeO

Temperature (K)	Conductivity (W/m·K)	Specific (J/kg·K)	Emissivity	Density (kg/m^3)
T < 473	58.17	495.0	0.50	7800
473 < T < 673	48.52	547.9	0.50	
673 < T < 873	36.14	557.9	0.50	
873 < T < 1073	31.15	872.1	0.55	
1073 < T < 1273	25.08	710.8	0.65	
T > 1273	23.81	652.0	0.65	
Oxide scale	3.2	725.0	0.7	7750

Table 9.2 Volume Compositions of gas in the furnace (%)

CO_2	H_2O	O_2	N_2
19.3	12.1	2.3	66.3

Table 9.3 Temperature at measurement location

Distance from the entrance [m]	7.40	11.10	12.21	15.73	16.65	18.00
Temperature (K)	775	876	946	1110	1197	1272
Distance from the entrance (m)	21.83	22.20	25.90	31.08	35.33	37.00
Temperature (K)	1517	1523	1578	1633	1663	1668

9.3 Mathematical Model

9.3.1 Heat Conduction Equation

We are selecting a single slab as research object and are following its movement when it is traveling from entrance to exit of the reheating furnace. According to geometry symmetry, only half of a single slab is taken as calculation zone. Unlike some of the other models [13, 14], the geometric model is not including the oxide scale thickness. The influence of oxide scale to heat transfer of the slab surface has been considered in the boundary conditions. Taking the center of the symmetry plane as origin point and the slab moving forward direction as positive in the x axis direction, a rectangular Cartesian coordinate system has been established as those shown in Fig. 9.2. Then, the 3-dimensional transient heat conduction equation (Eq. (9.1)) to calculate the temperature distribution is,

$$\rho c \frac{\partial T_c}{\partial t} = \frac{\partial}{\partial x}\left(\lambda_c \frac{\partial T_c}{\partial x}\right) + \frac{\partial}{\partial y}\left(\lambda_c \frac{\partial T_c}{\partial y}\right) + \frac{\partial}{\partial z}\left(\lambda_c \frac{\partial T_c}{\partial z}\right). \tag{9.1}$$

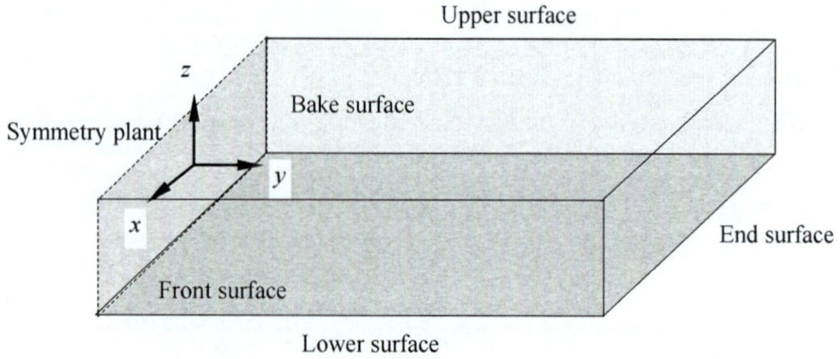

Fig. 9.2 Billet orientation considered in the numerical model

where x, y and z directions are considered as slab length, height and width directions respectively, such as those shown in Fig. 9.2, T_c is the temperature, ρ is the density, t is time, c is the specific heat, and λ_c is the conductivity.

9.3.2 *Initial Conditions*

For the characteristic of the object, only one parameter, i. e. temperature of the slab, need to be set up as initial condition. So, the slab is assumed to be at uniform temperature of 300 K, when charged to the reheating furnace.

9.3.3 *Boundary Conditions*

Oxide Scale. There are six faces, i. e., upper surface, lower surface, right surface, front surface, bake surface and symmetry face, for the studied geometry as those shown in Fig. 9.2. Apart from the symmetry face, all the surfaces are exposed to hot combustion gases environment during reheating process. The combustion gases consist of some oxidizing gases, such as H_2O, CO_2, and O_2. Under high temperature conditions, metal easily reacts with the combustion gases. As a result, an oxide layer, which consists of FeO, Fe_3O_4 and Fe_2O_3 from inside to outside, forms on these surfaces of the slab. But, the percentage thicknesses of the three oxide layers are not the same. And researcher [12] has suggested that FeO, which forms next to the metal, comprise about 95% of production. The second is Fe_3O_4, which is the intermediate phase, about 4%. The least is Fe_2O_3, which is the outermost layer of the scale, only about 1%, when the temperatures are higher than approximately 873 K. Therefore, in this work, it is assumed that the scale layer consists of only one component, FeO, like any other literature [13, 14]. The thermal properties of FeO are listed in Table 9.1. Meanwhile, in this case, the thickness of scale can be represented as Eq. (9.2) [12]:

$$\delta^2 = \delta_0^2 + k_0 t e^{-Q/RT}. \tag{9.2}$$

where δ is the scale thickness, δ_0 is initial scale thickness at time $t = 0$, t is the oxidation time, k_0 is a constant of 6.1, Q is the activation energy of iron oxidation of 169,452 J/mol, R is the gas constant, and T is absolute temperature.

Although, the oxidation reaction is exothermal reaction, the rate of heat release is almost negligible. So, in this work, the effect of the heat of oxidation reaction on temperature field within the slab is neglected [13, 14].

As mentioned previously, the oxide scale has poor thermal conductivity, large specific heat and less surface emission rate compared to that of slab, so the existence of the oxide layer on the slab surface can affect the heat transfer behavior. A thermal resistance calculation model which is a function of the thickness of oxide scale is adopted to solve the problems. The functions can be represented as Eq. (9.3):

$$R_t = \frac{\delta}{\lambda_t} \tag{9.3}$$

where R_t is the thermal resistance, δ is the oxide scale thickness, λ_t is the conductivity of the oxide scale as shown in Table 9.1.

Upper or End Surfaces. In Eq. (9.1), for the surfaces which are exposed to hot combustion gases environment, we adopt a comprehensive heat transfer flux of radiation and convection as the boundary conditions. The total heat flux of the surfaces (q_t) will include three parts: radiation heat transfer flux (q_{rg}) between furnace gas and slab surfaces, radiation heat transfer flux (q_{rw}) between furnace walls and slab surfaces, and convection heat transfer flux (q_{cg}) between furnace gas and slab surfaces. The total heat transfer flux can be obtained as Eq. (9.4) [14]:

$$q_t = \frac{\sigma \varepsilon_s \left(\varepsilon_g T_g^4 - a_{gs} T_s^4\right)}{\left[1 - (1 - \varepsilon_s)\left(1 - a_{gs}\right)\right]} + \sigma \varepsilon_{sw}\left(T_w^4 - T_s^4\right) + h_{cg}\left(T_g - T_s\right). \tag{9.4}$$

where σ is the Stefan-Boltzmann constant having a value of 5.67×10 W/m^2·k^4, ε_s is the emissivity of slab surfaces, T_g is the furnace gas temperature, T_s is slab surfaces temperature, ε_g is the emissivity and α_{gs} is the absorptivity of the furnace gas.

According to the relevant literature [15, 16], the gas emissivity (ε_g) and absorptivity (α_{gs}) can be obtained by the gray gas CO_2 and H_2O mixture models as Eqs. (9.5) and (9.6):

$$\varepsilon_g = \left(\varepsilon_{g,CO_2} + \varepsilon_{g,H_2O} - \Delta\varepsilon\right), \tag{9.5}$$

$$a_{gs} = \left[\varepsilon_{g,CO_2}\left(\frac{T_g}{T_s}\right)^{0.65} - \varepsilon_{g,H_2O}\left(\frac{T_g}{T_s}\right)^{0.45} - \Delta a_{gs}\right], \tag{9.6}$$

where ε_{g,co_2} and ε_{g,H_2O} is the emissivity of CO_2 and H_2O respectively, $\Delta\varepsilon$ is the correction factor for furnace gas emissivity because of the overlap of the spectra of CO_2 and H_2O, Δa_{gs} is the correction factor for furnace gas absorptivity which is based on the temperature of the furnace gas and slab surfaces.

In Eq. (9.6), ε_{sw} is the radiation heat exchange factor between furnace walls and slab surfaces which is obtained by Eq. (9.7):

$$\varepsilon_{sw} = \varepsilon_s \varepsilon_w \tau_{gm} \varphi_{sw}, \tag{9.7}$$

where τ_{gm} is mean transmissivity of the furnace gas which is expressed as Eq. (9.8):

$$\tau_{gm} = 1 - 0.5(a_{gs} + a_{gw}), \quad (9.8)$$

The absorptivity (α_{gw}) for radiation from the furnace walls in Eq. (9.6) can be obtained as Eq. (9.9):

$$a_{gw} = \left[\varepsilon_{g,CO_2} \left(\frac{T_g}{T_w} \right)^{0.65} - \varepsilon_{g,H_2O} \left(\frac{T_g}{T_w} \right)^{0.45} - \Delta a_{gw} \right]. \quad (9.9)$$

where Δa_{gw} is the correction factor for furnace gas absorptivity which is based on the furnace gas temperature and walls temperature.

The convection heat transfer in total heat transfer is very small. In the present study, the gas convective heat transfer coefficient between the furnace gas and slab surfaces (α_{cg}) in Eq. (9.7) is assumed to be a constant value of 8.5 W/m^2·K.

Lower Surface. The water cooled pipe system consists of 16 horizontal pipes and 4 longitudinal pipes in the furnace. All the skid buttons on which the slabs rest are welded on the longitudinal pipes. Although the insulating layer covering outside of the pipes may reduce heat transfer to the pipe surfaces, the temperature of the insulating layer surfaces is relatively low. Furthermore, a part of heat flux from the lower surface of the slab will transfer through skid buttons to the cooling water in the pipes. So, the water cooled pipe system will inevitably reduce the heat transfer on the lower surface of the slab [15, 16]. In the present work, the effect of the water cooled pipe system on heat transfer is considered using method of Dubey and Srinivasan [14].

For the lower surface which is not contact with skid buttons, the boundary condition is described as Eq. (9.10):

$$q_{b1} = Fq_t, \quad (9.10)$$

were F is radiation shielding factor.

For the lower surface which is contact with skid buttons, the boundary condition is expressed as Eq. (9.11):

$$q_{b2} = F(1 - \beta)q_t - \beta\left(\frac{T_s - T_h}{R_{bt}} \right). \quad (9.11)$$

whereβ is contact ratio, R_{bt} is overall thermal resistance for skid contact, T_h is cooling water temperature.

Front or Back Surface. In order to make the slabs can synchronously forward, the front and back surfaces of two adjacent slabs are very close together in a pusher type reheat furnace. Although it is possible that heat transfer will occur between the adjacent surfaces, in this work, the heat flux between them is neglected because of very small. Thus, the heat transfer flux on the front or back surface of the slab is assumed to zero.

Symmetry Plane. On the symmetry plane, the adiabatic boundary condition for Eq. (9.1) is adopted and can be expressed as Eq. (9.12):

$$q_t = 0. \tag{9.12}$$

9.3.4 Solution Method

According to Fig. 9.2 in Sect. 9.3.1, the calculation geometry was created and meshed with hexahedral grid using Gambit 2.3 software. A dense mesh was used in the surface zone to guarantee the computational efficiency and simulation accuracy. The oxide scale growth and the thermal resistance models were inputted using user-defined functions based on fluent software. At certain time step, on based of the slab surfaces temperature resulted at the previous time step, the oxide scale thickness and thermal resistance was solved by use of the oxide scale growth model. And then, the heat conduction equation was solved on base of the slab surfaces temperature resulted at current time step. The above process was repeated at each location in the furnace. The residual of energy convergence was set to 10^{-6}. The time step was set to 30 s.

9.4 Results and Discussion

9.4.1 Growth of Oxide Scale

The thermal resistance, which is caused by the oxide scale on the surfaces of the slab, is a significant factor to affect the comprehensive heat transfer coefficient between the surroundings and the slab's surfaces. The thickness of the oxide scale might determine the thermal resistance value. Thus, it is worth investigating that the variation regularity of the oxide scale thickness during the reheating process. Figure 9.3 shows the varying of average scale thickness on the upper and lower surface of the slab during the reheating process. It can be seen that the highest temperature zone of the slab is located at the intersecting point of the three surfaces, i.e. the front surface, the upper surface and the end surface. we take this point as investigated object. Figure 9.4 shows the varying of scale thickness and temperature of the investigated point during the reheating process. That as seen in Figs. 9.3 and 9.4, the surfaces temperature of the slab and the resident time in the high temperature zone in the furnace might be important factors for the oxide scale formation rate. When the temperature is lower than 833 K, the oxide scale formation rate is very small. So even prolonging the resident time here, the oxide scale will not become very thick. But, when the temperature is very high, it will increase rapidly with resident time.

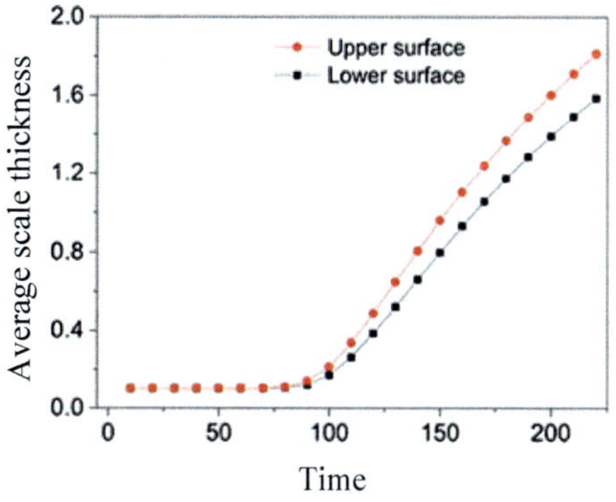

Fig. 9.3 Varying of average scale thickness on the upperand lower surface with resident time

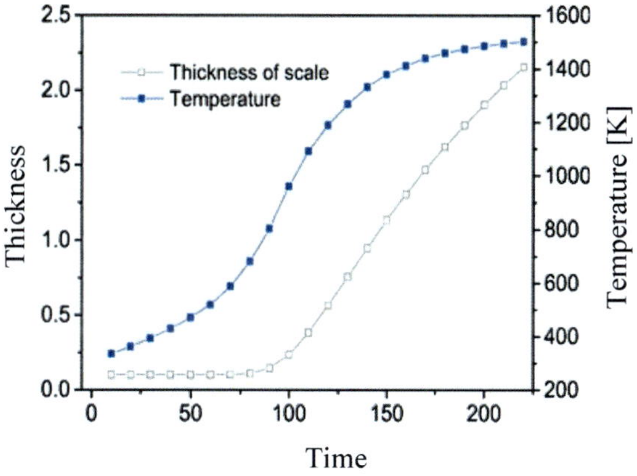

Fig. 9.4 Varying of oxide scale thickness and temperature of the investigated point with resident time

So, to shorten resident time in the high temperature zone in the furnace will help to decrease the oxide scale thickness on the slab surfaces.

9.4.2 Temperature Distribution Within the Slab

A pusher typed reheat furnace is one of the main steel slab reheating equipment, whose main task is reheating the slabs on the rolling line according to our process heating require. The reheating quality of steel slab will directly affect the final product quality and performance. Several basic indicators, such as slab discharge temperature, temperature difference of cross section, oxide scale thickness and so on, can be used to measure reheating quality. The slab discharge temperature mainly depends on some factors, such as sorts of slab steel, slab transporting temperature dropping, production specifications, rolling process, etc. Generally speaking, based on satisfying the technological demands, lower discharge temperature is favorable to reduce the per unit product energy consumption and to increase the product performance.

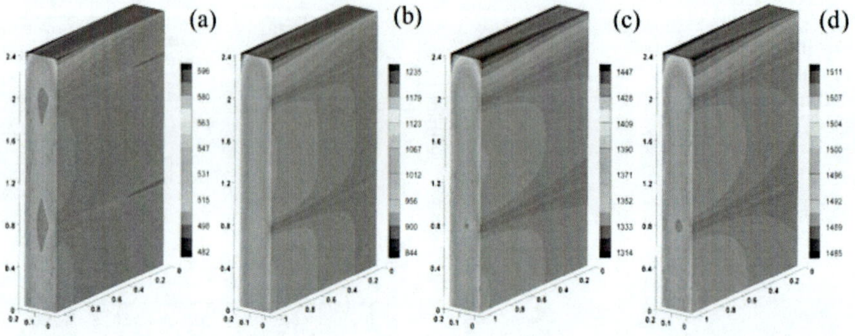

Fig. 9.5 Temperature field (temperature in K) of the slab: **a** after preheating zone, at t = 64 min. **b** during heating zone, at t = 112 min.**c** after heating zone, at t = 158 min. **d** after soaking zone, at t = 220 min

Figure 9.5 shows the temperature distribution within the slab after each zone of the furnace. It can be seen that the global temperature of the slab gradually rises along with the process of reheating. But, the temperature in different zone of the slab is different during reheating process. This can be attributed to difference of heat transfer flux on the different surfaces. Generally, the temperature of region near the end surface is higher than that of other regions in the slab. The maximum temperature appears at the interface between the upper and end surface, and is close to the front surface. Meanwhile, the low temperature regions in the slab lie in near the water cooled pipes, which are also known as skid mark, due to the shielding and heat transfer of the water cooled pipe system. Furthermore, the temperature in the core of the slab is also low because the direction of heat flux is from the surfaces to the core. We all know that the main problem of reheating slab is non uniform temperature and skid mark is an important influence factor. According to the results of simulation, the temperature difference in different regions will change along with the process of reheating because of varying heat transfer condition. The low temperature region near core and the skid mark has always been here, but with prolonging of the resident time, the smaller the inhomogeneity of the temperature in the slab. In a word, when the slab moves to the furnace outlet, the emperature of the front regions is higher than that of the rear of the slab. But, the maximum difference is no more than 35.8 K which meets the quality requirements.

9.4.3 Effect of Oxide Scale on Heat Transfer

In order to more closely investigate the effect of the oxide scale on heat transfer during the reheating process, the varying of average heat flux difference on the surfaces with and without oxide scale during the heating process is illustrated in Fig. 9.6. It can be seen from the figure that in the low temperature zone (preheating zone) in the furnace,

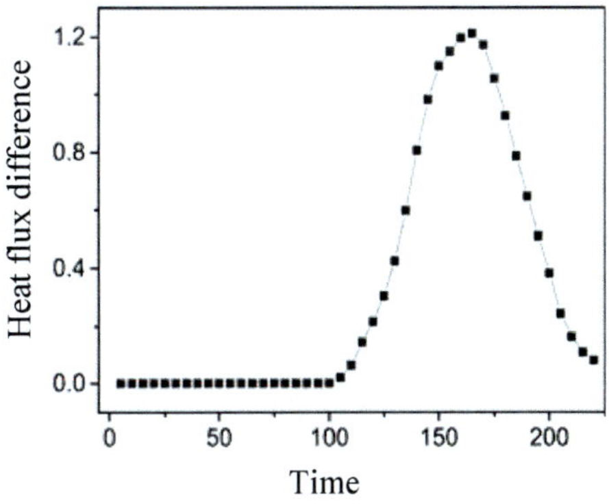

Fig. 9.6 Varying of heat flux difference on the surfaces with and without oxide

the oxide scale has almost no effect on the heat transfer flux on the slab surfaces because the oxide scale generated is very thin. When the slab moving into the heating zone, the effect of the oxide scale becomes more and more significant, Because the surfaces temperature increases quickly to cause the oxide scale increasing in thickness rapidly. The maximum average heat transfer flux difference is about 1.22 kw/m^2. Subsequently, although the oxide scale continuous to increase in thickness, its influence will decrease gradually because of the decreasing of the temperature difference between the environment and the surfaces of the slab.

9.4.4 Temperature Difference of Cross Section

To more directly illustrate the varying of temperature difference in the slab cross-section during the reheating process, a representative cross section (y = 0.8 m) has been taken as research subject. Figure 9.7 shows the varying of the surface average temperature (ST), the core average temperature (CT) and the temperature difference (DT) between them of the cross section with resident time. It can be seen from

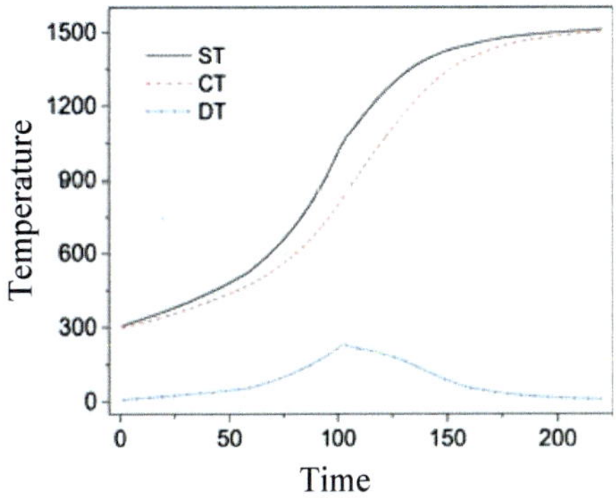

Fig. 9.7 The varying of ST, CT and DT with resident time

Fig. 9.7 that the relationship curve shape between the ST or CT and resident time are both similar to "S" shape. In the preheating zone, the heating rate of the slab is low at the start, then, gradually increases. After the slab moving to the heating zone, both of the surface and core temperature are rapidly increasing. In the soaking zone, the surface temperature is almost not increasing with resident time, and the core temperature still increases although the heating rate is somewhat lower. These cases are concert with the temperature distribution in the furnace. It can also be known from Fig. 9.7 that during the preheating zone and the first part of heating zone, the temperature difference between the ST and CT increases gradually with the slab gradually moving forward. And, it increases more quickly in the heating zone than that of in the preheating zone. When the slab moves to near the middle of the heating zone, the temperature difference reaches the peak value of 230.8 K. In the next zone, the temperature difference gradually decreases to about 21.9 K. That means the non-uniformity of temperature field in the slab increases firstly then decreases in the whole reheating process. As shown previously, the process of reheating slab in the furnace is a typical 3-D transient heat transfer process. In the process, the heating rate is mainly caused by the heat transfer flux on the surfaces of the slab, which mainly depends on the temperature difference between the gas and the surfaces. The higher temperature difference between the gas and the surfaces is favorable to higher heat transfer flux to increase the heating rate, but unfavorable to the uniformity of temperature field in the slab. This is exactly the reason why setting the soaking zone in the furnace. Generally, under the condition of keeping the size of the slab constant, the temperature difference in the cross section will be principally dependent on Bolt number (Bi) and Fourier number (Fo). The relationship can be expressed as Eq. 9.13:

$$\Delta T = f(Bi,\ Fo). \tag{9.13}$$

To decrease Bi or increase Fo is helpful for decreasing the temperature difference in the cross section of the slab and improving the quality of produce. By comparison, it can also be seen that the temperature of the surfaces with oxide scale is lower than that of without oxide scale during the reheating process, particularly in the high temperature zone. Because, the heat conductively of the oxide scale is very small. It can cause a great thermal resistance between the surroundings and slab surfaces to decrease the heat transfer flux.

9.5 Conclusions

In this work, a 3D transient mathematical heat transfer model for a pusher type reheating furnace was established. The heat transfer behavior of the slab was studied during the whole reheating process and the effect of oxide scale on the surfaces on heat transfer characteristics was full considered. Get the following conclusions:

1. The kinetic model of steel oxidation by Chen et al. can been used to predicted the growth of the oxide scale on the surfaces of the slab during the reheating process, and the simulation results is reasonable.
2. The surfaces temperature and the resident time in the high temperature zone is important factors for the oxide scale formation rate. To shorten resident time in the high temperature zone will help to decrease the oxide scale thickness on the slab surfaces.
3. In the preheating zone, the effect of oxide scale on the heat transfer was negligibly small. In the heating zone, its influence on the heat transfer of the slab cannot be ignored, and the maximum average heat transfer flux difference with and without oxide scale is about 1.22 kw/m^2. In the soaking zone, its influence will decrease gradually with resident time.
4. The temperature of the slab with oxide scale is lower than that of without oxide scale during the reheating process, particularly in the high temperature zone. The relationship curve shape between the ST or CT and resident time are both similar to "S" shape. The temperature difference within the cross section of the slab is first increasing and then decreasing during the whole reheating process.

Acknowledgements This work was supported by the National Key Research and Development Program of China (2017YFA0700300).

References

1. Kim, M.Y.: A Heat transfer model for the analysis of transient heating of the slab in a direct-fired walking beam type reheating furnace. Int. J. Heat Mass Transf. **50**, 3740–3748 (2007)
2. Hsieh, C.T., Huang, M.J., Lee, S.T., Wang, C.H.: Numerical modeling of a walking-beam-type slab reheating furnace. Numer. Heat Tr. A- App. **53**(9), 966–981 (2008)
3. Huang, M.J., Hsieh, C.T., Lee, S.T., Wang, C.H.: A coupled numerical study of slab temperature and gas temperature in the walking-beam-type slab reheating furnace. Numer. Heat Tr. A- App **54**(6), 625–646 (2008)
4. Karimi, H.J., Saidi, M.H.: Heat transfer and energy analysis of a pusher type reheating furnace using oxygen enhanced air for combustion. J. Iron Steel Res. Int. **17**(4), 12–17 (2010)
5. Emadi, A., Saboonchi, A., Taheri, M., Hassanpour, S.: Heating characteristics of billet in a walking hearth type reheating furnace. Appl. Therm. Eng. **63**(1), 396–405 (2014)
6. Han, S.H., Chang, D.: Optimum residence time analysis for a walking beam type reheating furnace. Int. J. Heat Mass Transf. **55**(15–16), 4079–4087 (2012)
7. Steinboek, A., Wild, D., Kugi, A.: Nonlinear model predictive control of a continuous slab reheating furnace. Control Eng. Pract. **21**(4), 495–508 (2013)
8. Morgado, T., Coelho, P.J., Talukdar, P.: Assessment of uniform temperature assumption in zoning on the numerical simulation of a walking beamreheating furnace. Appl. Therm. Eng. **76**, 496–508 (2015)
9. Lindholm, D.: A finite element method for solution of the three-dimensional time-dependent heat-conduction equation with application for heating of steels in reheating furnaces. Numer. Heat Tr. A- App **35**(2), 155–172 (1999)
10. Jaklic, A., Kolenko, T., Zupancic, B.: The influence of the space between the billets on the productivity of a continuous walking beam furnace. Appl. Therm. Eng. **25**(5–6), 783–795 (2005)

11. Torres, M., Colas, R.: A model for heat conduction through the oxide layer of steel during hot rolling. J. Mater. Process. Tech. **105**(3), 258–263 (2000)
12. Chen, R.Y., Yuen, W.Y.D.: Review of the high temperature oxidation of iron and carbon steels in air or oxygen. Oxid. Me. **59**, 433–468 (2003)
13. Jang, J.H., Lee, D.E., Kim, M.Y., Kim, H.G.: Investigation of the slab heating characteristics in a reheating furnace with the formation and growth of scale on the slab surface. Int. J. Heat Mass Transf. **53**(19–20), 4326–4332 (2010)
14. Dubey, S.K., Srinivasan, P.: Development of three dimensional transient numerical heat conduction model with growth of oxide scale for steel billet reheating simulation. Int. J. Therm. Sci. **84**, 214–227 (2014)
15. Jang, J.H., Lee, D.E., Kim, C., Kim, M.Y.: Prediction of furnace heat transfer and its influence on the steel slab heating and skid mark formation in a reheating furnace. ISIJ Int. **48**(10), 1325–1330 (2008)
16. Hsieh, C.T., Huang, M.J., Lee, S.T., Wang, C.H.: A numerical study of skid marks on the slabs in a walking-beam type slab reheating furnace. Numer. Heat Tr. A- App **57**(1), 1–17 (2010)

Chapter 10
Tube Hydro-Forming Process Design Based on Knowledge-Based Engineering

Yulong Ge and Zhiwei Ye

Abstract The principal objectives of a knowledge-based engineering (KBE) system are to solve a particular problem by reusing experience and knowledge in a flexible way meanwhile retain the new ones for future problems. This work presents KBE methodology for tube hydro-forming process design. Hydro-formed tubular parts are more complicated than common metal stamped parts because they are composed of groups bending, expansion and local forming features. Hydro-forming process design identifies and sequences the necessary operations and then producing appropriate die sets. Case-based reasoning (CBR) is employed into forming process planning to generate a hybrid KBE system for reuses existing cases to develop new designs. A part is defined using a feature-based representation composed of geometric and material parameters. Self-organized map algorithm is used as the retrieval engine in CBR. Combining the common rules of tube hydro-forming, the process route and the die layout can be determined quickly. This methodology can accelerate the design and manufacturing efficiently and assist engineers in preserving tacit knowledge of tube hydro-forming and accelerates the parts design and manufacturing.

Keywords Tube hydroforming · Process planning · Case-based reasoning · Self-organized map · Feature extraction

10.1 Introduction

The tube hydro-forming (THF) is a near net shape forming process technology which is widely adopted to produce tubular parts in aviation and automotive industries. It can raise the strength-weight ratio and rigidity with less post-processes [1].

Y. Ge (✉)
School of Vehicle and Mobility, Tsinghua University, Beijing 100003, China
e-mail: aaronge88@126.com

Z. Ye
Guangxi Automobile Holdings Limited, Liuzhou 545000, China

J. Xu and K. M. Pandey (eds.), *Mechanical Engineering and Materials*, Mechanisms and Machine Science 100,
https://doi.org/10.1007/978-3-030-68303-0_10

Since complicated shapes and mechanical material properties of different tubular parts (Fig. 10.1), the hydro-forming process can be considered as a combination of forming methods such as trimming, tube bending, hollow structure stamp, low pressure forming, high pressure forming and hole punching, etc. That makes the process planning of tube hydro-forming is more complicated than sheet forming and highly dependent on the former experience.

The forming quality of a particular tubular part is, in essence, depend on the necessary forming processes and their sequences and the knowledge based engineering (KBE) methodology can help make the decisions of the process planning with the ability of reuse the experience and knowledge.

Researches have made attempts to apply KBE in metal forming domain. Jinqiao Zheng introduced a process information model (PIM) using rule-based reasoning (RBR) for path planning in automobile panels stamping. The stamping features were represented as objective-oriented entities and the relationships as feature model network [2]. Neugebauer, R came up an advanced algorithm to deal with CAD data to extract information for KBE use and described the feature driven process for planning procedures which could overcome the disadvantages of fixed classification systems. And an approach to feature-based retrieval of sheet metal parts was presented by Dawei Wang [3]. The features frequently used in sheet metal part design were chosen as the "key words" and the relative position of features in sheet metal parts were expressed by four location description matrices to definite equivalent of relationships of two feature groups which was the basis to calculate the similarity of two sheet metal parts [4].

Due to the main drawback of rule-based KBE as it is hard to convert tacit knowledges (such as "experience") into explicit rules, the case-based reasoning (CBR) becomes a popular problem solving for decision support. The CBR is made of four

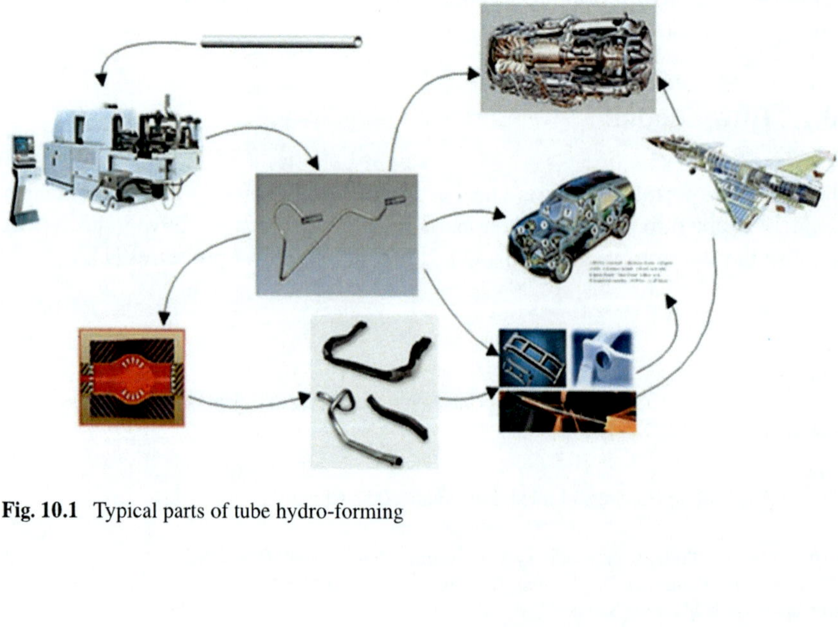

Fig. 10.1 Typical parts of tube hydro-forming

components: a case base for retrieving similar previous similar cases, reusing the solution for the new problem, revising the feasibility of the solution in this case and incorporating the solution into case base as a existed one for future problems [5].

Tor [6] represented a CBR approach with a new indexing and retrieval strategy for stamping die design. a feature relation graph representation scheme used for retrieval mechanism, w. A dual-step similarity analysis was proposed to shorten down the search time, and it performed well in multi-stage non-axisymmetric deep drawing [7]. Tsai [8] utilized the CBR methodology to plan stamping process and design stamping dies for automotive panels reusing existing designs to develop new designs. With the discussion mentioned above, however, few similar attempts have been made for the tube hydro-forming process planning. This paper is mainly concerned about: (1) a tube hydro-forming process design method with a KBE system using CBR; (2) a feature representation of tubular part; (3) using the self-organized map algorithm for CBR; (4) data preprocessing method for case retrieval; (5) and a case study to verify the effectiveness of this methodology.

10.2 Feature-Based Tube Hydro-Forming Process Design

The principal objectives of a KBE system are to solve a particular problem by reusing experience and knowledge in a flexible way meanwhile retain the new ones for future problems. In tube hydro-forming domain, the proposed KBE attempts to become an essential tool for improving design effectiveness and guiding the design engineers. The schematic diagram of it can be found in Fig. 10.2.

In a complete process sequence of tube hydro-forming, it contains bending, crushing, pre-forming, hydro-forming and calibrating as well as several minor intermediate and complementary operations [9]. Pre-forming is necessary when processes cannot be formed safely without preliminary operations which is most important in process planning. Especially, when the outer dimeter of the initial tubular blank is larger than the die allowance, there should be a crushing process. So that the blank can be loaded into the die cavity and the pinch must be avoided [10], seeing Fig. 10.3.

Accordingly, although bending and calibrating relate to the final part geometry, they are not very critical. Bending operations can be determined on the number and severity on the centerline. If there are mild bends that lie on a vertical plane or fall within expanded area, the hydro-forming process can make them formed in one step. Calibrating process is the foundation of fittability to the cavity surface and the pressure mainly depends on the smallest fillet radius on the part, as defined in Eq. (10.1) [11]:

$$p_{cal} = \frac{2}{\sqrt{3}} \frac{t_i \sigma_b}{RC_{\min} - t_i/2} \tag{10.1}$$

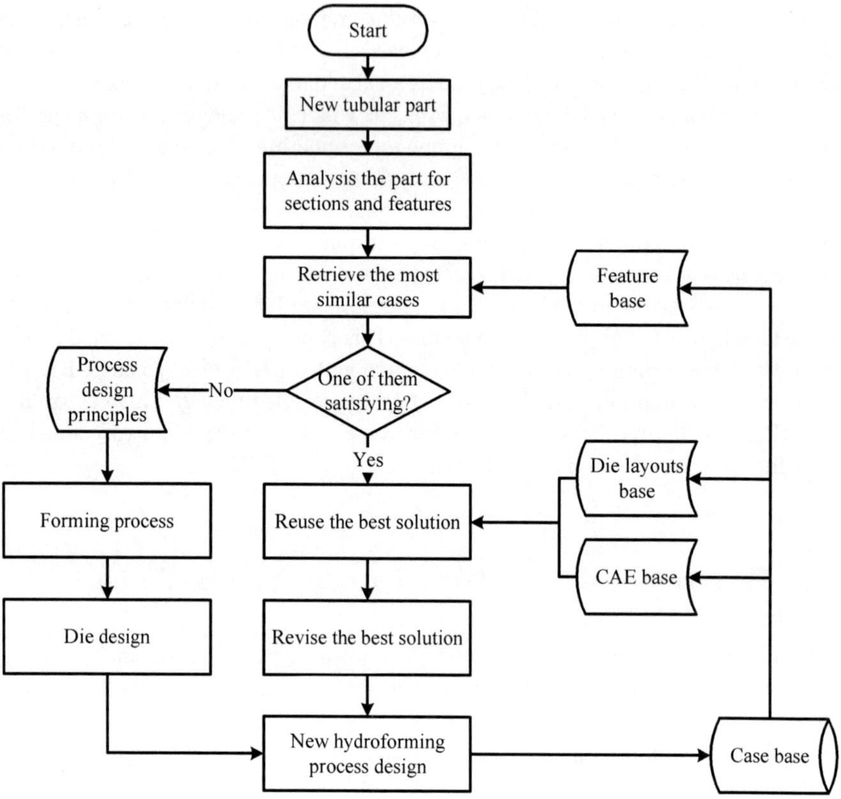

Fig. 10.2 Schematic diagram of the proposed KBE system

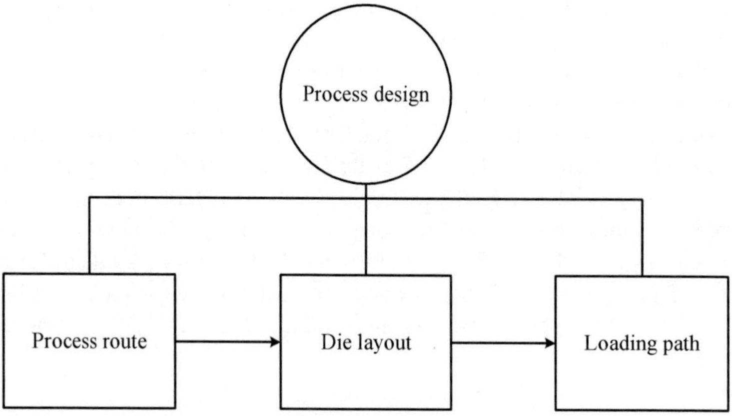

Fig. 10.3 Design path of tube hydro-forming

where t_i is the initial blank thickness, and RC_{min} is the minimum fillet radius of part and σ_b is the ultimate strength stress of material.

When the process and layout of die planning completed, the feasibility of the part should be evaluated to obtain the process parameters to which the maximum expansion and material parameters are particularly critical. The maximum expansion indicates the formability of the part and the axial feed amount. The initial pressure of hydro-forming process can be calculated by Eq. (10.2):

$$p = \frac{2t_i\sigma_s}{d_0 - t_i} \tag{10.2}$$

where d_0 the outer diameter and σ_s is the yielding stress. Then the loading path should be optimized for a better forming quality.

The purpose of the featured-based representation is to categorize a particular part into different segments to build the bridge between the geometry design and the manufacturing. The most important feature of a tubular part is the centerline with which it can be recognized with a list of segments with features on them. In this work, the centerline is represented in an incremental YBC coordinate which is widely used in tube NC bending process [12]. A spatial bent tube contains a list of three-tuples that feeding, rotating and bending whose bending radius is usually a constant. Thus, normal planes on the points of tangency can cut the part into segments naturally.

To obtain the centerline, two polylines on a pair of planes at right angle are drawn along the part trend and then stretched. The turning points of spatial bending process on two planes must be coincident to avoid tiny lines. The intersection of these stretching with spatial corner in a same radius can be considered as the centerline, seeing Fig. 10.4.

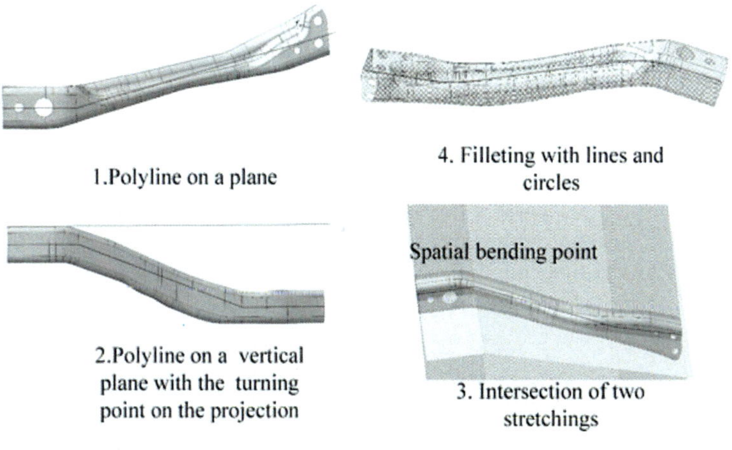

Fig. 10.4 Process to obtain the centerline

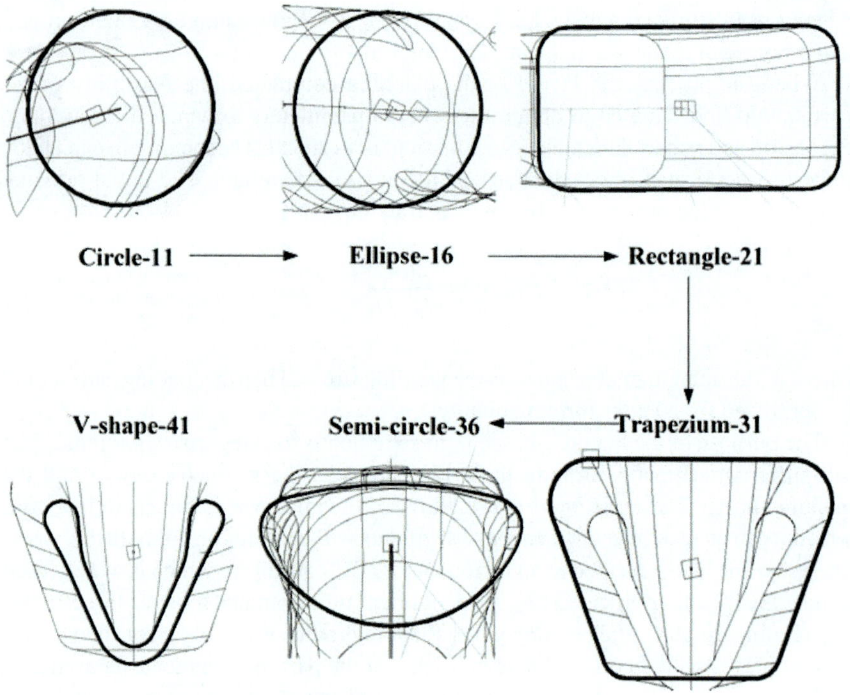

Fig. 10.5 Cross section shape quantitation

When the centerline is determined, the features on each segment should be defined. It is so complicated that it is variable in a segment, a part or between different parts and effected by many contributory factors. For simplicity, the features in this paper consist of cross section shapes and local forming features, which are the basis for the die cavity surface design. The cross section shapes are quantified according to the evolutionary process from a circle, seeing Fig. 10.5. The profile cut by the middle point's normal plane of each line segment is selected as the feature shape, the value of which can be chosen between typical values.

Local forming features are also quantified. Features that bulging up from the segment surface are assigned as positive values and depressed features are negative, as shown in Fig. 10.6. The inner values are available based on actual situations as well. For a segment that contains more than one local forming features, the feature with largest absolute value is used.

Thus, a segment can be described based on a five-tuple as Eq. (10.3):

$$seg_i = \{Y_i, B_i, C_i, cs_i, lf_i\} \tag{10.3}$$

where cs_i, lf_i represent the cross section shape and local forming features. This statement includes the information of a line and its following arc, so the local features

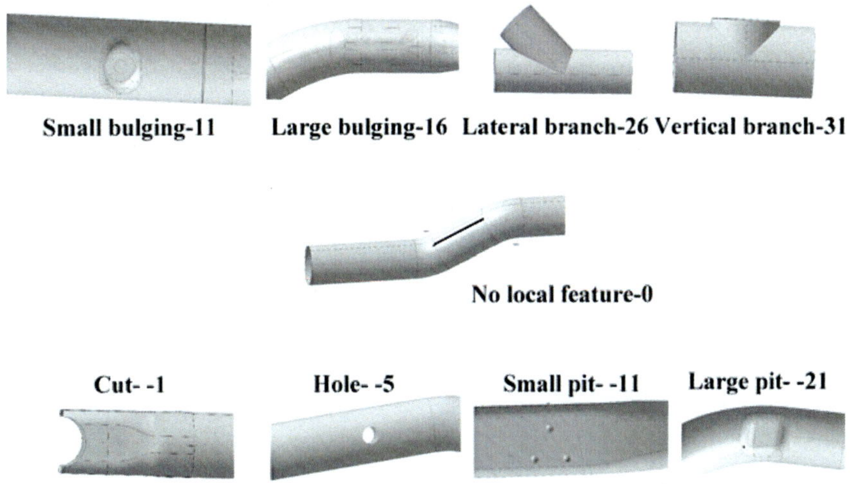

Fig. 10.6 Local forming features quantitation

of the bending areas should be considered when lf_i is assigned. In this work, the shorter end is placed in the front of the segment list uniformly.

Global parameters are used to describe the general geometry and material features of a tubular part that main decide the selection of blank and process design. The minimum diameter d_{min} is basis of the out diameter of the blank; the maximum diameter d_{max} determines the feeding supplement of hydro-forming; the calibrating pressure is mainly dependent on the wall thickness t and the minimum fillet corner radius rc_{min}; the feasibility of the forming process can be deduced with harden exponent n and yielding strength σ_s of the material; the radius R is the foundation of bending process, of which NC bending and pressurize bending are the alternation [9]. Never the less, if a part has a mirrored partner (mr), the die layout of two workpieces should be considered, shown as Fig. 10.7.

According to the YBC coordinate, B value of the first segment and last segment, C value of the last segment are always zeros, so they can be excluded. And since the number of segments is rarely more than six, a 35-bits feature vectors is used for the feature-based representation that contains both design and manufacturing information for potential process of a tubular part [7, 13].

Figure 10.8 can be represented in the feature vector shown in Table 10.1 for instance. Green lines mean feedings for straight lines and red arcs mean bendings.

10.3 Case-Based Reasoning for Tubular Parts

In a case based reasoning system, there are several critical points: the selection of features, organization and management of the case base, the method of case retrieval

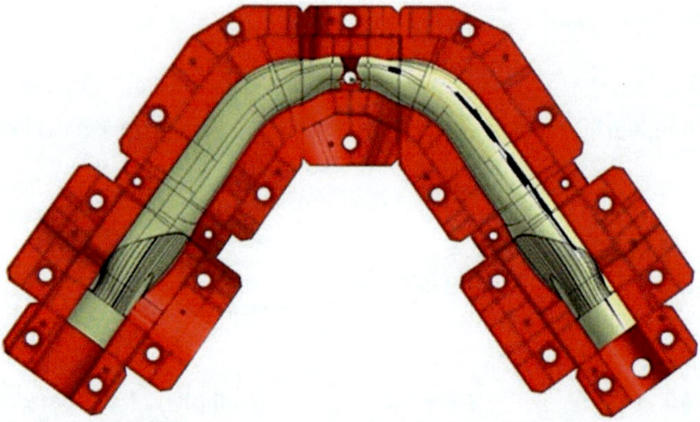

Fig. 10.7 Two workpieces die layout

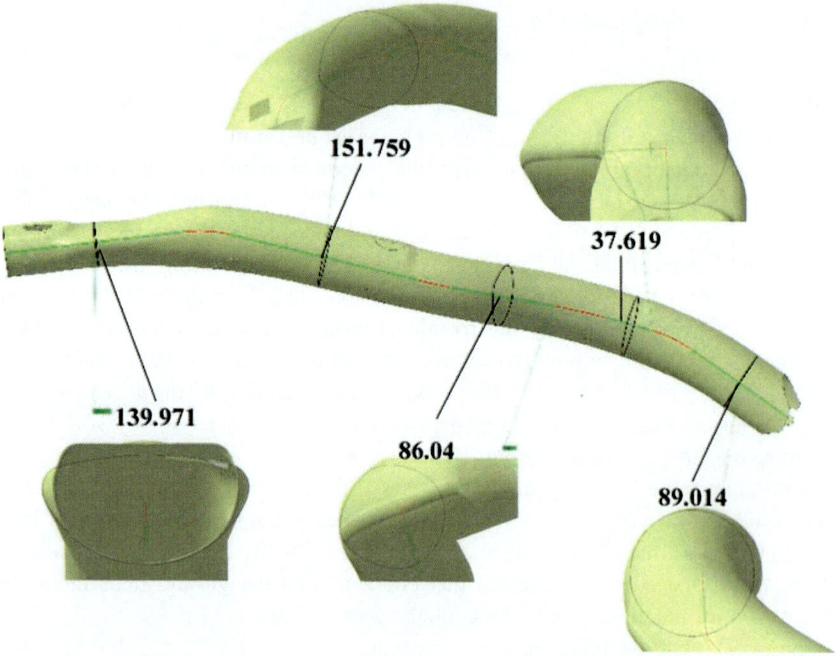

Fig. 10.8 Feature representation of a vehicle beam

and the adaption and learning in using and retain. A case can be expressed as below [14, 15]:

$$\text{casebase} = \{\text{case}_1, \text{case}_2, \ldots, \text{case}_\text{n}\}$$

Table 10.1 Feature vectors of the vehicle beam

Position	Symbol	Value	Position	Symbol	value	Position	Symbol	Value
1	MR	1	13	Y_2	37.619	25	C_4	20.287
2	D_{max}	48.16	14	B_2	86.656	26	cs_4	16
3	D_{min}	47.55	15	C_2	24.43	27	lf_4	12
4	t	2.0	16	cs_2	13	28	Y_5	139.97
5	rc_{min}	97.0	17	lf_2	12	29	B_5	0
6	n	0.23	18	Y_3	86.04	30	C_5	0
7	σ_s	346	19	B_3	0	31	cs_5	35
8	R	97	20	C_3	15.213	32	lf_5	−5
9	Y_1	89.014	21	cs_3	16	33	Y_6	0
10	C_1	20.138	22	lf_3	12	34	cs_6	0
11	cs_1	12	23	Y_4	151.7	35	lf_6	0
12	lf_1	−1	24	B_4	−90.00			

$$\mathrm{case_i} = \left\{\mathrm{problem_i}\right\} + \{\mathrm{solution_i}\}$$

$$\mathrm{problem_i} = \{\mathrm{feature_1}, \mathrm{feature_2}, \ldots, \mathrm{feature_m}\}$$

The features establishing a problem are essential to not only the management of cases but also the retrieval as they make up an index for a case. Proper feature selection can lead to a better performance in cluster detection in CBR.

The construction of the case base should be considered when the determination of representation of cases completed. Traditional case bases are built with the knowledge engineering and analysis of experts in specific domain that makes it a gap from satisfy result in case retrieval.

As the case of tube hydro-forming process is a high dimension problem, a popular algorithm is adopted as the CBR engine [16–18]. The self-organizing map (SOM) is an unsupervised machine learning algorithm that has been proved its high efficiency in CBR applications, comparing to the common nearest neighbor and clustering retrieval like KNN and K-means [19]. Meanwhile, SOM has the ability to handle missing values in samples [20]. It implements an orderly mapping of a high dimensional distribution onto a two-dimensional regular grid of units by competitive. Each unit in output grid is fully connected to all the neurons in the input layer. The SOM can preserve the most important topological and metric relationships of the primary data items in a visual way while compressed the data as kind of abstractions. Cases are automatically organized into a meaningful and ordered grid in which closer cases mean they share higher similarity, and neighboring units get similar weight vectors. Thus, the location of an output unit in the grid corresponds to a particular feature in the input space, shown in Fig. 10.9.

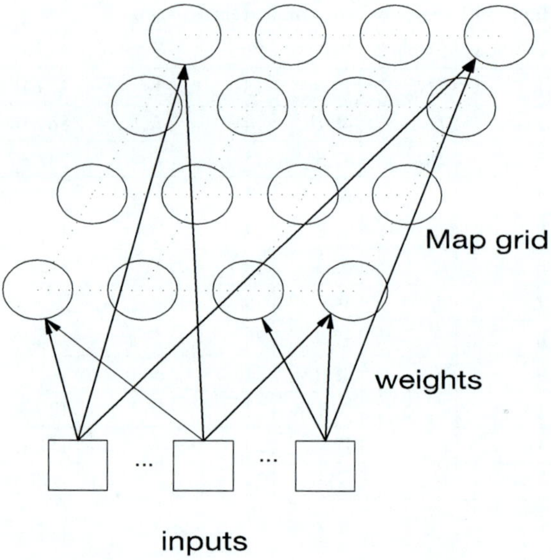

Fig. 10.9 Self-organizing map

Four major steps are required to set up a self-organizing map. The very first one is initializing every connection weight (w_{ji}) between an input layer unit (x_i) and an output layer unit (j) using a small random value. Then a competitive process is necessary for which a discriminant function provides the basis. The frequent selection is the squared Euclidean distance of weights which connect an output unit between all input neurons, as shown in Eq. (10.4):

$$d_j(x) = \sum_{i=1}^{D} \left(x_i - w_{ji}\right)^2 \tag{10.4}$$

where D is the number of input units and N is the number of output units. The inputs on continuous space are mapped to the discrete output space of units. And the unit whose indicator has the least value declares the winner in this competition.

Once there is a winner unit, the nearest ones tend to be more excited than the further ones and there is a lateral interaction among them. Thus, the distance of the topological neighborhood of the winner is decayed and the spatial location is changed along with time. A popular time dependence written as an exponential decay as Eq. (10.5):

$$\sigma(t) = \sigma_0 \exp\left(-t/\tau_\sigma\right) \tag{10.5}$$

It leads to the basis of the cooperation among the neighbor units. A topological neighborhood in SOM can be defined as Eq. (10.6):

$$T_{j,I(x)} = \alpha(t)\exp\left(-S_{j,I(x)}^2/2\sigma^2(t)\right) \tag{10.6}$$

where $\alpha(t)$ is a monotonically decreasing scaler function of t which could be a piece-wise linear one, S_{jk} is the lateral distance between unit j and k on the grid and $I(x)$ represents the winner. It has the advanced properties: the topological neighborhood is symmetrical about the grid and the maximum value is at the winner; the infinite distance is transferred to zero monotonically; and it is translation invariant. This step is called the cooperative process.

When the winner's weights updated, its neighbors will update their weights updated as well using Eq. (10.7):

$$\Delta w_{ji} = \eta(t)\cdot T_{j,I(x)}(t)\cdot\left(x_i - w_{ij}\right) \tag{10.7}$$

where $\eta(t)$ is a time dependent learning rate. It can be written as Eq. (10.8):

$$\eta(t) = \eta_0\exp\left(-t/\tau_\eta\right) \tag{10.8}$$

During the iteration of the learning process which consists of self-organizing phase and convergence phase, weight vectors w_j of the winner and its neighbors are move towards the input vector x. Thus, the output grid becomes self-organized and the feature map between inputs and outputs is formed topological orderly. The response of the best matching unit (BMU) to the similar cases to x is enhanced.

For all cases in base, the training process loops from competitive process until the feature map comes stable and constant. Thus, a SOM network can be used for a case retrieval for a high dimension problem.

Data preprocessing plays a very important in SOM algorithm. The dimensions of a feature vector are different in units and granularities, so that connective weights would be determined by larger values features. For example, thickness feature will be ignored practically comparing to feeding length. Scaling is a common method to rescaling the data along each dimension in the range (Eq. (10.9))[0, 1]:

$$x_i^{'} = \frac{x_i - x_{\min}}{x_{\max} - x_{\min}} \tag{10.9}$$

where x_i and $x_i^{'}$ represent the value before and after rescaling; x_{max}, x_{min} are the maximum and minimum value in this dimension. Considering the physical meaning of components in case vectors, all feedings, rotations and bendings in all parts and segments are stacked up for scaling.

Whitening is employed as the next preprocessing step to make SOM work better, in this work, which is principal component analysis (PCA). It is a dimensionality and noise reduction algorithm that can be used to speed up unsupervised feature learning algorithm like SOM. PCA will find a lower-dimensional subspace onto which to project the data using singular values decomposition (SVD) and the percentage of variance retained for different values decides the number of components that retained by Eq. (10.10):

$$\frac{\sum_{i=1}^{k} \lambda_i}{\sum_{i=1}^{n} \lambda_i} \tag{10.10}$$

where $\lambda_1, \ldots, \lambda_k$ are the eigenvalues of the covariance matrix of the dataset. As shown in Fig. 10.10, the tubular parts dataset is compressed into 13 components when the percentage of variance is 95%. Thus, the input vectors of cases are assigned with ten dimension arrays in which each element is in range [0, 1].

A map with hexagonal lattices is adopted in this paper, in which a unit has more neighbors in topology. The training is taken by a faster and safer batch algorithm [20] with 26 typical tubular part cases. The map owns 32 units in output layer in a 4×8 grid mesh. The topology and unified distance matrix are shown in Fig. 10.11. It visualizes the distances between the units. Darker color between the units signifies that the weight vectors are close to each other in the input space and lighter coloring corresponds to larger distance. So, Dark areas can be seen as clusters and light areas as separators.

The units best matched by cases are colored red, and the sizes indicate the matching number. It can be found that not every unit has matched cases. And the cases on the up of the map are much more similar than ones located below. The connection weights between each unit and the 13 neurons input layers are shown in Fig. 10.12. A light unit on a variable means the weight between them is closer to -1 and dark unit is closer to $+1$. According to the order of principle components, the largest weight decline in tune. The bars in each unit show the value of each component in that unit

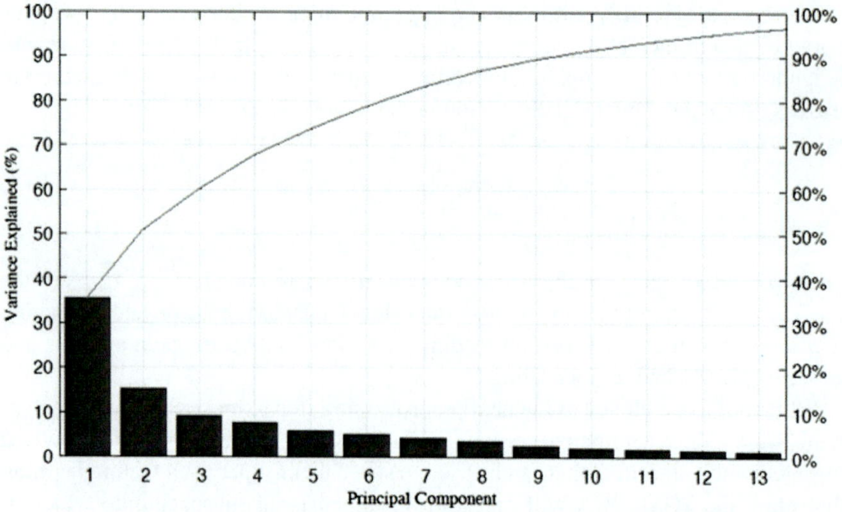

Fig. 10.10 Principle components (95%)

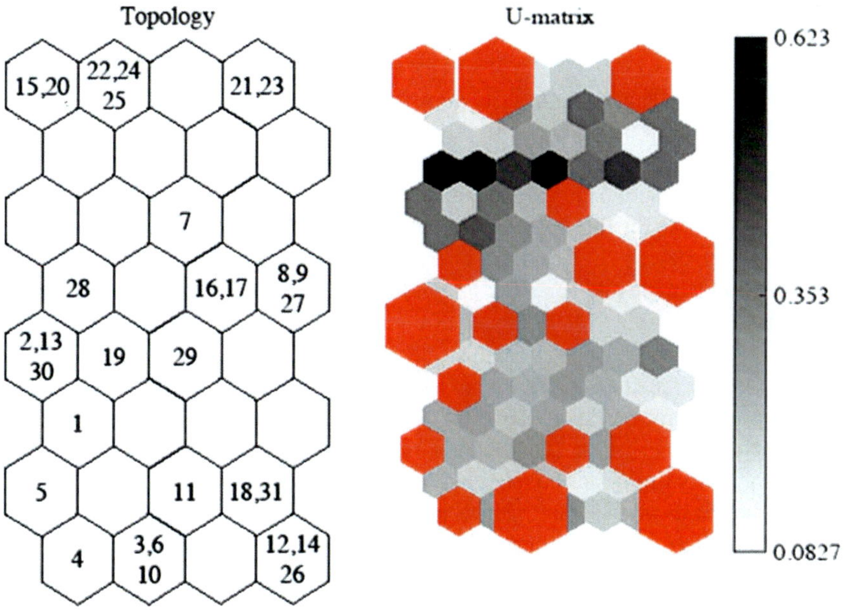

Fig. 10.11 Topology of SOM

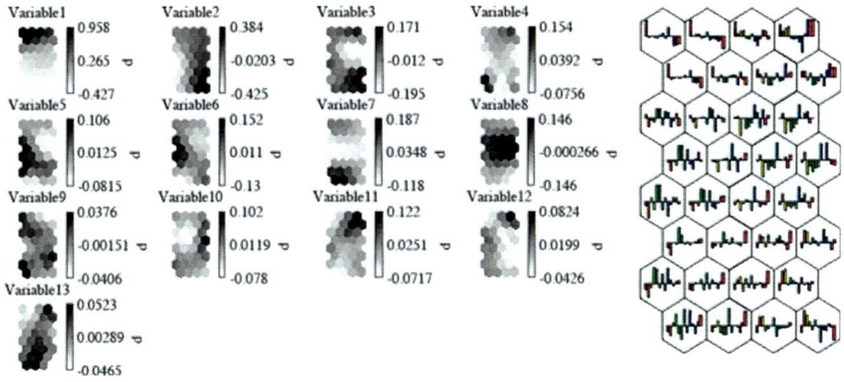

Fig. 10.12 Relationship between inputs and outputs

in relation to the overall maximum value (1) of that component. Associating them with the BMUs, it can be found that 6 cases of 26 has been clustered and classified by being strongly positive to the first component.

10.4 Case study

Five new cases are employed to verify the retrieval effectiveness of the representation and SOM algorithm for tubular parts, seeing Fig. 10.13.

The first step is converting them into feature-based representation, normalizing data and doing principle component analysis again using parameters from training. Secondly, by using the SOM, the new input vectors are guided to the corresponding units, to which cases match best. As the result of retrieval, all the test samples have matching cases in the map, and T2, T3, T4 have more than one retrievals according to the structure of map. The distances of the sample to similar cases are calculated and the case with least value should be selected to on this condition, as a one-target KNN algorithm. The best matched cases to the test samples are shown in Fig. 10.14. Regarding to the retrieval consistency, T1 is different from its best matching mainly in the cross section shapes; The direction of rectangle section distinguishes T2 pair; T3 is very much similar to the retrieved one and T4 can be considered as a portion of its partner and varies in cutting direction.

The third step is revising and reusing the retrieved result. Figure 10.15 shows an example of development using sample T2 as an instance, when the case (S2) in database is obtained using CBR, the process plan of it should be checked. Since the expansion ratio is slight, this plan consists of bending and calibrate forming stages and this suggested process plan is also suitable and can be reusable for T2. Considering the die design, bendings of the part S2 locate in the same plane (although the directions are opposite), so the splitting surface is exactly in this plane (Fig. 10.15a). as the most similar case, T2′s bendings are in the same plane as well. If the splitting surface is chosen as then former case, pinching will appear since the tube blank exceeds the die cavity (Fig. 10.15b). Thus, the retrieved die sets must be modified by changing the splitting surface direction to meet an appropriate design (Fig. 10.15c).

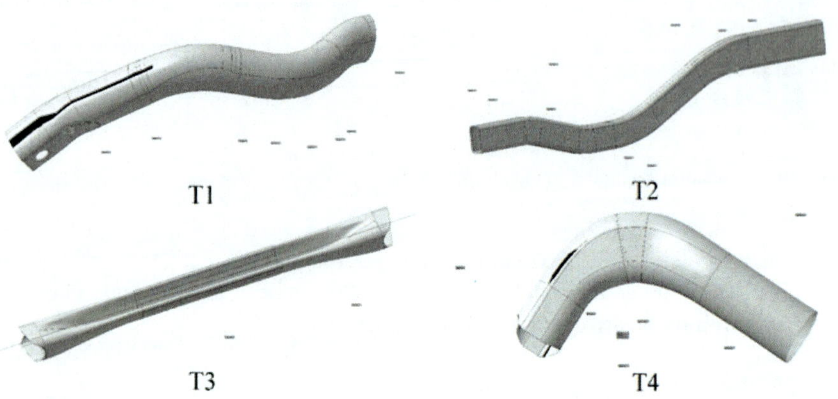

Fig. 10.13 Test cases

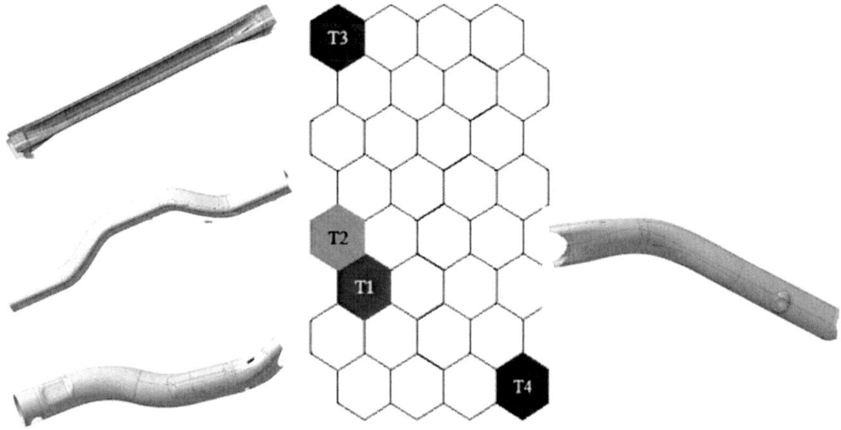

Fig. 10.14 Best matching cases in testing

10.5 Conclusions

As tubular parts for hydro-forming are composed of groups of bending, crushing and expansions, the process planning is very complicated. A new method to deal with the geometric and material data to extract information for KBE use as a feature-based representation and Based on this representation, a self-organized map CBR approach for process design are presented in this paper. The feature-based representation divides a tubular part into several segments with cross sections and local features so that the retrieval mechanism can index similar cases quickly and accurately. The SOM algorithm constructs similar cases more similar cases associating with nodes that are closer in a two-dimension grid, so it is very efficient in high-dimension data cluster. Using the hybrid methodology, process plans and dies can be designed by revising a previous case. It can assist engineers in preserving tacit knowledge of tube hydro-forming and accelerates the parts design and manufacturing.

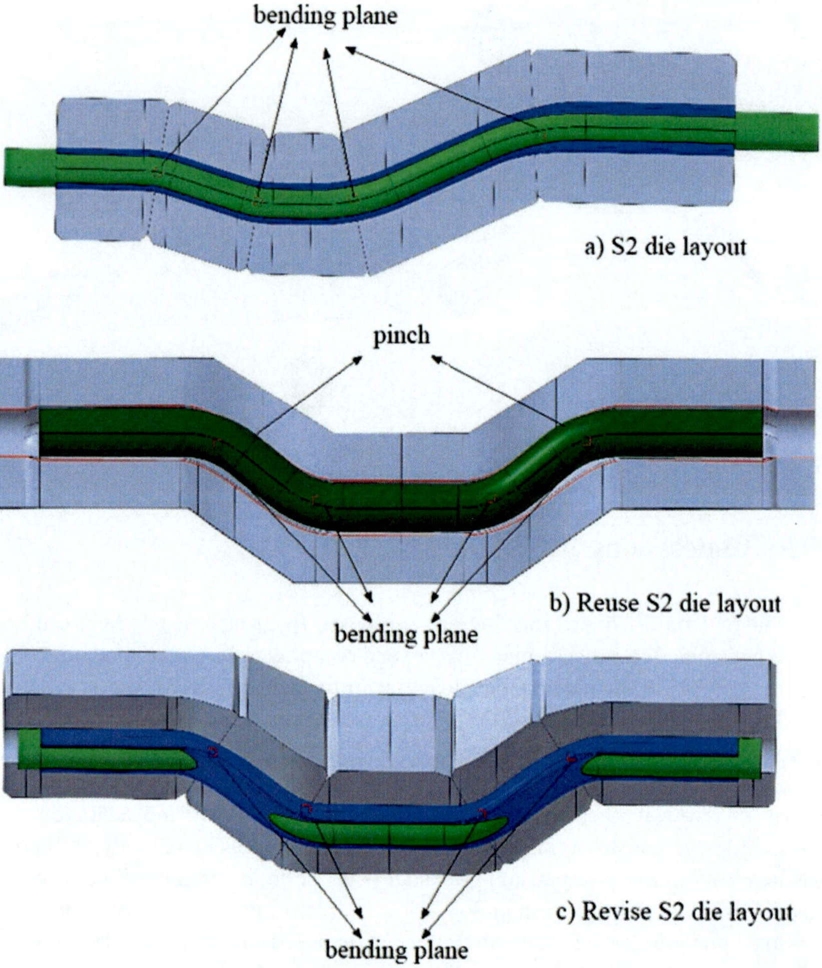

Fig. 10.15 Process planning and design based on the CBR methodology

Acknowledgements The presented work is sponsored by Program of High-End CNC Machine Tools and Basic Manufacturing Equipment under Contract No. 2019ZX04002030.

References

1. Zhang, Y., Chow, C.L., Wang, C.G., Wu, P.: Optimization for loading paths of tube hydroforming using a hybrid method. Mater. Manuf. Process.**24**(6), 700–708 (2009)
2. Zheng, J., Wang, Y., Li, Z.: KBE-based stamping process paths generated for automobile panels. Int. J. Adv. Manuf. Technol. **31**, 663–672 (2007)

3. Neugebauer, R., Werner, M., Pröhl, M., Brunnett, G., Kühnert, T.: New feature extraction and processing methods for the advanced knowledge based process planning of forming operations. Procedia CIRP **28**, 16–21 (2015)
4. Wang, D., Yan, G., Lei, Y., Zhang, J.: A Retrieval Algorithm of Sheet Metal Parts Based on Relationships of Features. Chin. J. Aeronaut. **25**(3), 453–472 (2012)
5. Guo, Y., Hu, J., Peng, Y.: Research of new strategies for improving CBR system. Artif. Intell. Rev. **42**, 1–20 (2014)
6. Tor, S.B., Britton, G., Zhang, W.: Indexing and retrieval in metal stamping die design using case-based reasoning. J. Comput. Inf. Sci. Eng. **3**(4), 353–362 (2003)
7. Zhang, W., Tor, S., Britton, G.: Indexing and retrieval in case-based process planning for multi-stage non-axisymmetric deep drawing. Int. J. Adv. Manuf. Technol. **28**, 12–22 (2006)
8. Tsai, Y.L., You, C.F., Lin, J.Y., Liu, K.Y.: Knowledge-based engineering for process planning and die design for automotive panels. Comput.-Aided Des. Appl. **7**(1), 75–87 (2010)
9. Koc, M. (eds.): Hydroforming for advanced manufacturing. Woodhead Publishing (2008)
10. Hoffmann A., Birkert A.: Design guidelines for hydroformed structural components of aluminium. In: Proceedings of the Hydroforming of Tubes, Extrusions and Sheet Metals, vol. 2, pp. 323–338. MAT-INFO Press, Fellbach (2001)
11. Yuan, S.J.: Modern Hydroforming Technology, 2nd edn. National Defense Industry Press, Beijing (2016)
12. Jia. M.H., Tang, C., Wang, W.G.: Spatial shape prediction of multi-bends metal tube bending. In: 2011 IEEE International Conference on Computer Science and Automation Engineering, pp. 770–775. IEEE, Shanghai (2011)
13. You, C.F., Jeng, C.R., Liu, K.Y.: Design for automotive panels supported by an expert system. INTECH Open Access Publisher (2013)
14. Shu-Hsien, L.: Expert system methodologies and applications—a decade review from 1995 to 2004. Expert Syst. Appl. **28**(1), 93–103 (2005)
15. Hou, J.S.: Applying AHP-based CBR to estimate pavement maintenance cost. Tsinghua Sci Technol **13**, 114–120 (2008)
16. Kohonen, T.: The self-organizing map. Proc IEEE **78**(9), 1464–1480 (1990). IEEE
17. Kolodner, J.L.: Improving human decision making through case-based decision aiding. Ai Mag. **12**(2), 52 (1991)
18. Kohonen, T.: Essentials of the self-organizing map. Neural Netw. **37**, 52–65 (2013)
19. Guo, Y., Hu, J., Peng, Y.: Research on CBR system based on data mining. Appl. Soft Comput. **11**(8), 5006–5014 (2011)
20. Vatanen, T., Osmala, M., Raiko, T., et al.: Self-organization and missing values in SOM and GTM. Neurocomputing **147**(1), 60–70 (2015)

Chapter 11
Numerical and Experimental Study on Orthogonal Cutting of GH4169 Superalloy by Micro-Pit Textured Cutter

Qian-Qian Cai, Jin-Kai Xu, Zhan-Jiang Yu, Zhong-Xu Lian, and Hua-Dong Yu

Abstract GH4169 superalloy has excellent properties such as high strength, high hardness and high temperature resistance, and is widely used in aerospace, automobile manufacturing, defense industry and biomedical fields. In the cutting process of superalloy, there are some problems such as serious tool wear and poor surface quality. The placement of micro-pit arrays on the rake face of cemented carbide tools can effectively improve the friction and wear resistance of the tools and improve the cutting conditions. In this article, ABAQUS software is used to simulate the micro-pit array tool orthogonal micro-cutting GH4169 superalloy. The cutting simulation of GH4169 superalloy was carried out by using micro-pit diameters of 35 μm, 30 μm, 25 μm, 20 μm and non-texture, respectively. The cutting-chip contact state, cutting force, cutting temperature and chip shape during the cutting process were analyzed. The study found that the placement of micro-pit texture can effectively reduce the cutting force during cutting and improve chip breaking. The micro-pit texture can intercept debris and hard particles on the bottom surface of the chip, reducing scratches on the rake face of the tool and the furrow effect of the hard particles, effectively improving the wear resistance of the tool.

Keywords GH4169 superalloy · Tool wear · Micro-pit array · ABAQUS · Cutting simulation

11.1 Introduction

With the rapid development of aviation manufacturing industry, the application of superalloys is more and more extensive [1–3], which is mainly due to its good wear resistance and excellent structure [4–6]. Under the condition of high temperature and high pressure, the superalloy can still maintain high tensile strength and yield strength, and has low creep and high corrosion resistance and oxidation resistance

Q.-Q. Cai · J.-K. Xu (✉) · Z.-J. Yu · Z.-X. Lian · H.-D. Yu
Ministry of Education Key Laboratory for Cross-Scale Micro and Nano Manufacturing, Changchun University of Science and Technology, Changchun 130022, China
e-mail: xujinkai2000@163.com

J. Xu and K. M. Pandey (eds.), *Mechanical Engineering and Materials*, Mechanisms and Machine Science 100,
https://doi.org/10.1007/978-3-030-68303-0_11

[7, 8]. During the cutting process, the microstructure of the superalloy material has high hardness, high shear stress, and low thermal conductivity, which is easy to cause problems such as high cutting force and high cutting temperature. It also has difficulties in chip breaking and serious wear and tear of the tool [9, 10]. Therefore, it is easy to affect the quality of the processed surface of the workpiece and reduce its service life. Therefore, it is necessary to quickly solve the problem of poor performance during the cutting process of superalloys [11, 12].

Based on the bionic tribology theory, texture processing on the tool surface can effectively improve the wear resistance of the tool [13, 14]. Research in the past few decades has confirmed that the placement of micro-texture on the tool surface has a positive effect on improving the friction contact state of the tool-chip interface, anti-adhesion, wear resistance and drag reduction, and storage of chips [15–17]. In this paper, the ABAQUS6.14–1 software is used to carry out the micro-array array orthogonal micro-cutting superalloy simulation test research [18–20]. The solid modeling of the tool and the workpiece were carried out respectively, the micro-pit arrays with different diameters were prepared on the surface of the micro turning tool, and the tool-chip contact state, chip stress distribution, cutting force, cutting temperature and chip shape were analyzed during the cutting process [21–23].

11.2 Creation of Finite Element Model

Solid modeling of the tool and the workpiece respectively, tool side length $s = 6.35$ mm, tool thickness $d = 2.38$ mm, tool rake angle $\gamma = 0°$, back angle $\alpha = 7°$, blade inclination angle $\lambda = 0°$, the tip angle $\theta = 80°$, the blade radius $r = 0.002$ mm, the tip radius is $R = 0.1$ mm; Work piece length $L = 10$ mm, width $k = 10$ mm. In order to be able to accurately carry out simulation experiments, local optimization of the model: tool side length $L = 0.280$ mm, width $d = 0.150$ mm, micro-pit diameter is 0.035 mm, 0.030 mm, 0.025 mm, 0.020 mm, depth is 0.01 mm, 6×10 micro-pit array, pit center distance is 0.06 mm; workpiece length $L_f = 0.600$ mm, width $k_f = 0.080$ mm, cutting depth $a_p = 0.01$ mm, cutting speed v = 400 mm/s, as shown in Fig. 11.1.

The cutting tool material is cemented carbide (92 wt. % WC, 8 wt. % Co). The detailed parameters of physical and mechanical properties are shown in Table 11.1. The workpiece material is superalloy GH4169. The chemical composition and physical mechanical properties of the superalloy are shown in Tables 11.2 and 11.3.

11.3 Analysis and Processing of Simulation Results

After the simulation model is constructed, the calculation is submitted, and the calculation results are post-processed in the ABAQUS6.14–1 Visualization module to output related graphic data.

Fig. 11.1 ABAQUS two-dimensional cutting model

Table 11.1 Physical and mechanical properties of cemented carbide tool materials

Physical and mechanical properties							
Density (g/cm^3)	Thermal conductivity (W/mK)	Hardness (HRA)	Bending strength (GPa)	Elastic modulus (GPa)	Young's modulus (GPa)	Linear expansivity (m/m°C)	Specific heat
14.5	75.36	89	1.5	600	580	4.5×10^{-6}	220

Table 11.2 Basic physical properties of GH4169 superalloy

Density ($Kg \cdot m^{-3}$)	Elastic modulus (GPa)	Poisson's ratio	Melting point (°C)
8240/8250	220	0.3	1260–1320

Table 11.3 J-C constitutive model and damage model parameters of GH4169 Superalloy

A (MPa)	B (MPa)	n	c	m	T_r (K)	Tm (K)	$\dot{\varepsilon}_0$
450	1700	0.65	0.017	0.71	293.15	1593.15	1

11.3.1 Derivative Cutting Effect

In the Visualization module, the tool-chip contact state during the simulation cutting process is observed. It can be seen from Figs. 11.2 and 11.3 that from the first 1.6×10^{-4} s, the chip and the micro-pit texture start to contact, and the chip stress around the micro-pit becomes larger. Figure 11.2b, e can be seen that when the chips flow into or out of the micro-pit, the edge of the micro-pit produces "derivative cutting" of the flowing chip, causing part of the chip on the bottom surface of the chip to fall off and be stored in the micro-pit. As the cutting continues, the micro-pit array performs "derivative cutting" of the chips passing through, and chips are continuously

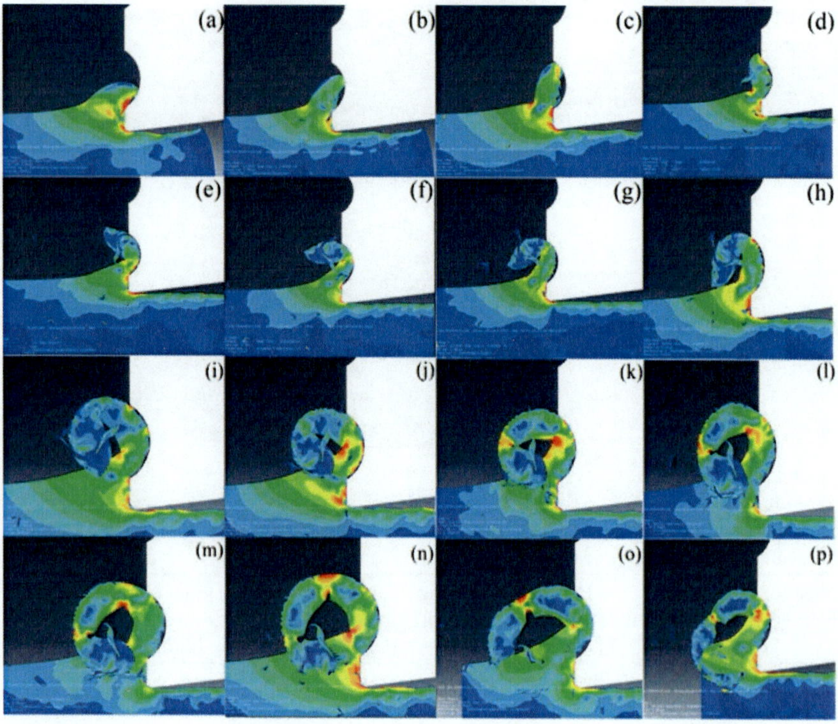

Fig. 11.2 Simulation process of "derivative cutting" with a tool with a diameter of 35 μm

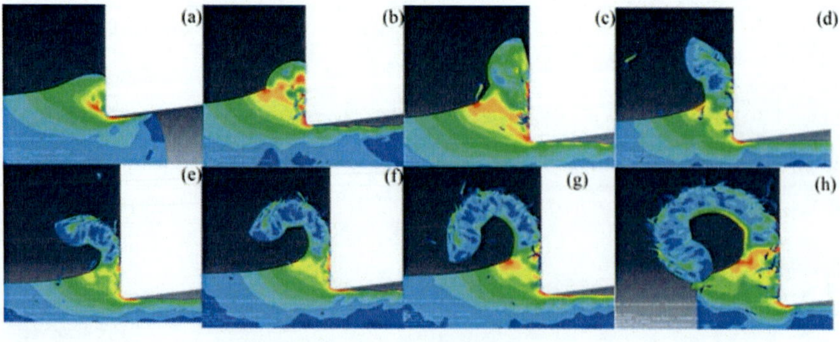

Fig. 11.3 Simulation cutting process of non-textured cutter

generated, and the "derivative cutting force" applied increases the strain of the chips. When the strain reaches a certain limit, the chip breaks (Fig. 11.2k) and "C" type chips are generated. Since this process is a finite element simulation, the stored debris cannot be kept in the pit all the time, and the debris will fall out. It can be seen from Fig. 11.3 that there is no "derivative cutting" phenomenon in non-textured tools.

It can be seen from the finite element simulation that the surface quality of the chip bottom surface after "derivative cutting" is relatively good. The micro-pit array can intercept debris and hard particles on the bottom surface of the chip, reduce the scratching and furrowing effect of the chip bottom on the rake face, and improve the surface quality of the chip bottom surface. It can be seen that the micro-pit texture is conducive to chip breaking, reduces the formation of spiral chips, improves the surface quality of the workpiece and the life of the cutting edge, the micro-pit intercepts a small amount of debris and hard particles on the bottom of the chip, and reduces the bottom face of the chip. The scratching of the surface and the furrow effect of hard particles improve the friction and wear resistance of the tool.

11.3.2 Comparative Analysis of Cutting Force

In the micro-cutting process, the cutting thickness ap is small, and the role of the cutting edge radius must be considered to analyze the force of the tool, as shown in Fig. 11.4. F_x, F_y and μ are defined in Eqs. (11.1), (11.2) and (11.3) respectively.

$$F_x = FR_1 = \frac{2\tau_s b\left[h_D - r + r\sin(\beta - \gamma_0)\right]\cos(\beta - \gamma_0)}{\sin\varphi\cos(\varphi + \beta - \gamma_0)} \tag{11.1}$$

$$F_y = FR_2 = \frac{2\tau_s b\left[h_D - r + r\sin(\beta - \gamma_0)\right]\sin(\beta - \gamma_0)}{\sin\varphi\cos(\varphi + \beta - \gamma_0)} \tag{11.2}$$

$$\mu = \tan\beta = \frac{F_f}{F_n} \tag{11.3}$$

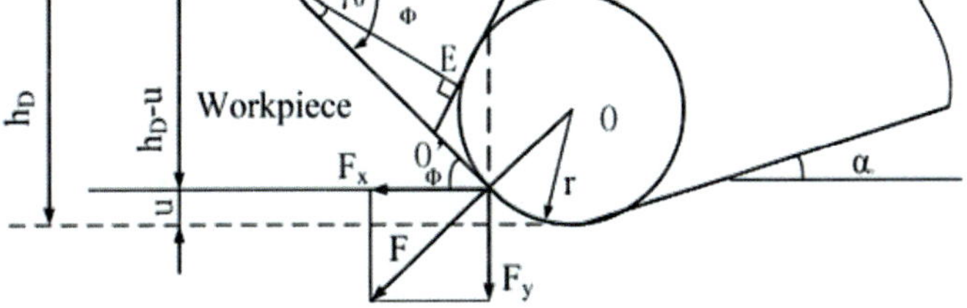

Fig. 11.4 Schematic diagram of micro-cutting cutting force

In above equations, τ_s is the shear stress on the shear plane, b is the cutting width, r is the cutting edge radius, γ_0 is the tool rake angle, β is the friction angle, andφ is the shear angle; F_n is the positive pressure on the rake face, and F_f is the friction force on the rake face. When $\gamma_0 = 0$, you can obtain Eq. (11.4):

$$\mu = \tan\beta = \frac{F_f}{F_n} = \frac{F_y}{F_x} = \frac{FR_2}{FR_1} \tag{11.4}$$

After the simulation results are post-processed, the cutting force components FR_1 and FR_2 acting on the tool are measured. The change of the cutting force is shown in Fig. 11.5. It can be seen that the changes of the main cutting force F_x and the feed force F_y of the micro-pit array tool and the non-textured tool during the micro-cutting simulation process are different. The main cutting force F_x is always greater than the feed force F_y. In the initial stage of cutting, the cutting force continued to increase and reached a certain value, and then decreased. This is because the initial tool continuously squeezed the cutting layer and the cutting force increased. When the equivalent plastic strain value at a certain point in the chip layer is greater than or equal to the fracture strain value of the workpiece material, the mesh fails, shear damage occurs, the chip separates from the workpiece, and the cutting force decreases. With the continuous advancement of cutting, the change of cutting force of non-textured tools tends to be stable, fluctuating around a certain value, which may be caused by the drastic change of the material's thermoplasticity or stress field in the

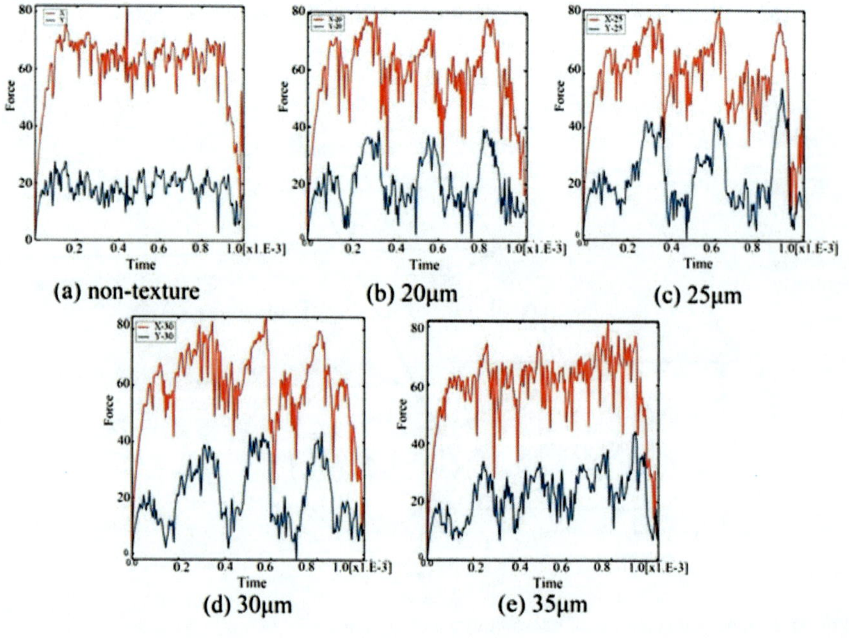

Fig. 11.5 Chip force distribution

micro-cutting test. During the cutting process of the micro-pit array tool, the cutting force changes greatly, and the cutting force waveform changes more violently in a sawtooth shape, which is caused by the increase in cutting force caused by "derivative cutting". When the chip flows through the surface of the micro-pit, the edge of the micro-pit exerts a certain force on the chip passing through to perform "derivative cutting" of the chip so that the cutting force increases.

Taking the average of the main cutting forces tending to be stable to obtain the main cutting force of the non-textured tool $F_x = 60.7$ N, the minimum cutting force of the micro-pit tool of Φ35 μm is $F_x = 51.8$ N, the cutting force of micro-pit array cutters is less than that of non-textured cutters. Taking the average of the main cutting forces tending to be stable to obtain the main cutting force of the non-textured tool $F_x = 60.7$ N, the minimum cutting force of the micro-pit tool of 35 μm is $F_x = 51.8$ N. The cutting force of micro-pit array cutters is less than that of non-textured cutters. The cutting force fluctuations of the pit cutters of 30, 25 and 20 μm are large. This is due to the increase in the strain of the chips when the "derivative cutting" of the flowing chips is performed at the edge of the micro-pits. When the strain is greater than the fracture limit of the chip, the chip fracture generates "C" type chips and the cutting force decreases. The chipping force of 35 μm micro-pit tool is relatively stable, and the cutting force is reduced by 14.7%. The fluctuation of the workpiece and tool caused by "derivative cutting" is small, which reduces the machining error and tool wear caused by the fluctuation of the workpiece or tool. It can be seen that the effect of the change in the diameter of the micro-pits on the cutting force is also different, and the size parameters of the micro-pits need to be properly designed when the tool micro-pit array is placed. Under the premise of ensuring the storage of chips and chip breaking in "derivative cutting", it is also necessary to reasonably control the size and change of cutting force to reduce cutting fluctuations.

11.3.3 Comparative Analysis of Cutting Temperature

Figure 11.6 shows the change in tool temperature during the cutting process. The average cutting temperature of the non-textured tool is $T_1 = 159.7$ °C. The cutting temperature of 35 μm micro-pit tool is $T_2 = 159.9$ °C. The cutting temperature of 30 μm micro-pit tool is $T_3 = 162.2$ °C. The cutting temperature of 25 μm micro-pit cutter is $T_4 = 160.9$ °C. The cutting temperature of 20 μm micro-pit cutter is $T_5 = 161.5$ °C. Tool temperature change is not obvious. This is due to the "derivative cutting" effect of the micro-pit texture tool during the cutting process. The interaction of the chip and the bottom surface increases the friction force, so that the cutting temperature is not significantly reduced. It can be seen that, when placing the tool micro-pits, it is necessary to design the size parameters of the micro-pits reasonably to reduce the cutting temperature changes caused by "derivative cutting".

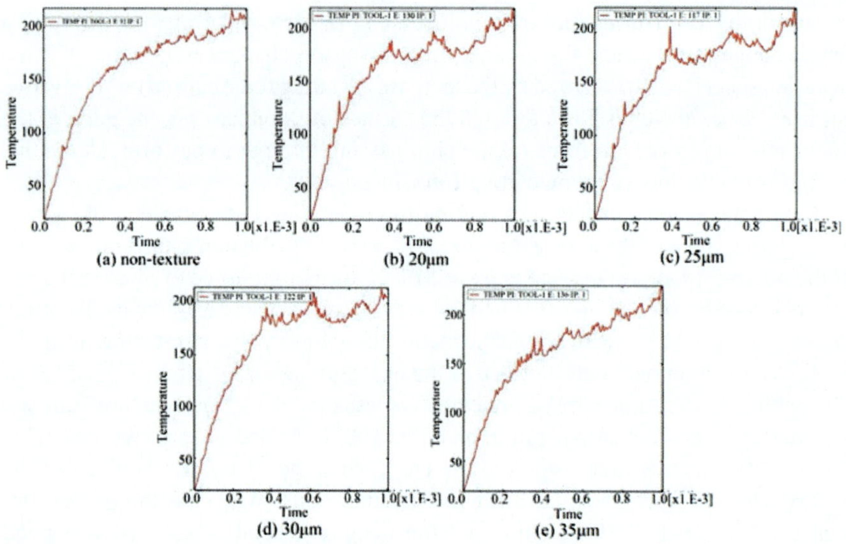

Fig. 11.6 Cutting temperature distribution

11.4 Conclusions

The finite element simulation test of orthogonal cutting of GH4169 superalloy with cemented carbide tools is carried out is this study. The experiment creates the components, materials and other modules, and calculates and analyzes the results to obtain the required data. The contact area, cutting force and cutting temperature of tool chip are compared and analyzed:

The micro-pit texture is conducive to chip breaking, reducing the formation of swarf accumulation chips affecting the surface quality of the workpiece and the life of the cutting edge, reducing the generation of spiral chips. The micro-pits intercept a small amount of debris and hard particles on the lower surface of the chip, reducing the scratching of the rake face and the furrow effect of the hard particles, and improving the wear resistance of the tool.

During the cutting process of GH4169 superalloy, the chip force of the tool with a pit diameter of $\Phi 35\mu m$ is reduced by 14.7%, the stress concentration at the tool tip is reduced, and the generation of crescents at the tool tip during cutting is effectively reduced, improve the wear resistance and durability of the tool. The influence of different micro-pit diameter on cutting force is also different. In the process of tool micro-pit placement, the size parameters of micro-pit array need to be designed reasonably. On the premise that "derivative cutting" can store chip and chip breaking, the size and change of cutting force should be reasonably controlled to reduce cutting fluctuation.

The effect of the placement of micro-pits on the cutting temperature during the cutting process is not obvious. Therefore, it is necessary to properly design the size parameters of the micro-pits when placing the micro-pits in the tool to reduce the impact of "derivative cutting" on the cutting temperature.

The placement of the micro-pit array has a certain influence on the contact area of the tool-chip, the stress distribution of the chip, the generation of chip accumulation, the surface quality of the workpiece, the cutting force and the cutting temperature. By rationally designing the shape and size parameters of the micro-pit array, it is possible to achieve reasonable control of the stress level of the chip contact surface and reduce the stress concentration phenomenon, improve the chip breaking rate, reduce the formation of chip accumulation, improve the surface quality of the workpiece, and reduce cutting force.

Acknowledgements This work was supported by the National Natural Science Foundation of China (No. U19A20103), the China Postdoctoral Science Foundation (No. 2019M661184), and the Jilin Province Scientific and Technological Development Program (No. Z20190101005JH)(No. 20190302076GX).

References

1. Wu, L.H., Jiang, C.H.: Effect of shot peening on residual stress and microstructure in the deformed layer of Inconel 625. Mater. Trans. **58**, 164–166 (2017)
2. Wu, L.H., Jiang, C.H.: Effect of thermal relaxation on residual stress and microstructure in the near surface layers of dual shot peened Inconel 625. Adv. Mech. Eng. **10**, 1–6 (2018)
3. Kong, T., Kang, B., Ryu, H.J., Hong, S.H.: Microstructures and enhanced mechanical properties of an oxide dispersion-strengthened Ni-rich high entropy superalloy fabricated by a powder metallurgical process. J. Alloy. Compd. **839**, 155724 (2020)
4. Song, X., Lei, J., Gu, Z., Zhou, S.: Boosting wear properties of Inconel 718 superalloy by uniform dispersing graphene nanoplatelets through laser melting deposition. J. Alloy. Compd. **834**, 155086 (2020)
5. Zhao, L., Tan, Y., Shi, S., You, X., Li, P., Cui, C.: Microsegregation behavior of inconel 718 superalloy prepared by electron beam smelting layered solidification technology. J. Alloy. Compd. **833**, 155019 (2020)
6. Hu, Z., Qin, C., Chen, X., Tang, A., Fang, T., Yang, Z., Luo, S., Mao, M.: Chemical-mechanical polishing of cemented carbide insert surface for extended tool life in turning of GH4169 nickel-based superalloy. Int. J. Precis. Eng. Man. **21**, 1421–1435 (2020)
7. Shi, P.Y., Wan, S.H., Yi, G.W., Sun, H.W., Yu, Y., Xie, E.Q., Wang, Q.H., Shen, S.Z., Alam, N.: TiO_2–ZnO/Ni–5wt.%Al composite coatings on GH4169 superalloys by atmospheric plasma spray techniques and theirs elevated-temperature tribological behavior. Ceram. Int. **46**(9), 13527–13538 (2020).
8. Yin, Y., He, W., Xie, H., Wu, L.: High-temperature fatigue crack propagation study of superalloy GH4169 by single-lens 3D digital image correlation. Sci. China Technol. Sc. **63**, 693–704 (2020)
9. Yang, J., Liu, D., Zhang, X., Liu, M., Zhao, W., Liu, C.: The effect of ultrasonic surface rolling process on the fretting fatigue property of GH4169 superalloy. Int. J. Fatigue **133**, 105373 (2020)
10. Miao, Q., Ding, W., Kuang, W., Yang, C.: Comparison on grindability and surface integrity in creep feed grinding of GH4169, K403, DZ408 and DD6 nickel-based superalloys. J. Manuf. Process. **49**, 175–186 (2020)

11. Shi, L., Zhang, C.: Experimental research on turning of superalloy GH4169 under high pressure cooling condition. Integr. Ferroelectr. **207**, 75–85 (2020)
12. Pan, L., Wu, Z., Fang, L., Song, Y.: Investigation of surface damage and roughness for nickel-based superalloy GH4169 under hard turning processing. P. I. Mech. Eng. B-J Eng. **234**(4), 679–691 (2020)
13. Pang, K., Wang, D.: Study on the performances of the drilling process of nickel-based superalloy Inconel 718 with differently micro-textured drilling tools. Int. J. Mech. Sci. **180**, 105658 (2020)
14. Pratap, A., Patra, K.: Combined effects of tool surface texturing, cutting parameters and minimum quantity lubrication (MQL) pressure on micro-grinding of BK7 glass. J. Manuf. Process. **54**, 374–392 (2020)
15. Wu, Z., Bao, H., Liu, L., Xing, Y., Huang, P., Zhao, G.: Numerical investigation of the performance of micro-textured cutting tools in cutting of Ti-6Al-4V alloys. Int. J. Adv. Manuf. Tech. **108**(10), 1–2 (2020)
16. Chen, P., Tong, J., Zhao, J., Zhang, Z., Zhao, B.: A study of the surface microstructure and tool wear of titanium alloys after ultrasonic longitudinal-torsional milling. J. Manuf. Process. **53**, 1–11 (2020)
17. Pena-Paras, L., Maldonado-Cortes, D., Rodriguez-Villalobos, M., Romero-Cantu, A.G., Montemayor, O.E.: Enhancing tool life, and reducing power consumption and surface roughness in milling processes by nanolubricants and laser surface texturing. J. Clean. Prod. **253**, 119836 (2020)
18. Wu, Z., Bao, H., Liu, L., Xing, Y., Huang, P., Zhao, G.: Numerical investigation of the performance of micro-textured cutting tools in cutting of Ti-6Al-4V alloys. Int. J. Adv. Manuf. Tech. **108**(10), 463–474 (2020)
19. Daniyan, I., Fameso, F., Ale, F., Bello, K., Tlhabadira, I.: Modelling, simulation and experimental validation of the milling operation of titanium alloy (Ti6Al4V). Int. J. Adv. Manuf. Tech. **109**, 1853–1866 (2020)
20. Liu, X., Xiong, R., Xiong, Z., Zhang, S., Zhao, L.: Simulation and experimental study on surface residual stress of ultra-precision turned 2024 aluminum alloy. J. Braz. Soc. Mech. Sci. **42**, 386 (2020)
21. Mane, S., Joshi, S.S., Karagadde, S., Kapoor, S.G.: Modeling of variable friction and heat partition ratio at the chip-tool interface during orthogonal cutting of Ti-6Al-4V. J. Manuf. Process. **55**, 254–267 (2020)
22. Gao, M.Y., Zhang, K., Zhou, Q., Zhou, H.F., Liu, B.L., Zheng, G.J.: Numerical investigations on the effect of ultra-high cutting speed on the cutting heat and rock-breaking performance of a single cutter. J. Petrol. Sci. Eng. **190**, 107120 (2020)
23. Zhang, W., Zhuang, K., Pu, D.: A novel finite element investigation of cutting force in orthogonal cutting considering plough mechanism with rounded edge tool. Int. J. Adv. Manuf. Tech. **108**, 3323–3334 (2020)

Chapter 12
Strength Analysis of a Plate Heat Exchanger with Two Design-By-Analysis Methods

Zhenning Liu, Huifang Li, and Caifu Qian

Abstract Plate heat exchangers are widely used equipment in engineering. As the structure of the equipment is complicated, it is hard to perform accurate design with the traditional method or so-called design-by-rules. In this paper, two design-by-analysis methods, namely Elastic Stress Analysis Method and Limit-load Analysis Method were used to perform stress assessment of a plate heat exchanger. Results show that the plate heat exchanger meets the strength requirements according to ASME VIII-2. It is also found that if applying Elastic Stress Analysis Method, a lot of SCLs should be specified on components of the plate heat exchanger and stresses have to be categorized according to their effects on the strength failure of the equipment. If applying Limit-load Analysis Method, the materials are assumed to be ideal elastic-plastic, and no Stress Classification Lines (SCLs) and stress categorization are needed. Strength assessment is simply to find convergent solutions for the maximum loading. So if the equipment is complicate in structure like the plate heat exchanger, it is suggested to apply Limit-load Analysis Method to perform strength assessment.

Keywords Plate heat exchanger · Finite element analysis · Strength assessment · Design-by-analysis

12.1 Introduction

Heat exchangers are devices that transfer heat from hot fluid to cold fluid to meet specific process requirements or to recover energy. Based on the structures, heat exchangers can be divided into shell-and-tube heat exchangers, plate heat exchangers, etc. Plate heat exchangers are composed of multiple plates that have different corrugations to enhance the heat transfer between the two fluids [1]. Plate heat exchangers have higher heat transfer coefficient than shell-and-tube heat exchangers [2] and are considered to be the ideal equipment for liquid-liquid heat exchange and vapor-liquid exchange. A lot of studies were addressed on the performance of plate heat

Z. Liu (✉) · H. Li · C. Qian
Beijing University of Chemical Technology, Beijing 100029, China
e-mail: 1758609342@qq.com

J. Xu and K. M. Pandey (eds.), *Mechanical Engineering and Materials*,
Mechanisms and Machine Science 100,
https://doi.org/10.1007/978-3-030-68303-0_12

exchangers. Shokouhmand and Hasanpour [3] found that reducing the uneven distribution can improve the thermal efficiency of the plate heat exchangers. Corrugation has a great influence on the heat transfer efficiency of plate heat exchangers. Islamoglu and Parmaksizoglu [4] found that increasing in the height of the corrugation leads to an increase in the coefficient of friction and a reducing in heat transfer efficiency. Grijspeerdt et al. [5] used CFD (Computational Fluid Dynamics) software to build optimized corrugated model of plate heat exchanger. Galeazzo et al. [6] used CFD to perform simulation calculations and found that turbulence in the flat flow channel is only partially generated, and the main state is laminar. The flow state of non-Newtonian fluid in the plate heat exchanger was also studied by Femandes et al. [7]. The sequence of chemical reactions also has a certain impact on fluid flow [8]. Barbaryan et al. [9] used numerical analysis methods to design low fluid pressure safety valves. Nagaraju et al. [10, 11] studied the flow state of magnetic fluid in pipes.

Under the combined action of the plate-side pressure, shell-side pressure and temperature field, a possible failure mode of the heat exchangers is cracking owing to insufficient structural strength [12]. So it is needed to perform strength assessment for the heat exchangers. Yang Xirong simulated the stress distribution on U-shaped heat exchangers [13]. Hoseinzadeh et al. [14] conducted life fatigue analysis on heat exchangers in power plants. Fen Xiao [15] performed stress analysis on fixed tube-plate heat exchangers. Regarding the influences of the working environment on the heat exchangers, Gagliardia [16] studied the plate heat exchangers working in the river and found that its failure reason was related to the upstream water treatment process.

As the structure of plate heat exchangers is complicated, it is hard to perform accurate design with the traditional method or so-called design-by rules. So in this paper, finite element method was used to perform stress analysis for a plate heat exchanger under different load cases. Two methods, namely stress category method and limit analysis method, are applied based on ASME VIII-2 to perform strength assessment of the heat exchanger.

12.2 Finite Element Modelling

12.2.1 Geometry and Grid Models of the Plate Heat Exchanger

The basic design parameters of the plate heat exchanger are listed in Table 12.1. As shown in Fig. 12.1 for the structure, the heat exchanger is composed of curved plate, pipes, shell, supports, etc.

Solid186 with the software ANSYS was used to mesh the plate heat exchanger. The total number of grids is 1526053. The mesh quality was checked in terms of skewness, and the average value was 0.14, meaning that the mesh quality is good.

Table 12.1 Design parameters

Parameter/Unit	Plate-side	Shell-side
Design pressure (MPa)	0.1	0.4
Design temperature (°C)	170	50
Working pressure (MPa)	0.01	0.3
Medium	Air	Water
Corrosion allowance (mm)	0	0
Material	S30408	S30408

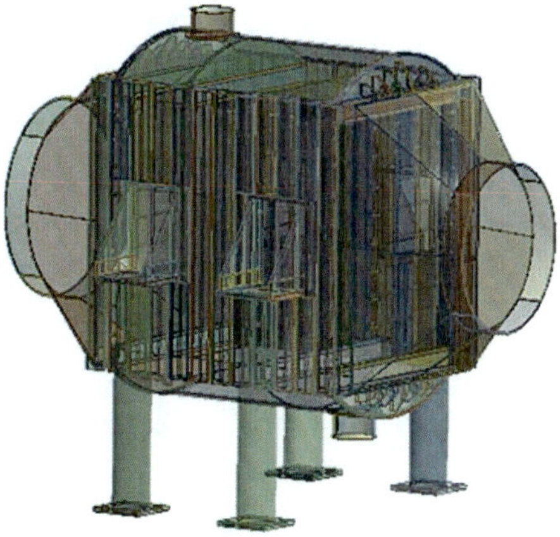

Fig. 12.1 Geometric model of the plate heat exchanger

12.2.2 Load and Constraint Conditions

Design pressures, dead load, seismic load, external nozzle load, and equivalent nozzle force are considered in this study. Among them, seismic load is assumed to be static with the acceleration being equal to 0.2 times of the gravitational acceleration. Each nozzle is acted by the equivalent nozzle force with magnitude being calculated by the Eq. (12.1):

$$p_e = \frac{d_i^2}{d_o^2 - d_i^2} p \tag{12.1}$$

where p is the pressure on the inner wall of nozzles.

On the air inlet and outlet nozzles of the equipment, there are also external loads with the magnitudes listed in Table 12.2.

In this study, two load cases are specified for static strength assessment as listed in Table 12.3 for the included pressures.

As shown in Fig. 12.2, in case 1, only the plate-side surface of the heat exchanger was loaded with the pressure of 0.1 MPa. The equivalent force applied on the air outlet and inlet nozzles is 2.5 MPa. Fixed constraint was imposed on the supporters.

Table 12.2 External loads on nozzles

	Force (N)			Moment (N·m)		
	Fx	Fy	Fz	Mx	My	Mz
Air inlet nozzle	−66,100	18,800	−32,600	−23,100	79,200	−5000
Air outlet nozzle	5000	50,700	20,200	35,000	13,700	−25,200

Table 12.3 Pressures in load cases for static strength assessment

	Static load cases	
	Case 1	Case 2
Plate-side pressure(MPa)	0.1	0
Shell-side pressure(MPa)	0	0.4

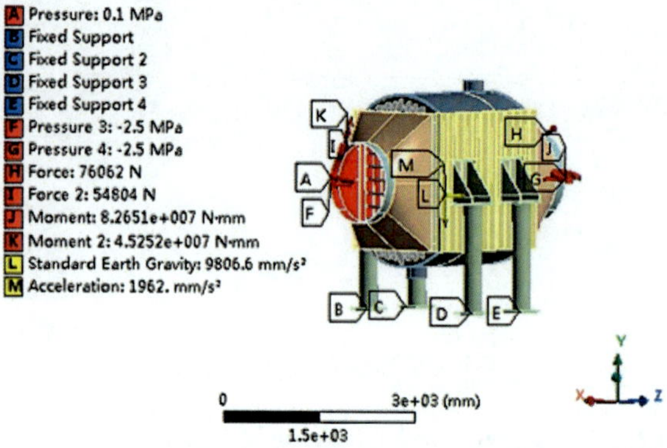

Fig. 12.2 Loads and constraints in case 1

As shown in Fig. 12.3, in case 2, only the shell-side surface of the heat exchanger was loaded with the pressure of 0.4 MPa. The equivalent force applied to the water outlet and inlet nozzles is 3.765 MPa. Fixed constraint was imposed on the supporters.

12.3 Strength Assessment Based on Elastic Stress Analysis Method

Elastic Stress Analysis Method assumes the material of the equipment is linear elastic. Stresses on the equipment will be classified into different categories in terms of Mises equivalent stress. Different categories of stresses will be limited with different allowable values according to their effects on the possible strength failure of the equipment. The allowable stress S for S30408 at the design temperature is 134.2 MPa.

The Mises equivalent stress distributions at the plate heat exchanger under different working cases were shown in Figs. 12.4–12.5.

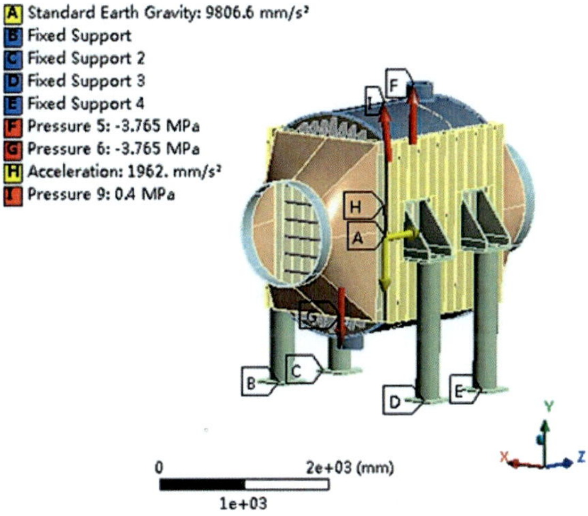

Fig. 12.3 Loads and constraints in case 2

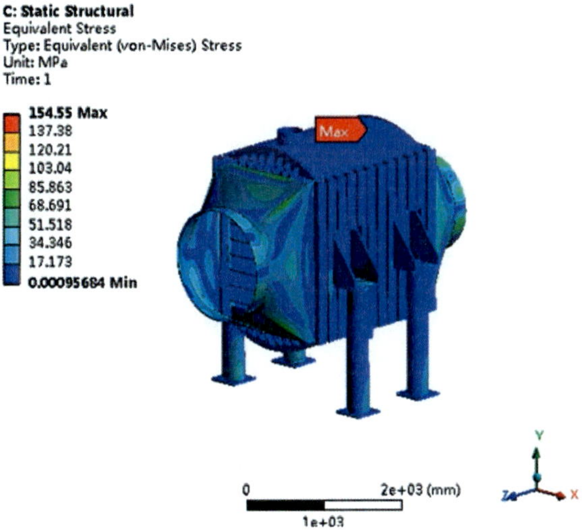

Fig. 12.4 Mises equivalent stress distribution under load case 1

The Mises equivalent stress in case 1 is 154.55 MPa which is located at the connection zone between the square-circle transition shell and connecting plate. The maximum Mises equivalent stress in case 2 is 189.15 MPa which is located at the T-shaped ribs.

According to Elastic Stress Analysis Method, Mises equivalent stresses should be categorized into General Primary Membrane Equivalent Stress (P_m), Local Primary Membrane Equivalent Stress (P_L), Primary Membrane (General or Local) Plus Primary Bending Equivalent Stress ($P_L + P_b$) and Primary Membrane (General or Local) Plus Primary Bending Equivalent Stress Plus secondary Equivalent Stress ($P_L + P_b$ + Q). For calculating the equivalent stresses, the total stress shall be

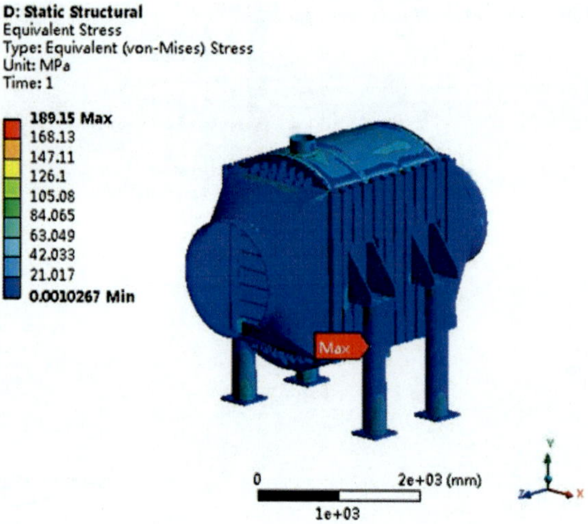

Fig. 12.5 Mises equivalent stress distribution under load case 2

linearized on a stress component basis to produce membrane and bending stresses. Membrane and bending stresses are developed on cross sections through the thickness of a component. Therefore, Stress Classification Lines (SCLs) should be specified orienting normal to contour lines of the stress component of highest magnitude.

In this study, SCLs or paths are specified at components of the equipment. As an example, Fig. 12.6 shows the paths at the connection zone between the square-circle transition shell and air inlet nozzle under load case 1. Table 12.4 lists stress categorization and assessment for each path. It should be noted, like at the small end of a conic shell, the local primary membrane equivalent stress P_L at the round end of square-circle transition shell is limited by 1.1 S.

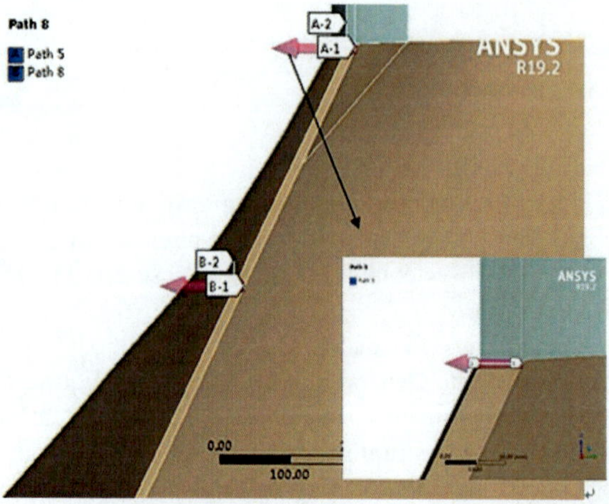

Fig. 12.6 Paths at the connection zone between the square-circle transition shell and air inlet nozzle under load case 1

Table 12.4 Stress categorization and assessment at the connection zone between the square-circle transition shell and air inlet nozzle under load case 1

Paths	Stress category	Calculated stress, MPa	Allowable stress, MPa	Assessment results
Path 8	P_m	19.20	134.2	Pass
Path 5	P_L	43.60	147.6	Pass
	$P_L + P_b + Q$	58.29	402.6	Pass

Similarly, as an example, Fig. 12.7 shows the paths at the connection zone between the curved plate and water inlet nozzle under load case 2. Table 12.5 lists stress categorization and assessment for each path.

It turns out that all types of Mises equivalent stresses in the whole equipment are less than the corresponding allowable values according to ASME VIII-2, meaning the plate heat exchanger meets the strength requirements based on Elastic Stress Analysis Method.

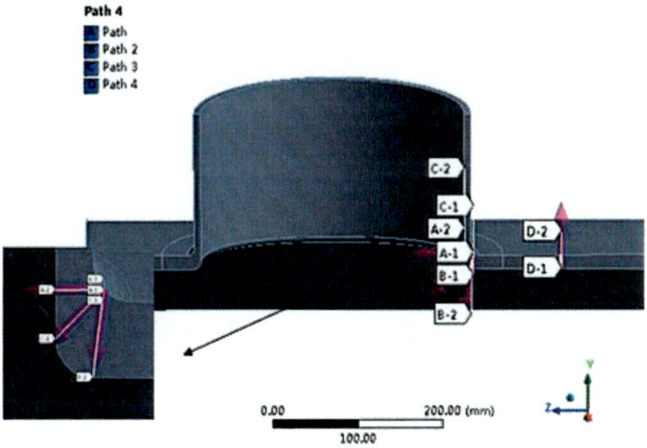

Fig. 12.7 Paths at the connection zone between the curved plate and water inlet nozzle under load case 2

Table 12.5 Stress categorization and assessment at the connection zone between the curved plate and water inlet nozzle under load case 2

Paths	Stress category	Calculated stress, MPa	Allowable stress, MPa	Assessment results
Path 4	P_m	21.76	134.2	Pass
Path 1	P_L	12.61	147.6	Pass
	$P_L + P_b + Q$	88.92	402.6	Pass
Path 2	P_L	24.17	147.6	Pass
	$P_L + P_b + Q$	111.32	402.6	Pass
Path 3	P_L	23.68	147.6	Pass
	$P_L + P_b + Q$	110.8	402.6	Pass

12.4 Strength Assessment Based on Limit-Load Analysis Method

When the maximum stress reaches the yields limit, the actual structure usually does not lose its load-carrying capability immediately. With the increase of the load, the stress in the high stress zone first reaches the yield limit, and the material enters the yield state. Then the stress will be redistributed and the range of plastic zone will be expanded. The ultimate load refers to the load corresponding to the plastic state when the component changes from elastic state to plastic state under the action of external load. When the ultimate load is reached, the component will enter the instability state. The results of the Limit-load Analysis Method are closer to the real failure conditions.

The ideal elastic-plastic material mode is adopted in the Limit-load Analysis Method as shown in Fig. 12.8 for the material used in the plate heat exchanger studied here.

For the plate heat exchanger studied here, two load cases are specified in the limit-load analysis. The first is only applying pressure on the plate-side and finding the maximum pressure by performing limit-load analysis. The second is only applying pressure on the shell-side and finding the maximum pressure by performing limit-load analysis.

Dead load, seismic load, external nozzle load, equivalent nozzle force and constraints for the limit-load analysis are the same as those for stress category analysis discussed above. Gravity acceleration, seismic load, and external nozzle load do not change with time.

Limit-load analysis is a non-linear analysis, so load-steps and sub-steps are required. Based on the limit-load analysis theory, the limit load, i.e. the maximum

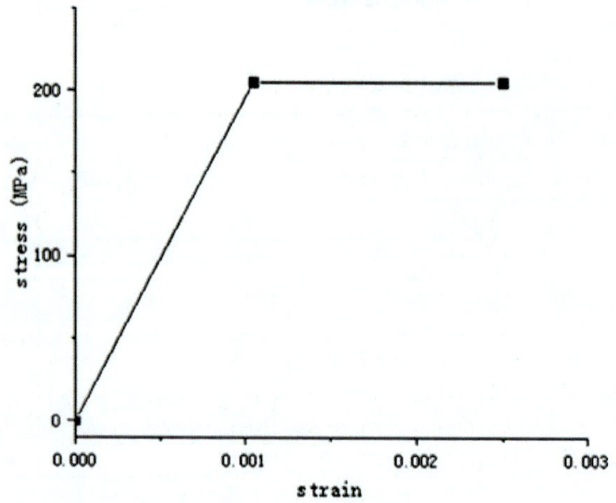

Fig. 12.8 Ideal elastic-plastic stress-strain curve

pressure is the one corresponding to the last sub-step before the calculation becomes divergent. In this study, 1 load-step and 100 sub-steps are set for the computation. About 48hrs are required to finish a computation to get the limit load.

For the first load case where the pressure only acts on the plate-side, it is found that the calculation becomes divergence after the 68th sub-step. Figure 12.9 shows the load-displacement relationship corresponding to each sub-step before divergence. When the deformation is not large, the displacement increment of each sub-step is stable. But the slope of the curve gradually decreases to zero, which means that the structure could generate a very large displacement increment under a very small load increment. From the last step results before divergence, it is found the maximum pressure or the limit pressure is 0.53 MPa. If taking 1.5 as the safety factor, the allowable pressure is 0.35 MPa, much larger than the design pressure (0.1 MPa) on the plate-side, meaning the strength on this side meets the strength requirement. Figure 12.10 shows the von Mises distribution under the maximum pressure on the plate-side. It is clear that the yielding zone is at air outlet nozzle connecting area.

For the second load case where the pressure only acts on the shell-side, it is found that the calculation becomes divergence after the 66th sub-step. Figure 12.11 shows the load-displacement relationship corresponding to each sub-step before divergence and it turns out the maximum pressure or the limit pressure is 2.25 MPa. If taking 1.5 as the safety factor, the allowable pressure is 1.5 MPa, much larger than the design pressure (0.4 MPa) on the shell-side, meaning the strength on this side meets the strength requirement. Figure 12.12 shows the von Mises distribution under the maximum pressure on the shell-side. It is found that the yielding zone is at water outlet nozzle connecting area.

As all allowable pressures are larger than the corresponding design pressures, the plate heat exchanger meets the strength requirements based on Limit-load Analysis Method according to ASME VIII-2.

Fig. 12.9 Load-displacement relationship when only plate-side pressure was applied

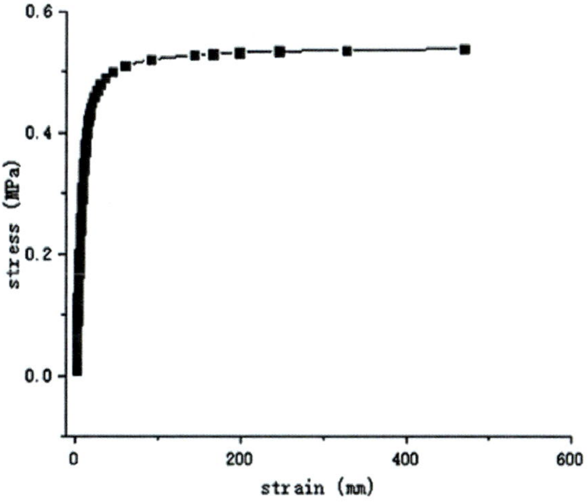

Fig. 12.10 von Mises distribution under the maximum pressure on the plate-side

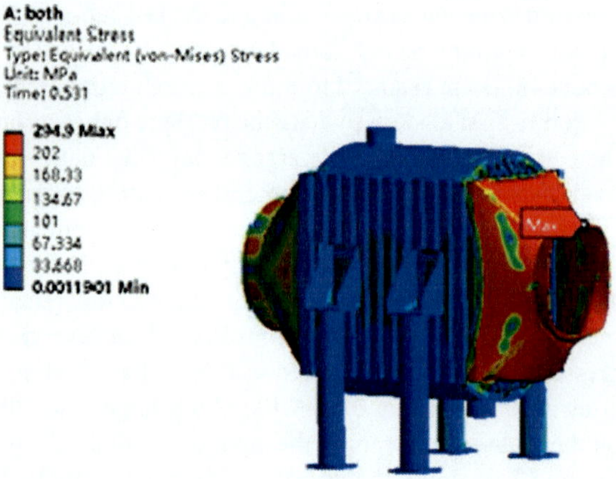

Fig. 12.11 Load-displacement relationship when only shell-side pressure was applied

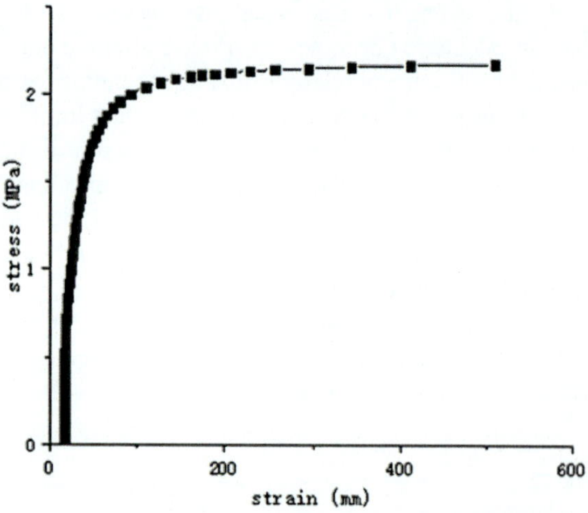

12.5 Conclusions

In order to make accurate and effective strength design of plate heat exchangers, finite element analysis was conducted to evaluate the static strength of a plate heat exchanger under the different load cases. Two design-by-analysis methods, namely Elastic Stress Analysis Method and Limit-load Analysis Method were used to perform stress assessment. Results show that the plate heat exchanger meets the strength requirements according to ASME VIII-2. It is also found that if applying Elastic Stress Analysis Method, a lot of SCLs should be specified on components of the plate heat exchanger and the stresses have to be categorized according to their effects on the strength failure of the equipment, which significantly complicates the computation and strength assessment. If applying Limit-load Analysis Method, the

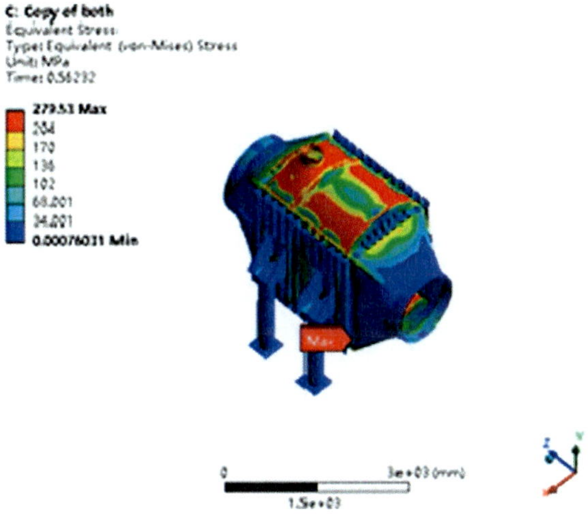

Fig. 12.12 von Mises distribution under the maximum pressure on the shell-side

materials are assumed to be ideal elastic-plastic, which is conservative since real materials are of strain-hardening and can still undertake some more loadings after yielding. In addition, no SCLs and stress categorization are needed and the strength assessment is simply to find convergent solutions for the maximum loading. So if the equipment is complicate in structure like the plate heat exchanger, it is suggested to apply Limit-load Analysis Method to perform strength assessment.

References

1. Numan, A., Mohd, I., Sayyed, H.: Study and analysis of plate type heat exchanger. Invertis J. Renew. Energy **7**(3), 137–141 (2017)
2. Pelliccione, A.S., SantAnna, R., Siqueira, M.H.S., Ribeiro, A.F., Ramos, J.E., Silva, O.P., Pimentel, M.F.: Failure analysis of a titanium plate heat exchanger–Mechanical fatigue. Eng. Fail. Anal. **105**, 1172–1188 (2019)
3. Shokouhmand, H., Hasanpour, M.: Effect of flow maldistribution on the optimal design of plate heat exchanger using constrained multi objective genetic algorithm. Case Stud. Thermal Eng. **18**, 100570 (2020)
4. Islamoglu, Y., Parmaksizoglu, C.: The effect of channel height on the enhanced heat transfer characteristics in a corrugated heat exchanger channel. Appl. Therm. Eng. **23**(8), 979–987 (2003)
5. Grijspeerdt, K., Hazarika, B., Vucinic, D.: Application of computational fluid dynamics to model the hydrodynamics of plate heat exchangers for milk processing. J. Food Eng. **57**(3), 237–242 (2003)
6. Galeazzo, F.C.C., Miura, R.Y., Gut, J.A.W., Tadini, C.C.: Experimental and numerical heat transfer in a plate heat exchanger. Chem. Eng. Sci. **61**(21), 7133–7138 (2006)
7. Femandes, C.S., Dias, R., Nobrega, J.M., et al.: Simulation of stirred yoghurt processing in plate exchangers. J. Food Eng. **69**(3), 281–290 (2005)
8. Science - Applied Physical Science; Research Data from Department of Mathematics Update Understanding of Applied Physical Science (Order of chemical reaction and convective

boundary condition effects on micropolar fluid flow over a stretching sheet). J. Technol. Sci. (2018)
9. Heat and Fluid Flow; Study Results from University of Pretoria in the Area of Heat and Fluid Flow Reported (Developing a low-fluid pressure safety valve design through a numerical analysis approach). Mathematics Week (2020)
10. Nagaraju, G., Garvandha, M.: Magnetohydrodynamic viscous fluid flow and heat transfer in a circular pipe under an externally applied constant suction. Heliyon 5(2) (2019)
11. Murthy, J.V.R., Nagaraju, G., Sai, K.S.: Numerical solution for MHD flow of Micro polar fluid between two concentric rotating cylinders with porous lining. Int. J. Nonlinear Sci. **13**(2), 183–193 (2012)
12. Xu, L., Qian, C.F., Liu, J.Y., Liu, Z.S.: Thermal stress analysis of tube sheet of floating head heat exchanger. Pressure Vessel **32**(6), 55–60 (2015). (In Chinese)
13. Yang, X.R., Wang, Z.W., Meng, X.Y.: U-tube heat exchanger tube sheet and tube stress calculation and fatigue analysis. Contemporary Chem. Ind. **48**(04), 855–858 (2019). (In Chinese)
14. Hoseinzadeh, S., Heyns, P.S.: Thermo-structural fatigue and lifetime analysis of a heat exchanger as a feedwater heater in power plant. Eng. Fail. Anal. **113**, 104548 (2020)
15. Xiao, F., Chen, Z.W.: Stress analysis and fatigue analysis of fixed tube-plate heat exchanger. Chem. Equip. Pipel. **47**(5), 5–7 (2010). (In Chinese)
16. Gagliardi, A., Lanzutti, A., Simonato, M., Furlanetto, R., Magnan, M., Andreatta, F., Fedrizzi. L.: Failure analysis of a plate heat exchanger used in a blast chiller. Eng. Failure Anal. **92**, 289–300 (2018)

Chapter 13
Study on Bearing Mechanism of Composite Stiffened Panel with Delamination Under Shear Load

Xue Bi, Peng Zou, and Xiangming Chen

Abstract As a typical thin wall structure, the composite stiffened panel not only has the splendid properties of composite material, but also shows the excellent performance of the composite wall structure with low quality and high bearing capacity. For the structure, the buckling and post-buckling failure of the panel and the effect of damage on the performance of the stiffened panel will happen. Thus, in this paper, the effect of damage on post-buckling bearing capacity of composite stiffened panels under shear load is researched. Three finite element models are built, including no damage shear model, delamination embedded shear model and shear after impact model. The shear failure, effect of delamination and impact damage on panel bearing strength is discussed. Furthermore, the effects of the size of delamination and position of impact on structure performance are also researched. These researches provided abundant theoretical supports for the experiment design and hopeful to be applied in industry manufacturing.

Keywords Shear load · Delamination · Stiffened panel

13.1 Introduction

Due to the excellent specific strength, specific stiffness and fatigue resistance, carbon fiber reinforced polymer (CFRP), is widely used in the aviation and aerospace industries. The composite stiffened panel, as a typical thin wall structure with low quality and high bearing capacity, plays a significant role in aircraft design because of its outstanding performance, such as fuselages, horizontal/vertical tail wings and wing. For the research on typical failure modes of composite stiffened panel, more attention should be paid to the buckling and post-buckling performance failure and damage defect. The bearing capacity of composite stiffened panel under different loads have been studied by many famous relevant institutions, such as COCOMAT [1] (Improved

X. Bi (✉) · P. Zou · X. Chen
Full Scale Aircraft Structural Static/Fatigue Laboratory, AVIC Aircraft Strength Research Institute of China, Xi'an 710065, China
e-mail: 441811017@qq.com

J. Xu and K. M. Pandey (eds.), *Mechanical Engineering and Materials*, Mechanisms and Machine Science 100,
https://doi.org/10.1007/978-3-030-68303-0_13

Material Exploitation at Safe Design of Composite Airframe Structures by Accurate Simulation of Collapse) and POSICOSS (Improved Post-buckling Simulation for Design of Fiber Composite stiffened Fuselage Structures) [2] etc.

The bearing capacity of composite stiffened panel is easily affected by low velocity impact damage that may occur during the working process of tool dropping, runway gravel and hail [3, 4]. In the impact damage, both barely visible impact damage (BVID) and visible impact damage (VID) can be encountered and cause a significant decrease in structural strength and endanger flight safety. In addition, reinforced composite materials often bear in-plane compression load and shear load during aircraft service, so relevant researchers have carried out a lot of research work on composite stiffened panels after impact [5–22]. Buckling and post buckling behavior of Glare laminates containing splices and doublers were researched from both instrumented tests [5] and numerical modelling [6]. Yu Feng et al. investigated effect of impact damage positions on the buckling and post-buckling behaviors of stiffened composite panel under compression load [7], shear load [8], fatigue load and compression-after-impact- fatigue (SAIF) load [9] and shear-after-impact- fatigue (SAIF) behaviors [10]. Sun Wei et al. [11, 12] made an experimental and finite element research on effect of stiffener damage caused by low velocity impact on compressive buckling and failure modes of T-stiffened composite panels. In addition, the edge impact damage, which shows different damage mechanism on Compression-After-Impact (CAI) behavior of stiffened composite panels, is discussed in [13]. The panel style is also a research point, except T-stiffened panel [14–16], J-stiffened panel [17–19] and hat-stiffened panel [20–22]. It is found that the research on the bearing capacity of stiffened panels after impact is mainly focused on compression load, while the research on in-plane shear load is limited, especially for the research on the influence of delamination position and size on the post buckling performance of structures, there is still a lack of targeted research work. Therefore, this paper will focus on the failure mechanism of composite stiffened panels under in-plane shear loading.

13.2 Question Description

13.2.1 Configuration and Size of Specimen

In this project, the effect of embedded damage on load-bearing failure and post buckling performance of three stringer stiffened panels under shear load is studied. Two kinds of simulation models are studied: nondestructive model and embedded delamination damage model. The type of stringer is type I. The configuration and detailed dimensions of the test piece are shown in Fig. 13.1. The configuration size and section of type I stringer are shown in Fig. 13.2. The paving of each part of the stiffened plate is shown in Table 13.1. Among them, the skin and the stringer are glued together, and the adhesive film is J-116b with a thickness of 0.2 mm.

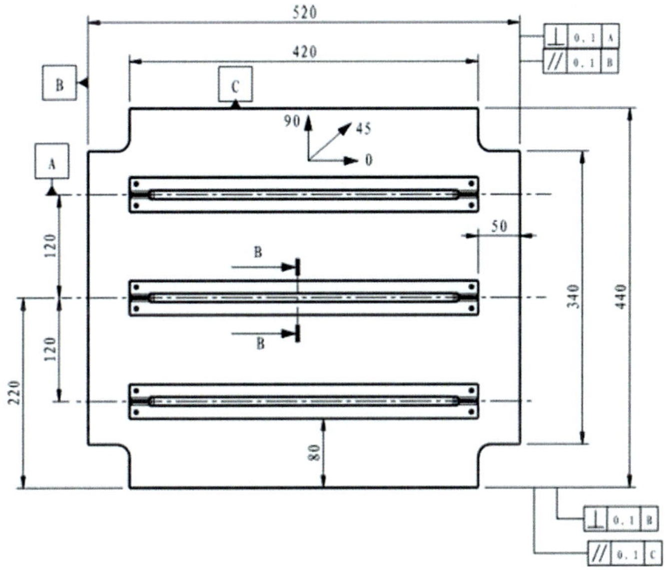

Fig. 13.1 Dimensions of shear test pieces for type I stiffened plates (mm)

13.2.2 Material System

The laminate material system is CCF300/BA9916-II, which is manufactured by conventional prepreg method. The nominal thickness of the single layer is 0.155 mm, the nominal fiber volume content is about 62%, and the porosity is within 2.0%. The material properties are shown in Table 13.2 below.

13.2.3 Size and Location of Precast Layers

This study mainly simulates the delamination prefabricated by polytetrafluoroethylene film between the truss and the skin. The position of the simulated delamination in the finite element method is shown in Fig. 13.3. The delamination dimensions are 40 × 40 mm, 80 × 40 mm and 120 × 40 mm.

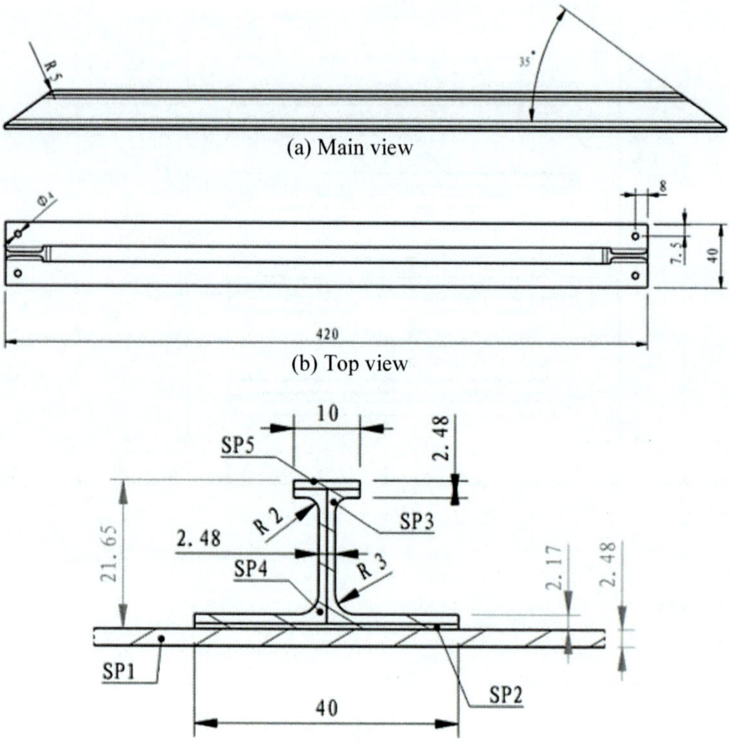

(a) Main view

(b) Top view

(c) Section

Fig. 13.2 Configuration and size of I-type stringer (mm)

13.3 Numerical Simulation Method

13.3.1 Finite Element Model

According to the specimen configuration and size in the previous section, two types of finite element models are established in this study: (a) the non-destructive shear loading model; (b) the prefabricated layered shear loading model between the truss and skin, as shown in Fig. 13.4 below.

The main components of various models are described as follows: for (a) model, it is mainly composed of skin, bottom plate of flange, flange-web-upper flange (right), flange-web-upper flange (left) and upper flange roof, which is mainly composed of traditional deformable shell elements. Meanwhile, in order to keep the basic conditions of the other two models consistent, and to compare and analyze the effects of damage and delamination on the post buckling capacity under shear load, cohesive elements are preset between 1–2, 8–9 and 15–16 layers of the skin, and between the skin and the bottom flange plate. In model (b), based on model (a), the corresponding

Table 13.1 Laying sequence of each part of stiffened plate

Ply code	Positon	Ply ratio (0/ ± 45/90)	Laying sequence	Number of layers	Nominal thickness [mm]
SP1	Skin	(37.5/50/12.5)	$[45/-45/0/45/90/0/-45/0]_S$	16	2.48
SP2	Bottom plate of flange	(33/50/17)	[0/45/90/0/-45/45]	6	0.93
SP3	Flange, web and upper flange (right)	(37.5/50/12.5)	[0/-45/0/90/45/0/-45/45]	8	1.24
SP4	Flange, web and upper flange (left)	(37.5/50/12.5)	[0/45/0/90/-45/0/45/-45]	8	1.24
SP5	Upper flange roof	(37.5/50/12.5)	[0/-45/0/90/45/0/-45/45]	8	1.24

Table 13.2 CCF300/BA9916-II, unidirectional board performance

E_1 [Gpa]	E_2 [Gpa]	G_{12} [Gpa]	12	X^T [Mpa]	X^C [Mpa]	Y^T [Mpa]	Y^C [Mpa]	S [Mpa]
124.8	9.755	5.26	0.312	1871	1291	59.7	215	102

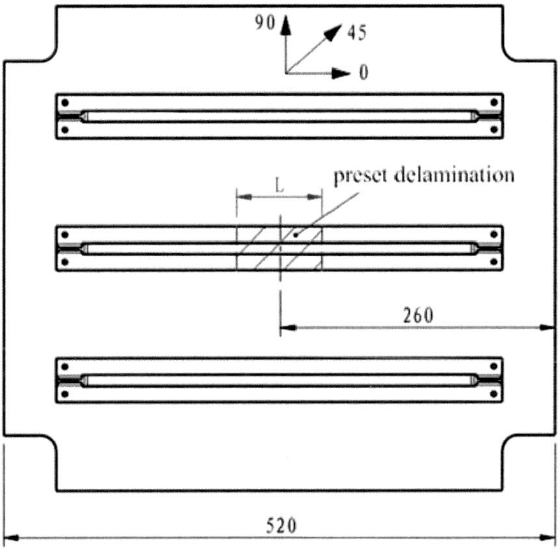

Fig. 13.3 Location of preset delamination of shear test piece of type I stiffened plate (mm)

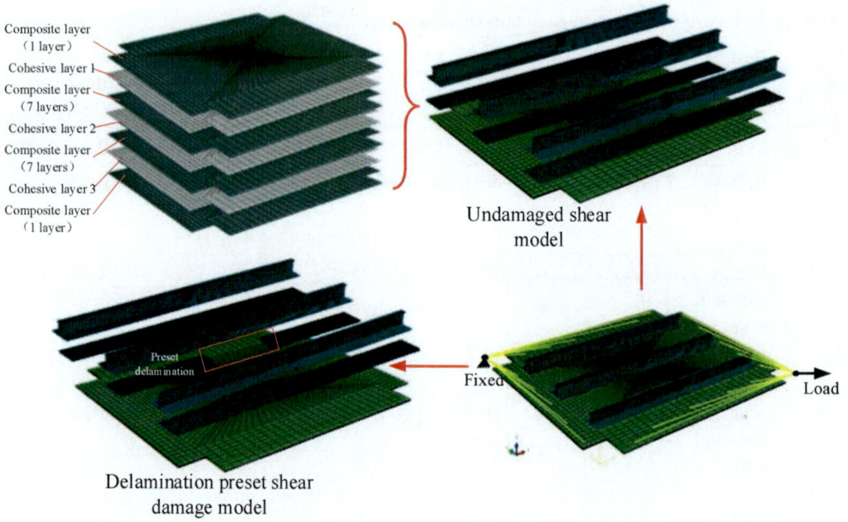

Fig. 13.4 Finite element model

delamination size is preset between the bottom plate of the flange and the skin of the middle stringer, that is, the cohesive element layer is divided into corresponding space. The load and boundary conditions for both models are set as follows: two adjacent sides of the panel are coupled to the reference point, which is set fixed. The other two adjacent sides are also coupled to another reference point, where a displacement load is applied. In addition, the four sides are restrained in the direction perpendicular to the panel. Detailed boundary conditions are shown in Fig. 13.4.

13.3.2 Constitutive Model

For the prediction of the in-plane damage of composite laminates, the Hashin failure criterion of ABAQUS is used to judge the initial failure of in-plane damage. The discriminant formulas of various failure forms are provided in Eqs. (13.1)–(13.4):

$$\text{Fiber tensile failure}: F_f^t = \left(\frac{\sigma_{11}}{X^T}\right)^2 + \alpha\left(\frac{\tau_{12}}{S^L}\right)^2 \quad \sigma_{11} \geq 0 \tag{13.1}$$

$$\text{Fiber compression failure}: F_f^c = \left(\frac{\sigma_{11}}{X^C}\right)^2 \quad \sigma_{11} < 0 \tag{13.2}$$

$$\text{Tensile failure of matrix}: F_m^t = \left(\frac{\sigma_{22}}{Y^T}\right)^2 + \left(\frac{\tau_{12}}{S^L}\right)^2 \quad \sigma_{22} \geq 0 \tag{13.3}$$

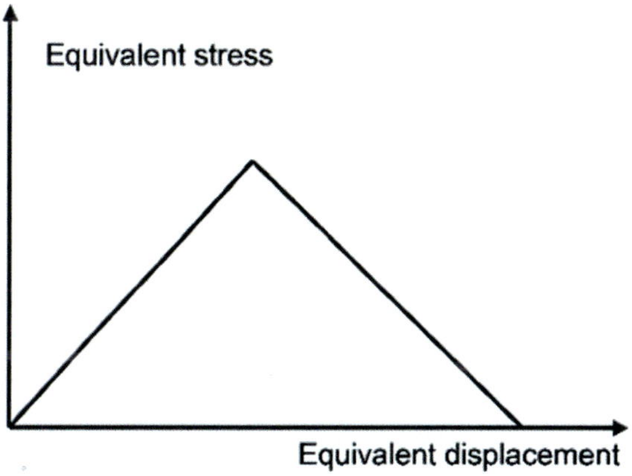

Fig. 13.5 Linear damage evolution constitutive model

Compression failure of matrix :

$$F_m^c = \left(\frac{\sigma_{22}}{2S^T}\right)^2 + \left[\left(\frac{Y^C}{2S^T}\right)^2 - 1\right]\frac{\sigma_{22}}{Y^C} + \left(\frac{\tau_{12}}{S^L}\right)^2 \quad \sigma_{22} < 0 \tag{13.4}$$

The linear degradation criterion based on energy is used for damage evolution, and the final failure condition of the element is determined by inputting the energy requirements under various failure modes (the area of the triangle in Fig. 13.5 below).

The bilinear constitutive model based on cohesive element is used to predict the delamination damage of composite laminates. The constitutive equation of the interface represents the mechanical relationship between the interfacial traction force and the relative separation displacement of the interface, and the curve of the traction separation constitutive relationship is shown in Fig. 13.6.

Figure 13.6a describes the normal behavior of the interface, Fig. 13.6b describes the behavior of two tangent directions of the interface, and Fig. 13.6c is a bilinear constitutive model of mixed mode under the combined action of tensile delamination and shear delamination. The mixed pattern in Fig. 13.6c graph is the key to express the hierarchical coupling, and its corresponding initiation and expansion criteria also change. ABAQUS provides an elastic relationship that directly defines traction and separation, which can be expressed as Eq. (13.5):

$$\begin{Bmatrix} t_n \\ t_s \\ t_t \end{Bmatrix} = \begin{bmatrix} K_n & & \\ & K_s & \\ & & K_t \end{bmatrix} \begin{Bmatrix} \delta_n \\ \delta_s \\ \delta_t \end{Bmatrix} \tag{13.5}$$

t_n, t_s and t_t: nominal traction in normal and two tangential directions;
δ_n, δ_s and δ_t: corresponding displacement value;
K_n,K_s and K_t: respective interface stiffness.
$K_i(i = n, s, t)$ can be expressed by Eq. (13.6):

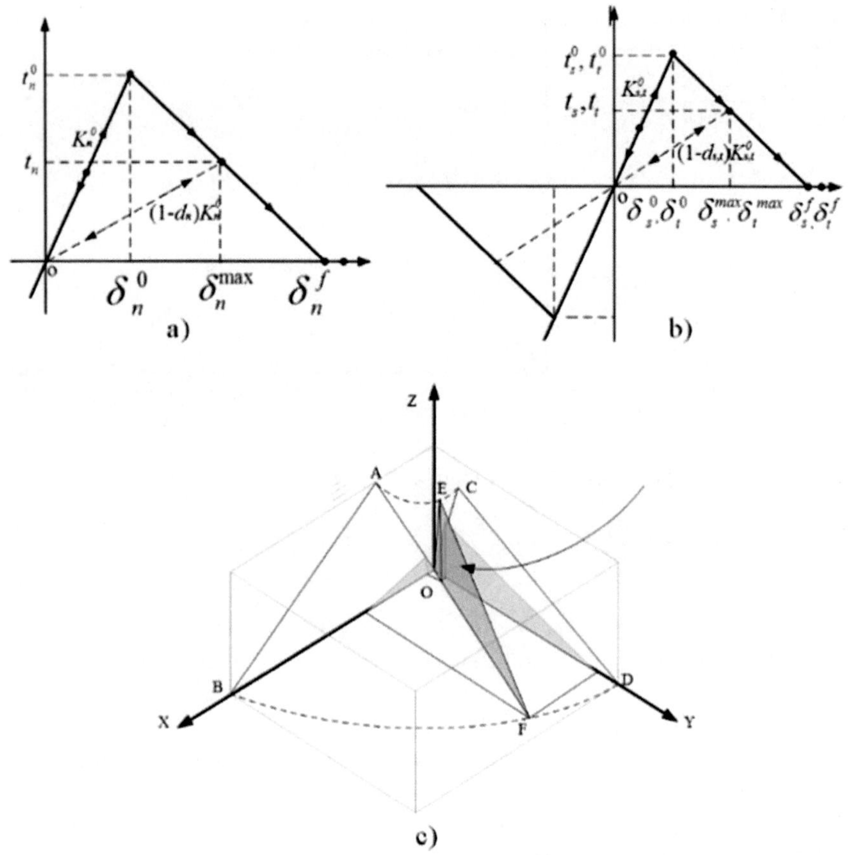

Fig. 13.6 Cohesive bilinear constitutive model: **a** normal behavior, **b** tangential behavior, **c** mode mixing

$$K_i = \begin{cases} K_i^0 & \delta_i \le \delta_i^0 \\ (1 - d_i)K_i^0 & \delta_i^0 \le \delta_i \le \delta_i^f \quad i = n, s, t \\ 0 & \delta_i \ge \delta_i^f \end{cases} \tag{13.6}$$

K_i^0: initial stiffness of interface;

δ_i^0 and δ_i^f: in single mode, the nominal strain values when stratified initiation and stratification are completely formed.

d_i: damage variables that controlling delamination propagation expansion.

After the initiation of delamination, the delamination behavior of the interface layer is controlled by the delamination softening law. The secondary nominal stress criterion is used to predict the initiation of mixed mode delamination, as shown in the following formula, which allows each traction component to interact to produce

Table 13.3 Cohesion model parameters

Parameters	Value
Mode I delamination critical strain energy release rate [mJ/mm^2]	0.434
Mode delamination critical strain energy release rate [mJ/mm^2]	0.744
Mode delamination critical strain energy release rate [mJ/mm^2]	0.744
Interface strength $t_{0n}/t_s^0/t_t^0$ [MPa]	20/35/35
Initial interface stiffness K [N/mm^3]	5000/1923/1923

delamination together. When the traction stress rate meets the following criteria (Eq. (13.7)), delamination begins.

$$\left\{\frac{\langle t_n\rangle}{t_n^0}\right\}^2+\left\{\frac{t_s}{t_s^0}\right\}^2+\left\{\frac{t_t}{t_t^0}\right\}^2=1 \tag{13.7}$$

Hierarchical expansion criterion is used to judge the occurrence of delamination. In this paper, energy based hierarchical evolution method is adopted. At present, there are two energy criteria: energy law criterion and B-K criterion. In combination with relevant research literature, this paper uses energy law to define mixed mode delamination failure, as shown in the Eq. (13.8):

$$\left[\frac{G_n}{G_n^C}\right]^\beta+\left[\frac{G_s}{G_s^C}\right]^\beta+\left[\frac{G_t}{G_t^C}\right]^\beta=1 \tag{13.8}$$

G_s and G_t: instantaneous fracture energy in three directions;

G_n^C, G_s^C and G_t^C: critical fracture energy causing normal and two tangential delamination in single mode.

β: empirical parameters, it is used to characterize the coupling degree of the three layered modes, and the value of β in this paper is taken as 1.

The material properties used in the simulation are shown in Table 13.3.

13.4 Shear Failure Analysis of Intact Stiffened Panel

13.4.1 Skin Damage Analysis

Figure 13.7 shows the location of the initial delamination damage. It can be seen from Fig. 13.7 that the initial delamination of intact stiffened panels mainly occurs on the bottom skin of the lower flange of the middle stringer. Three cohesive layers are set up in the simulation, the stratification degree of each layer is basically the

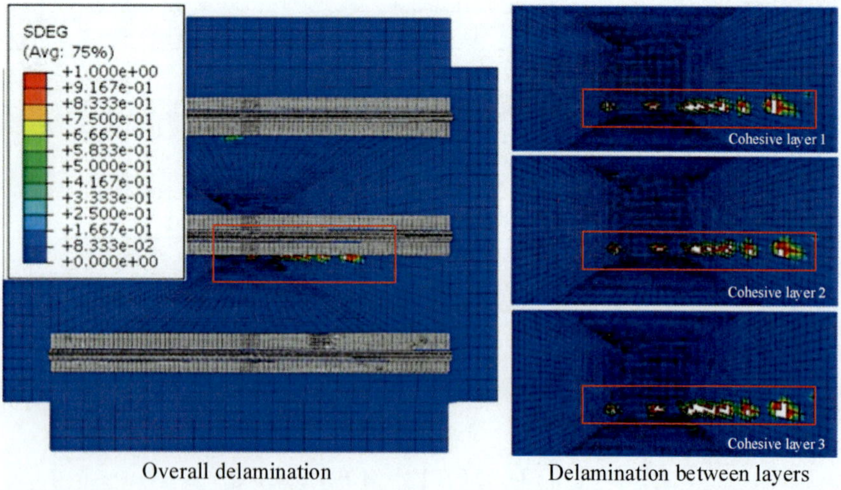

Fig. 13.7 Analysis of initial delamination position of skin

same, and the stratification degree of upper and lower layers is slightly larger than that of the middle layer, which is mainly related to the deformation degree of different delamination positions along the thickness direction. Because of being close to the loading end, the skin of the upper stringer also has a certain tendency of delamination, while there is no delamination at the lowest stringer near the fixed end.

With the load increases, the delamination will extend vertically and horizontally. It can be seen from Fig. 13.8 that the delamination will gradually expand along the longitudinal axis of the middle stringer after the delamination occurs at the bottom of the flange under the middle stringer, but it is difficult to extend to the other two stringers, which is different from the case of stringer debonding in the following paper. In addition, the law is consistent in the three-layer cohesive unit

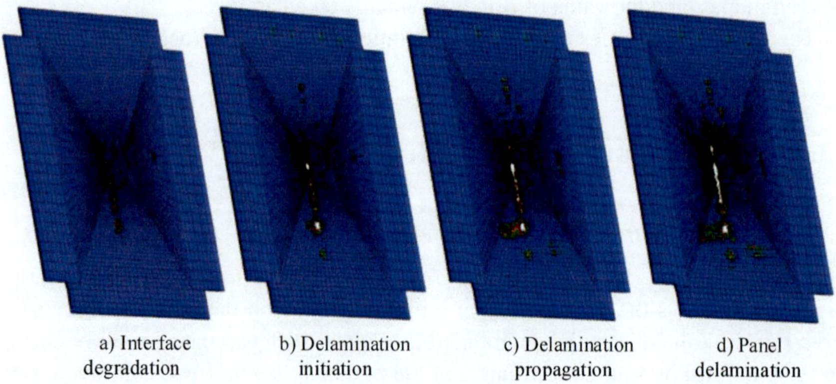

Fig. 13.8 Analysis of delamination growth

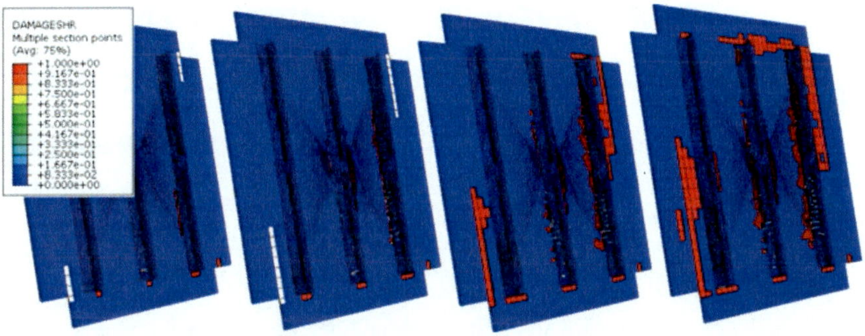

a) Intralaminar damage initiation b) Intralaminar damage local propagation c) Intralaminar damage global propagation d) Intralaminar failure

Fig. 13.9 Damage analysis in skin layer

layer, indicating that the ply angle near the ply has little effect on the delamination propagation.

Figure 13.9 shows the in-plane damage of the skin. The in-plane damage first occurs at the bottom edge of the lower flange of the truss near the loading end, and the damage is mainly based on the matrix tension, and there is no other form of damage. This is mainly due to the lower properties of matrix and fiber.

At the moment, slight matrix damage will not affect the bearing capacity of the structure, and the bearing capacity is still in the rising stage. With the increase of load, the damage extends under the three stringers, especially at the lower flange of the two stringers near the loading point, and there are also a lot of damage at the edge of the skin, which is mainly caused by the boundary constraints in the simulation process, which may be different from the actual test.

13.4.2 Analysis of Skin-Flange Debonding

Figure 13.10 shows the debonding of the stringer during the loading process. It can be seen from the figure that the debonding starts at the stringer near the loading end

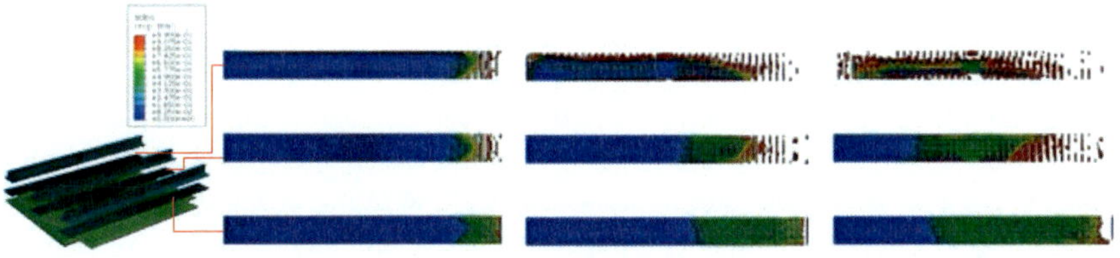

a) Debonding initiation b) Debonding propagation c) Stringer debonded

Fig. 13.10 Analysis on debonding of flange

Fig. 13.11 Damage propagation analysis of stringer

and gradually expands inward along the diagonal line. Meanwhile, with the increase of load, debonding also occurs at the other two stringers.

13.4.3 Damage Analysis of Stringer

Figure 13.11 shows the damage propagation of the stringer. It can be seen from the figure that the damage first occurs on the stringer near the loading point, and then the three stringers have different degrees of in-plane damage, and mainly distributed in the lower half of the lower flange. Compared with skin damage and skin-stringer debonding damage, it can be concluded that due to the relatively low property setting of adhesive film in the simulation process, the premature debonding between the skin and the stringer reduces the load transferred to the stringer. Therefore, the damage of the stringer is relatively light under the condition of large area damage of the skin under three stringers.

13.5 Analysis of the Influence of Embedded Delamination on Shear Failure

13.5.1 Analysis of the Influence of Embedded Delamination on Skin

In this study, three kinds of delamination sizes, 40 mm × 40 mm, 80 mm × 40 mm and 120 mm × 40 mm, are taken into account. Figures 13.12 and 13.13 show the in-plane damage and delamination damage propagation under different sizes of embedded defects, and it can be seen from the figure that for in-plane damage, the influence of embedded defects with different sizes on in-plane damage at each stage is very small. Compared with the delamination damage at the same stage (corresponding to the last stage of in-plane damage propagation), the delamination damage of 80 mm × 40 mm embedded defects is relatively larger, while that of 120 mm × 40 mm is the

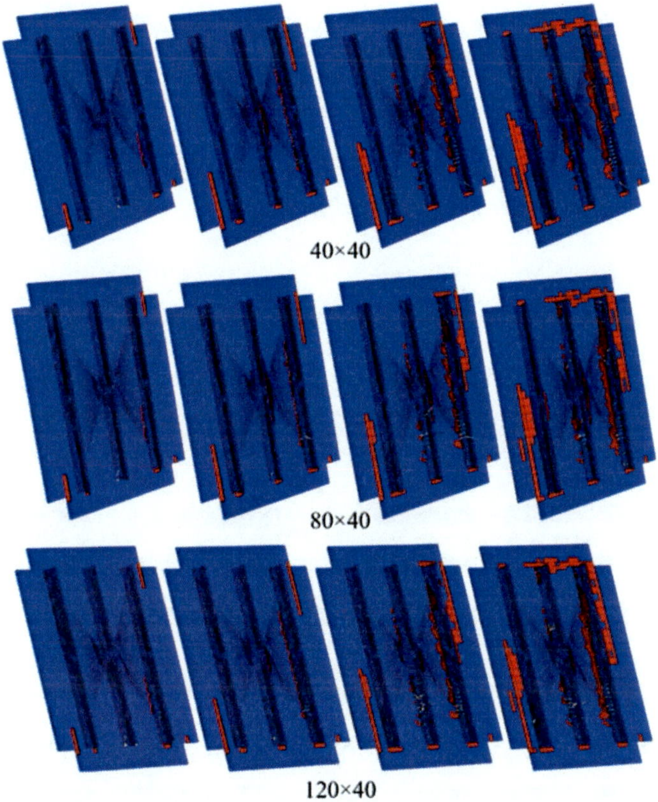

Fig. 13.12 Effect of embedded delamination on in-plane damage of skin

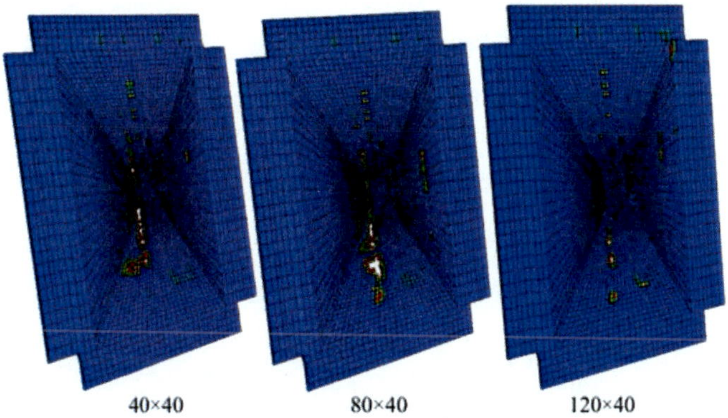

Fig. 13.13 Effect of embedded delamination on skin delamination damage

smallest. In general, the size of delamination damage is smaller than that of in-plane damage (mainly in-plane matrix damage).

13.5.2 *Analysis of the Influence of Embedded Delamination on Skin-Stringer Debonding*

Figure 13.14 shows the effect of different size embedded defects on skin-stringer debonding. It can be studied that the delamination propagation law is basically consistent. For the middle truss with embedded defects, delamination mainly extends from the right half to the left, and the left side begins to gradually occur when the right side is almost completely layered.

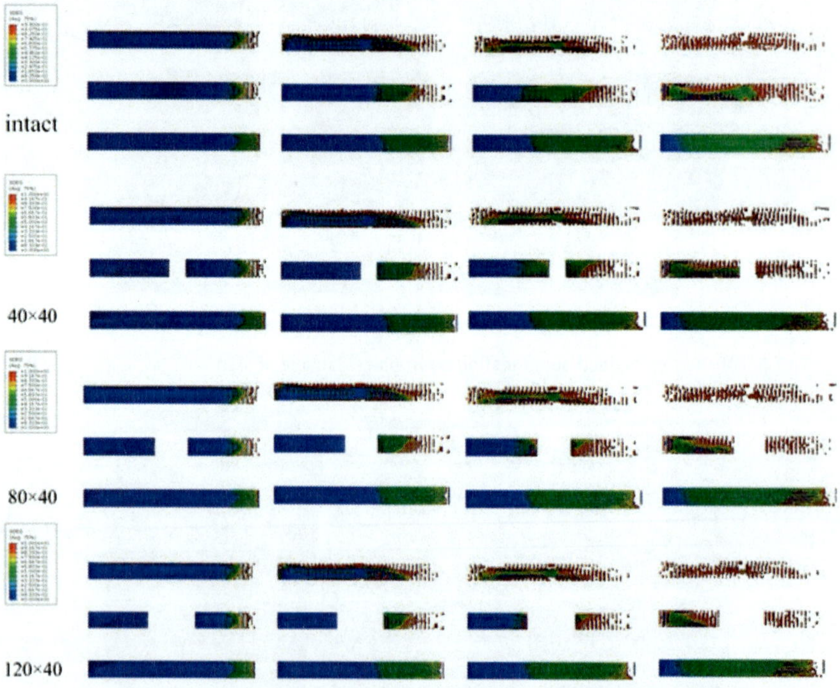

Fig. 13.14 Effect of embedded delamination on skin-stringer debonding

13.5.3 Analysis of the Influence of Embedded Delamination on the Damage of Stringer

Similarly, due to the relatively weak performance of the adhesive film, the premature debonding between the skin and the stringer reduces the load transferred to the stringer. Therefore, the influence of different sizes of embedded delamination on the damage propagation law of the stringer is also weak. It can be seen from Fig. 13.15 below that there is no obvious difference in the damage degree and distribution of the stringer.

From the above analysis, it can be seen that different sizes of embedded delamination have limited influence on the damage propagation of skin and stringer. The main reason is that the performance of the adhesive film is relatively weak. Before the damage of the skin and the stringer, they will be debonding prematurely, which reduces the load transferred to the stringer.

Taking the embedded delamination of 80 mm × 40 mm as an example, the damage propagation of each part is shown in Fig. 13.16 below. It can be found that there is a large degree of debonding when the in-plane initial damage occurs to the skin,

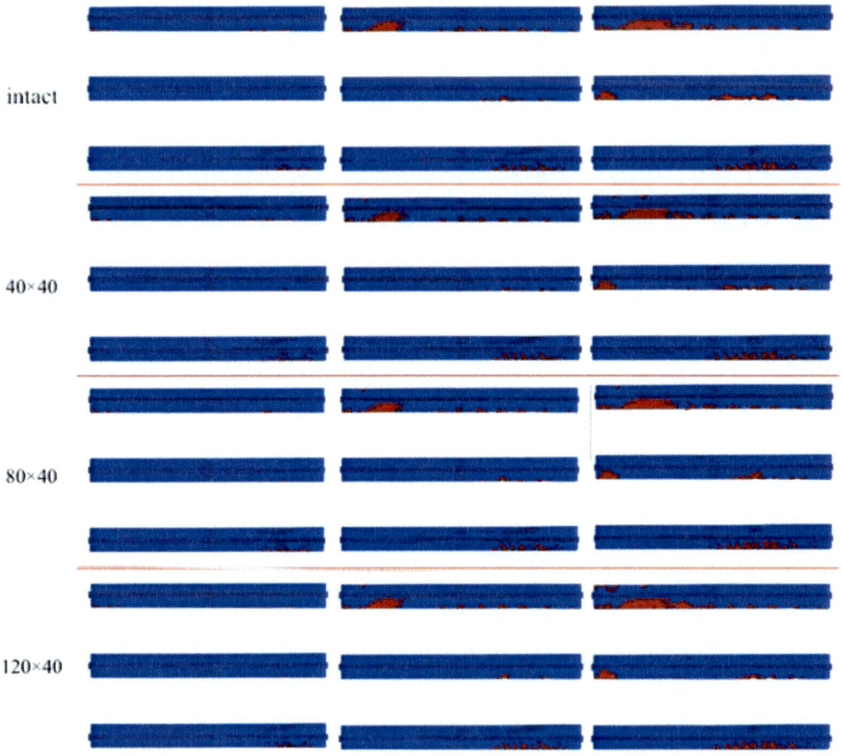

Fig. 13.15 Effect of embedded delamination on damage of stringer

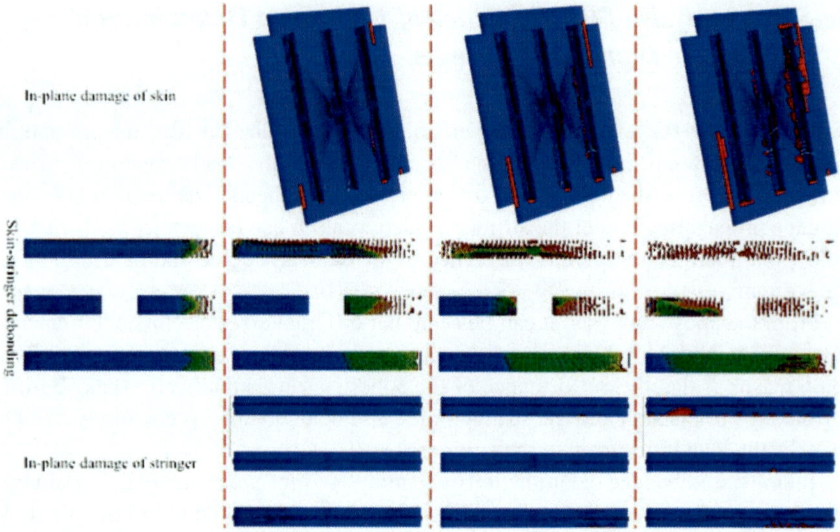

Fig. 13.16. Damage extension

while the stringer is not damaged at this time. When the damage occurs, the degree of debonding is more serious. As a result, both skin and stringer are less affected by the size of embedded layer.

13.6 Conclusions

In this report, the influence of defect damage on the post buckling capacity of composite stiffened panels under shear loading is studied. The damage free shear model and the prefabricated delamination damage shear model are established. The influence of the damage of the prefabricated delamination between the skin and the truss on the mechanical properties of the stiffened panel under shear load is analyzed, and the influence of the size of the prefabricated delamination on the results is further studied, which provides theoretical support for the experimental design. The main conclusions are as follows:

(a) For intact stiffened panels, the initial delamination occurs at the skin of the bottom flange of the middle stringer, and the in-plane damage first occurs at the bottom skin of the lower flange of the stringer near the loading end, and the damage is mainly caused by matrix tension. The results show that the debonding starts from the stringer near the loading end and gradually expands inward along the diagonal line. The damage of the stringer occurs on the stringer close to the loading point, and then the three stringers have different degrees of in-plane damage, and mainly distributed in the position where the lower flange is far away from the loading point.

(b) The main reason of different size of embedded delamination on the damage propagation of skin and stringer is limited is that the performance of the adhesive film is relatively low, and the debonding of the skin and the stringer occurs prematurely before the damage occurs, which reduces the load transferred to the stringer.

References

1. Degenhardt, R., Rolfes, R., Zimmermann, R., Rohwer, K.: COCOMAT—improved material exploitation of composite airframe structures by accurate simulation of postbuckling and collapse. Compos. Struct. **73**(2), 175–178 (2006)
2. Zimmermann, R., Rolfes, R.: POSICOSS—improved postbuckling simulation for design of fibre composite stiffened fuselage structures. Compos. Struct. **73**(2), 171–174 (2006)
3. Mcelroy, M.: Use of an enriched shell finite element to simulate delamination-migration in a composite laminate. Compos. Struct. **167**, 88–95 (2017)
4. Park, H., Kong, C.: Experimental study on barely visible impact damage andvisible impact damage for repair of small aircraft composite structure. Aerosp. Sci. Technol. **29**, 363–372 (2013)
5. Al-Azzawi, A.S.M., McCrory, J., Kawashita, L.F., Featherston, C.A., Pullin, R., Holford, K.M.: Buckling and postbuckling behaviour of Glare laminates containing splices and doublers. Part 1: Instrumented tests. Compos. Struct. **176**, 1158–1169 (2017)
6. Al-Azzawi, A.S.M., Kawashita, L.F., Featherston, C.A.: Buckling and postbuckling behaviour of Glare laminates containing splices and doublers. Part 2: Numerical modelling. Compos. Struct. **176**, 1170–1187 (2017)
7. Feng, Y., Zhang, H., Tan, X., He, Y., An, T., Zheng, J.: Effect of impact damage positions on the buckling and post-buckling behaviors of stiffened composite panel. Compos. Struct. **155**, 184–196 (2016)
8. Feng, Y., He, Y., Tan, X., An, T., Zheng, J.: Experimental investigation on different positional impact damages and shear-after-impact (SAI) behaviors of stiffened composite panels. Compos. Struct. **178**, 232–245 (2017)
9. Feng, Y., He, Y., Zhang, H., Tan, X., An, T., Zheng, J.: Effect of fatigue loading on impact damage and buckling/post-buckling behaviors of stiffened composite panels under axial compression. Compos. Struct. **164**, 248–262 (2017)
10. Feng, Y., He, Y., Tan, X., An, T., Zheng, J.: Investigation on impact damage evolution under fatigue load and shear-after-impact- fatigue (SAIF) behaviors of stiffened composite panels. Int. J. Fatigue **100**, 308–321 (2017)
11. Sun, W., Guan, Z., Ouyang, T., Tan, R., Zhong, X.: Effect of stiffener damage caused by low velocity impact on compressive buckling and failure modes of T-stiffened composite panels. Compos. Struct. **184**, 198–210 (2018)
12. Sun, W., Guan, Z., Li, Z., Ouyang, T., Jiang, Y.: Modelling and simulating of the compressive behavior of T-stiffened composite panels subjected to stiffener impact. Compos. Struct. **186**, 221–232 (2018)
13. Li, N., Chen, P.: Experimental investigation on edge impact damage and Compression-After-Impact (CAI) behavior of stiffened composite panels. Compos. Struct. **138**, 134–150 (2016)
14. Boni, L., Fanteria, D., Lanciotti, A.: Post-buckling behaviour of flat stiffened composite panels: Experiments vs. analysis. Compos. Struct. **94**(12), 3421–3433 (2012)

15. Wang, X., Cao, W., Deng, C., Wang, P., Yue, Z.: Experimental and numerical analysis for the post-buckling behavior of stiffened composite panels with impact damage. Compos. Struct. **133**, 840–846 (2015)
16. Masood, S.N., Vishakh, R., Viswamurthy, S.R., Gaddikeri, K.M., Sridhar, I.: Influence of stiffener configuration on post-buckled response of composite panels with impact damages. Compos. Struct. **194**, 433–444 (2018)
17. Bai, R., Lei, Z., Wei, X., Tao, W., Yan, C.: Numerical and experimental study of dynamic buckling behavior of a J-stiffened composite panel under in-plane shear. Compos. Struct. **166**, 96–103 (2017)
18. Lei, Z., Bai, Ru., Tao, W., Wei, X., Leng, R.: Optical measurement on dynamic buckling behavior of stiffened composite panels under in-plane shear. Opt. Lasers Eng. **87**, 111–119 (2016)
19. Chen, Q., Qiao, P.: Post-buckling analysis of composite plates under combined compression and shear loading using finite strip method. Finite Elem. Anal. Des. **83**, 33–42 (2014)
20. Raju, G., Wu, Z., Weaver, P.M.: Buckling and postbuckling of variable angle tow composite plates under in-plane shear loading. Int. J. Solids Struct. **58**, 270–287 (2015)
21. Zhang, T., Li, S., Chang, F., Shi, X., Li, L.: An experimental and numerical analysis for stiffened composite panel subjected to shear loading in hygrothermal environment. Compos. Struct. **138**, 107–115 (2016)
22. Tsai, S.W.: Composite design, 4th edn. Think Composites, Dayton (1988)

Chapter 14
PSD Random Vibration Strength and Fatigue Analysis of a CNG Tube Trailer

Zhiqian Liu, Caifu Qian, and Zuzhi Chen

Abstract Tube trailer is a kind of gas transportation equipment, which is widely used in transportation of various industrial gases including compressed natural gas (CNG). However, due to its particularity of carrying high-pressure gas and operating and parking on urban roads, the requirements for its safety performance are especially concerned. In this paper, finite element random vibration analysis on a tube trailer was performed according to the standard GB/T 4857.23-2012, random vibration analysis under the "ISTA 3A vehicle power spectral density (PSD) curve". The first 9 orders natural frequencies are obtained from modal analysis and it is found that only the first natural frequency is covered in the speed limitation of 80 km/h for the tube trailer. The maximum deformations in the first 3 orders all occur in the middle of the frame of the tube trailer; Stress analysis under random vibration excited by the specified PSD found that the maximum stresses corresponding the probability values at 1σ, 2σ and 3σ are all less than the yield stress of the material, meaning that the structure meets the strength requirements. Within the fatigue life under PSD random vibration, the trailer's fatigue damage cycle ratio D is less than 1, meaning the structure meets the fatigue design requirements.

Keywords Tube trailer · Finite element analysis · Random vibration analysis · PSD

14.1 Introduction

Compared with pipeline transportation, tube trailers have the advantages of lower investment and higher suitable for short and medium distance transportation.

Z. Liu · C. Qian (✉)
Beijing University of Chemical Technology, Beijing 100029, China
e-mail: qiancf@mail.buct.edu.cn

Z. Chen
China Special Equipment Inspection and Research Institute, Beijing 100029, China

J. Xu and K. M. Pandey (eds.), *Mechanical Engineering and Materials*,
Mechanisms and Machine Science 100,
https://doi.org/10.1007/978-3-030-68303-0_14

However, since tube trailers are mostly used on highways during land transportation, they sometimes shuttle or stop in densely populated areas. As the working pressure is high and the medium are flammable and explosive, it will cause huge human life and property losses once an accident occurs. Therefore, comprehensive analysis and safety assessment of tube trailers is of very practical significance.

In recent years, there have been more and more relevant studies on tube trailers in China and abroad. Yu et al. [1] studied the fatigue life of tube trailer gas cylinders, and proposed to control the inertial load during the transportation. Liu et al. [2] discussed and analyzed the reasons for the wear of the external thread of the cylinder bottle, the external thread structure of the bottle mouth, and the arrangement position of the cylinder. Wang [3] analyzed the failure modes of CNG tube trailer gas cylinders comprehensively, and believed that to reduce the failure damage, it is necessary to improve the fixing structure of the bottle ends. Gao et al. [4] analyzed the actual problems of tube trailer gas cylinder under different working conditions. Dong et al. [5] proposed that the standardization of design and manufacture is a prerequisite for safe use, and perfecting and improving technical requirements, forming technical standards are inevitable for industry development. Zhang et al. [6] put forward suggestions on risk prevention for tube trailers from the aspects of material selection, shape optimization, compliance with standards and anti-corrosion methods to ensure the safety of the product under normal road conditions and correct operation.

As a kind of transportable equipment, tube trailers will inevitably experience bumps and vibrations during road operation, which places more stringent requirements on the strength and fatigue life of the structure under random vibration conditions. It is quite difficult to perform accurate vibration analysis of tube trailers with the conventional design method. Instead, it is effective to conduct random vibration analysis using finite element method.

Different methods of random vibration analysis are available. Hao et al. [7] performed random vibration analysis of the converter cabinet based on the ASD (Acceleration Spectrum Density) spectrum values applied to all directions of the structure. Yang et al. [8] conducted their analysis of the frame of Radix Ophiopogonis harvester by applying the displacement spectral density of road roughness.

In this paper, ANSYS Workbench 19.0 was applied to overall model the key structures of tube trailers (i.e. frames and gas cylinders). Random vibration analysis was carried out under the excitation of ISTA 3A random vibration PSD curve from the standard GB / T 4857.23-2012 "Basic Experiments of Transport Package Part 23: Random Vibration Test Method". Stress analysis was conducted on the structure based on 3σ theory and the strength and fatigue life of the structure under the random vibration is evaluated and studied.

14.2 Random Vibration Analysis Concept

Random vibration analysis is also called power spectral density analysis [9], which is a statistical method based on probability [10]. The power spectral density spectrum represents a measure of the mean square value of random variables and is mostly used for random vibration analysis. In many situations in reality, the load on the structure and the corresponding time course are uncertain, so the method of transient analysis is not applicable in the simulation calculation under this premise [11], continuous transient response can only be described by a probability distribution function, i.e. the probability corresponding to a certain level of response. The power spectral density is the statistical result of the response of the structure under random dynamic load excitation, and is a relationship curve between the power spectral density value and the frequency value.

The traditional method is to repeat the design and experiment multiple times. This method will lead to repeated design and excessive optimization, which is time-consuming and inefficient [12]. In order to solve this problem regarding the shortcomings of traditional design methods, the products that have preliminarily designed are analyzed using ANSYS simulation software which has been widely used in various analyses including modal and random vibration analysis and then the weak parts of the products are reinforced [13].

Random vibration analysis belongs to frequency domain analysis, and modal analysis must be performed to obtain the natural frequency of the structure before performing random vibration analysis.

14.3 Finite Element Modeling of Tube Trailer

For simplification but without affecting the results a lot, the flanges of the connection between the gas cylinder and the baffle and the details of the individual support plates were simplified. The finite element model is shown in Fig. 14.1. The model size is 12192 × 2438 × 1400, the design temperature is −40–60 °C, and the fully loaded

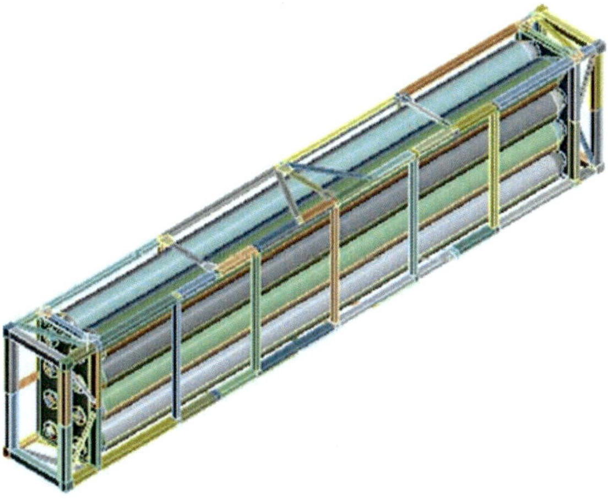

Fig. 14.1 Finite element geometry model of the tube trailer

Table 14.1 Main properties of the material

Material	Elastic modulus [MPa]	Poisson's ratio	Density [kg/m^3]
Q345R	2.09×10^8	0.3	7850

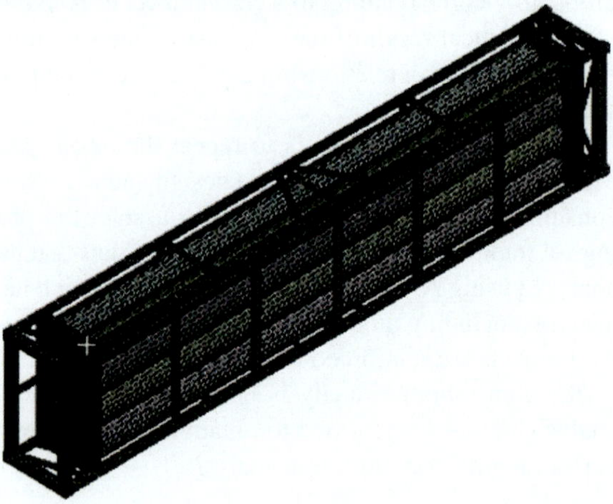

Fig. 14.2 Finite element grid model of the tube trailer

mass is 28755 kg with the net weight being 25,500 kg and the loading capacity being 3255 kg.

The material used in the equipment is Q345R, and its main properties are listed in the Table 14.1.

Element Solid186 is used to mesh the model. The grid is hexahedral with mostly a size of 50 mm after performing irrelevant testing. The whole grid model is shown in Fig. 14.2.

For the boundary conditions, the four corner supports at the bottom of the frame are fully restrained for each analysis.

It should be pointed out that there are stresses at the tubes due to internal pressure, but this paper mainly concerns the strength of trailer frames which is not affected by the pressure. So the pressure inside the tubes is not considered in this analysis.

14.4 Modal Analysis

Performing modal analysis, the natural frequency results of the first 9 orders are obtained and listed in Table 14.2.

Tire 12R22.5PR is widely used in tube trailers manufactured in China. According to "GB/T 2977-2016 specifications, size, pressure and load of truck tires", the diameter of the tire is 1085 mm.

Based on Eq. (14.1):

Table 14.2 The first 9 order modal analysis results

Order	Natural frequency [Hz]
1	6.1198
2	7.9711
3	9.5705
4	11.199
5	11.621
6	11.954
7	12.74
8	12.903
9	12.983

$$v = \omega r = \frac{2\pi}{T} r = 2\pi f r \tag{14.1}$$

where: v is the linear speed of the tube trailer, m/s;ω is the angular speed of the tube trailer, rad/s; r is the radius of the tire, mm; T is the period, s; f is the frequency, Hz.

According to the notice issued by the National Development and Reform Commission (NDRC) on soliciting comments on the national standard "Technical Requirements for Vehicle Speed Limiting Systems", the maximum speed of dangerous goods transport vehicles shall not be higher than 80 km/h. Substituting the speed limit and tire radius values into Eq. (14.1) yields Eq. (14.2):

$$80 \times 3.6 = 2 \times 3.14 \times f \times \frac{1}{2} \times 1085 \times 10^{-3} \tag{14.2}$$

Therefore, $f = 6.52$ Hz.

So the maximum frequency that can be allowed for a tube trailer with the tire 12R22.5PR under safe driving conditions is 6.52 Hz.

The above calculations show that in normal road operation, only the 1st order natural frequency of 6.1198 Hz is covered in the speed limitation. But in order to ensure the safety of the long-tube trailer in case of possible emergencies during driving or speeding, the analysis is extended to cover the 3rd order, where the calculated driving speed is 117.38 km/h.

Deformation contours for the first three order vibration modes are shown in Fig. 14.3. It is seen that the maximum deformations for the first 3 orders are all located in the middle of the frame of the tube trailer.

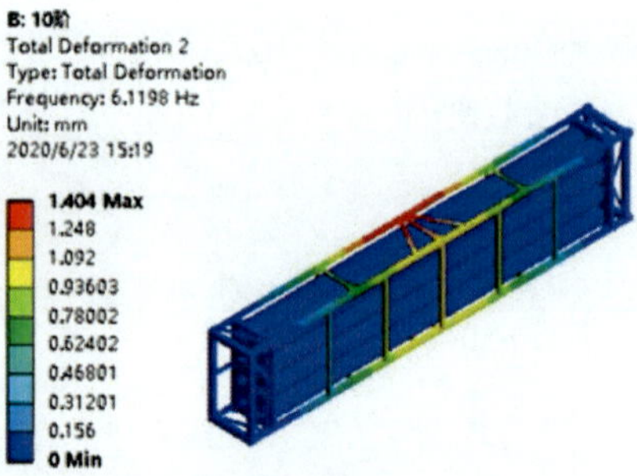

(a) 1st order vibration mode

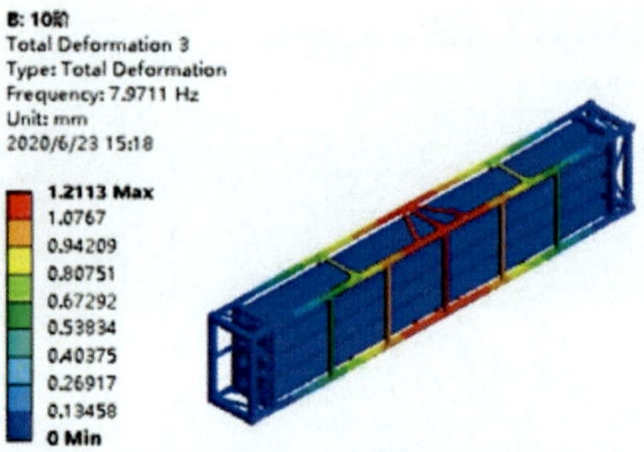

(b) 2nd order vibration mode

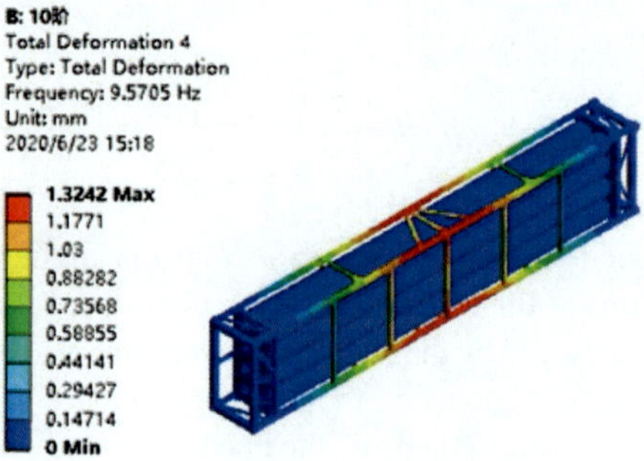

(c) 3rd order vibration mode

Fig. 14.3 Deformation contours at the vibration modes for the first 3 orders

14.5 PSD Random Vibration Analysis

14.5.1 Selection of PSD Loads

In practice, random vibration loads applied to the vehicle are mainly caused by uneven road surfaces. The standard GB/T 4857.23-2012 "Basic Experiments on Transport Packages Part 23. Random Vibration Test Methods" provides the experimental acceleration power spectral density curves and data in the relevant international standards for random vibration. In this study, the acceleration power spectral density curve for the vehicle random vibration according to ISTA 3A-2008 was selected as shown in Fig. 14.4.

Figure 14.4 gives a PSD excitation, which is not representative of a particular transport situation, but averaged from different weather, different road conditions, different driving speeds, etc. and used as an input excitation in a real vibration test.

ANSYS Workbench 19.0 is applied to perform random vibration simulations of the model. Using the Random Vibration module, the coordinates of the corresponding points within Fig. 14.4 were entered as PSD G Acceleration for loading with the value

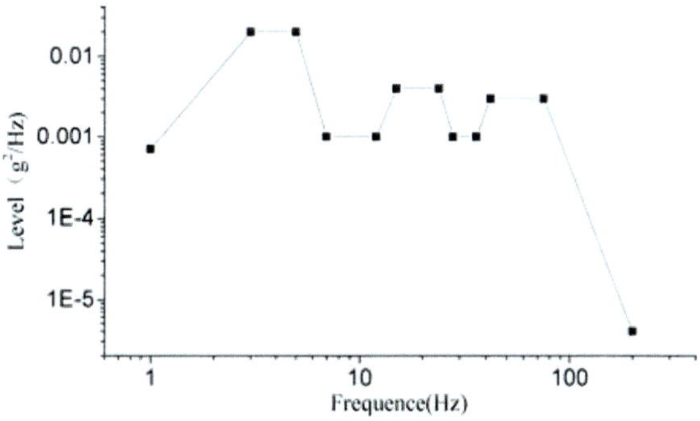

Fig. 14.4 Random vibration PSD curves for model loading

of the modal damping ratio being taken as 0.01. von Mises stress was obtained for the results.

14.5.2 Strength Analysis

From the random vibration calculations, stress solutions under random loads can be obtained according to different probability described by 1σ, 2σ and 3σ as shown in Fig. 14.5 for the distribution. The most dangerous case is the solution occurring at the 3σ case.

Stress distributions are shown in Fig. 14.6. The maximum stresses at 1σ, 2σ and 3σ are 93.256 MPa, 186.51 MPa and 279.77 MPa, respectively. Figure 14.7 shows that the location of maximum stress of 279.77 MPa is located at the end of the frame, where the crossbeam joins the longitudinal beam.

As the maximum stress value at 3σ of this structure is 279.77 MPa, which is less than the yield stress of material Q345R (345 MPa), it is reached that the structure meets the strength requirements under random vibration excited by the specified PSD.

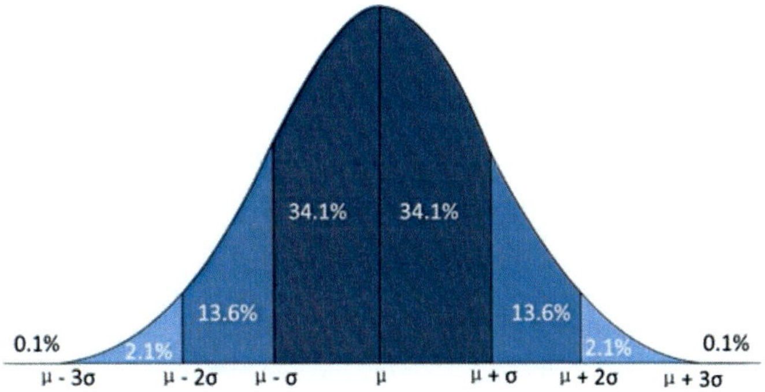

Fig. 14.5. 3σ normal distribution

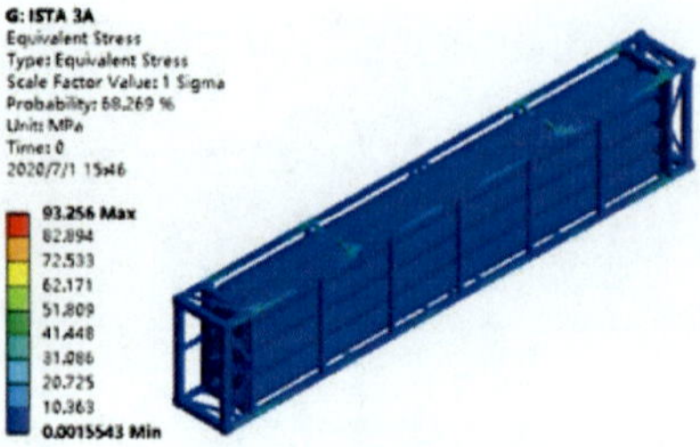

(a) Stress distribution at 1σ

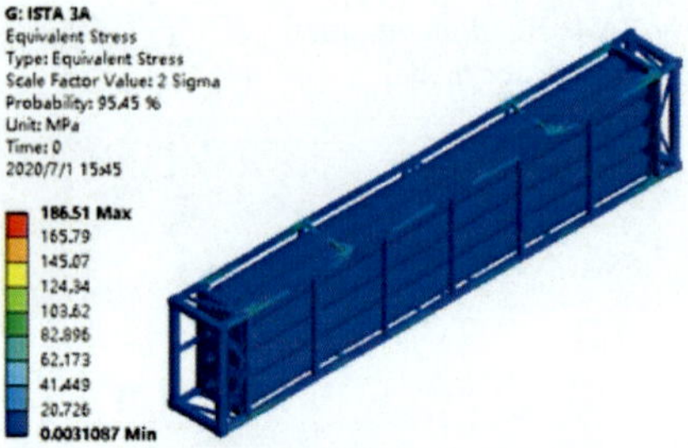

(b) Stress distribution at 2σ

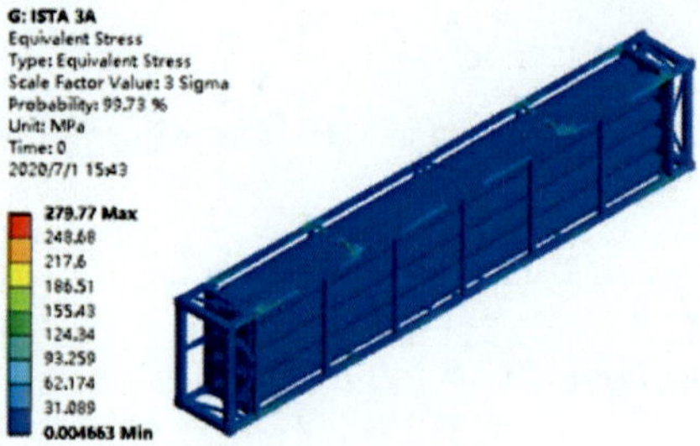

(c) Stress distribution at 3σ

Fig. 14.6 Stress solutions at 1σ, 2σ and 3σ

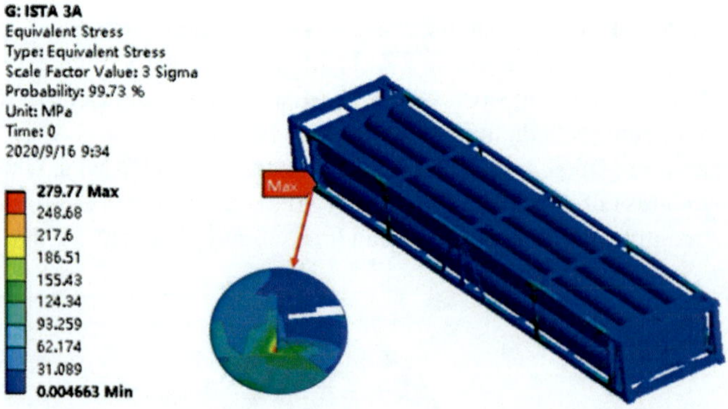

Fig. 14.7 Position of the maximum stress

14.5.3 *Fatigue Analysis*

Regarding fatigue calculations of structures under random vibration loads, for structures whose stresses obey a Gaussian distribution, it is usual to calculate the fatigue of the structure by the Steinberg's proposed three-interval method [14], that is, the

Mises stress is divided into three intervals according to a 3σ distribution, with each interval corresponding to an occurrence probability of 68.3%, 95.4% and 99.7%.

As the stresses for the case greater than 3σ occur only 0.27% of the time, it is assumed that they do not cause any damage.

From Miner's law, the formula for the fatigue life under PSD random vibration can be obtained as Eq. (14.3) [15]:

$$D = \frac{n_{1\sigma}}{N_{1\sigma}} + \frac{n_{2\sigma}}{N_{2\sigma}} + \frac{n_{3\sigma}}{N_{3\sigma}} \tag{14.3}$$

In Eq. (14.3), D is the fatigue damage cycle ratio; n1σ, n2σ, n3σ is the actual number of cycles under the corresponding 1σ, 2σ, 3σ level; N1σ, N2σ, N3σ is the number of cycles under the corresponding 1σ, 2σ, 3σ solution, which can be found in JB 4732. Since $n_{1\sigma} = 0.6831 v_0^+ T$,$n_{2\sigma} = 0.272 v_0^+ T$,$n_{3\sigma} = 0.4331 v_0^+ T$[16], it is easy to transform Eq. (14.3) into Eq. (14.4):

$$D = \frac{0.6831 v_0^+ T}{N_{1\sigma}} + \frac{0.272 v_0^+ T}{N_{2\sigma}} + \frac{0.04331 v_0^+ T}{N_{3\sigma}} \tag{14.4}$$

In Eq. (14.3), v_0^+ is the average frequency of vibration, Hz; T is the expected life of the structure, s.

Based on relevant requirement, assuming that the value of expected life T is $1.08 \times 10^4 s$, the average frequency v_0^+ is 5 Hz, and the maximum stresses at 1σ, 2σ and 3σ are respectively 93.256 MPa, 186.51 MPa and 279.77 MPa. Based on JB 4732 C-1, the corresponding N1σ, N2σ, and N3σ can be calculated as 4.96×10^5, 3.06×10^4 and 8.23×10^3, respectively.

Substituting these values into Eq. (14.4) yields Eqs. (14.5) and (14.6):

$$D = \left(\frac{0.6831}{4.96 \times 10^5} + \frac{0.272}{3.06 \times 10^4} + \frac{0.04331}{8.23 \times 10^3} \right) \times 1.08 \times 10^4 \times 5 \tag{14.5}$$

$$D = 0.834 < 1 \tag{14.6}$$

Therefore, the structure can meet the fatigue design requirements.

14.6 Conclusions

In this paper, finite element simulation was performed to analyze the strength of a tube trailer under PSD stochastic excitation. The main conclusions are drawn as follows.

(1) The first 9 orders natural frequencies are obtained from modal analysis and only the first natural frequency is covered by the speed limitation of 80 km/h for the

trailer. The maximum deformation at the first 3 orders is all located in the middle of the frame of the tube trailer.

(2) Stress analysis under random vibration excited by the specified PSD found that the maximum stresses corresponding the probability values at 1σ, 2σ and 3σ are respectively 93.256 MPa, 186.51 MPa and 279.77 MPa, respectively. They are less than the yield stress of the material, meaning that the structure meets the strength requirements.

(3) Fatigue analysis found that in the fatigue life under PSD random vibration, the trailer's fatigue damage cycle ratio D is 0.834, less than 1, meaning the structure meets the fatigue design requirements.

References

1. Yu, X., Song, B., Zhang, Z., Liu, Q.: Analysis of fatigue life for tube trailer cylinders. J. Hebei Univ. Sci. Technol. **36**(4), 407–412 (2015)
2. Liu, Z., Li, X., Zhu, B., Yue, Z.: Random vibration analysis of two-component vehicle hydrogen supply system based on power spectral density. Packag. Eng. **40**(1), 69–74 (2019)
3. Wang, Q.: Analysis and discussion of potential risks in application of natural gas tube trailer. Chem. Eng. Machinery. **46**(4), 372–374 (2019)
4. Gao, H., Lin, S., Li, D., Wei, H., Tao, F., Wu, B.: The road running research of tube trailer gas cylinder based on finite element method. Fujian Chem. Ind. **12**, 11–15 (2016)
5. Dong, H., Li, B., Bo, K., Chen, Z.: Review of the development of tube trailer cylinder on safety technology. China Special Equipm. Saf. **30**(8), 1–5 (2014)
6. Zhou, R., Zhang, J., Xu, G., Zhao, W.: Developing of safety accessories application technology on long tube trailer. Shandong Chem. Ind. **44**(24), 103–105 (2015)
7. Hao, Y., Feng, J., Hu, J.: Probabilistic distribution of Von Mises stress and strength evaluation of structures under static-random vibration combined condition. J. Vibrat. Shock **39**(5), 188–193 (2020)
8. Yang, Y., Liang, S., He, Y.: Random vibration analysis of an ophiopogon japonicus harvester frame. J. Xihua Univ. (Natural Science Edition) **38**(6), 87–91 (2019)
9. Woodcock, D.L.: An Introduction to Random Vibration. Robsinm J.D. Cambridge University Press, Edinburgh (1963)
10. Zheng, M., Shen, F., Luo, P.: Vibration fatigue analysis of the structure under thermal loading. J. Adv. Mater. Res. **853**, 559–564 (2014)
11. Bai, Y., Qiu, E., Wang, H.: Random vibration fatigue life analysis and optimization design of connector based on ANSYS. J. Sichuan Ordnance **40**(11), 178–182 (2019)
12. Zhao, W., Zhang, Y.: Reliability analysis of random vibration transmission path systems. Mech. Syst. Signal Process. **113**, 77–89 (2018)
13. Xia, J., Li, G., Li, B.: Fatigue life prediction of Package-on-Package stacking assembly under random vibration loading. Microelectron. Reliab. **71**, 111–118 (2017)
14. Steinberg, D.S.: Vibration analysis for electronic equipment, 2nd edn. Wiley, New York (1988)
15. Sun, Y., Deng, J., Liu, Z.: Random vibration fatigue analysis of body parts. Internal Combustion Engine & Parts **1**, 49–51 (2019)
16. Wang, K., Shi, Y., Yan, G.: Random vibration analysis of unmanned electric vehicle frame based on pavement power spectrum. Agric. Equip. Vehicle Eng. **58**(6), 129–133 (2020)

Chapter 15
Simulation of Multi-step Tube Hot Gas Forming Process of a UHSS Torsion Beam

Shangwen Ruan, Jianping Zhao, Zhenli Mi, Lan Su, and Pengzhi Cheng

Abstract Ultra-high strength steel can significantly improve the safety and lightweight of vehicles, especially for the closed-profile components. In this study, a novel process simulation of multi-step tube hot metal gas forming (HMGF) of a V-shape torsion beam is conducted. The aim of this paper is to provide the insight into the influence of process variables, such as pre-heating condition, gas pressure and tool temperature, on the formability, hardening and component behavior. The press hardening steel used is hot-rolled BR1500 steel and the stress-strain relationships under different temperatures are investigated for the thermo-mechanical simulation. A consistent simulation model with all relevant sub-stages is then established in LS-DYNA software. Component geometry, wall thinning, hardness and microstructure distributions are achieved in the simulation and the forming parameter dependency is analyzed. And the forming equipment is introduced. The hot gas forming step is the most complicated process, and is carried out in a nitrogen gas environment, which prevents oxidation of the parts after heating.

Keywords Ultra-high strength steel · Tube hot gas forming · Torsion beam · Multi-step forming

S. Ruan · Z. Mi (✉) · L. Su
Institute of Engineering Technology, University of Science and Technology Beijing, Beijing 100083, China
e-mail: mizl@nercar.ustb.edu.cn

J. Zhao
Guangxi Automobile Holdings Limited, Liuzhou 545000, China

P. Cheng
Hangyu Intelligent Manufacture Limited, Beijing 100089, China

J. Xu and K. M. Pandey (eds.), *Mechanical Engineering and Materials*, Mechanisms and Machine Science 100,
https://doi.org/10.1007/978-3-030-68303-0_15

15.1 Introduction

The use of high strength steel sheets increases mainly for body-in-white parts for the reduction of weight and improvement of strength. Hot stamping is common practice in automotive industry for cover panels [1].

On the other hand, the rigidity takes advantages of hollow structures. Tube hydroforming is increasingly employed to manufacture hollow parts of high strength steels, but it is not easy for severe requirement of equipment, small ductility of materials and the inevitable spring-back [2].

Warm tube hydroforming is then employed to heighten the formability of high strength steels. But using oil and water as pressure media limits the heating temperature below 300 °C [3].

Hot metal gas forming (HMGF) is an innovative forming technology whose concepts are based on tube hydroforming and hot to remove the limitation of the heating temperature. HMGF process is coupled thermo-mechanical problem process. The heating transferring could alter the flow-stress of the materials and a large cooling rate is also necessary for martensite formation [4, 5].

Ferritic stainless steel showed at least 55% expansion at high temperatures during bugling [6]. Then, the tube hot gas forming was introduced into the automotive industries by Fraunhofer Institute for Machine Tools and Forming Technology IWU [7–9]. Induction heating and additional heating elements was used for uniform temperature distribution [10].

However, it is difficult to control the internal pressure in HMGF process since the charging time is less than 1 s [11]. Thus, a simplified controlling strategy using resistance heating and filling the air into seal tubes was introduced by Maeno et al. [12], so that the tube could be bulged at low flow stress and the controlling of the internal pressure is avoided.

The torsion beam is an important part in the rear suspension of the chassis. Its torsion resistance and stiffness have always been crucial to the system. The tube hydroformed torsion beams have been widely employed in the industry. A new gas forming process of ultra-high strength steel hollow parts using air filled into sealed tubes and resistance heating to omit the subsequent heat treatment was developed [13].

Since direct HMGF process may induce a flange on the torsion beam, Zhang et al. conducted a simulation study on an indirect HMGF process with furnace heating. When the bugling finished, it showed the lowest temperature on the part was about 500 °C, which is below the martensite start temperature [14].

In this paper, an indirect hot gas metal forming process for a V-shape torsion beam is introduced. An in-situ resistance heating method is employed to avoid the heat loss during transition. The hot-rolled BR1500 press-harden steel is used in the process. The consistent simulation model with all relevant sub-stages is then established in LS-DYNA software and the reasonable pieces with high precision and low thinning rate of 12% could be obtained.

15.2 Indirect Hot Gas Forming of a Torsion Beam

In the present study, the ultimate state of a V-shape torsion beam is depicted in Fig. 15.1. It has a bent axial line and complex section shapes along it. The desired diameter in the ends is much larger than that in the middle, which leads to a large expansion. The initial thickness of the tube is 3.5 mm to match the requirement of large stress level during severe conditions. The geometry parameters of the tube blank are listed in Table 15.1. A multi-step indirect HMGF process is necessary in the study. The work-flow scheme is shown in Fig. 15.2.

Firstly, end expansion is conducted to enlarge the diameters of the ends on the machine tools with two hydraulic cylinders, two expansion punches and slide rails. Secondly, the press bending is conducted to obtain the bent axial line on a 3000 kN CNC hydraulic press. In addition, the pre-forming process aims to achieve the V-shape on the component as that the flange could be prevented. It is conducted

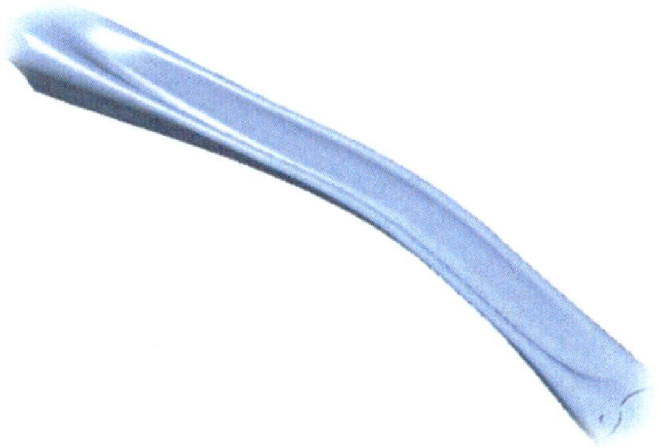

Fig. 15.1 Desired V-shape torsion beam

Table 15.1 Geometry of the tube

Thickness	Length	Outer diameter
3.5 mm	160 mm	98 mm

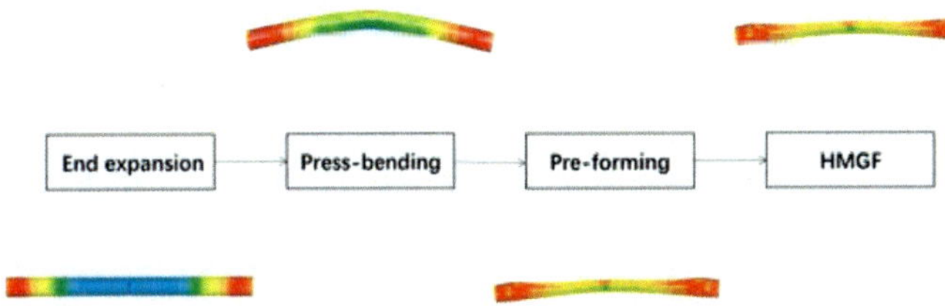

Fig. 15.2 Process scheme of indirect HMGF

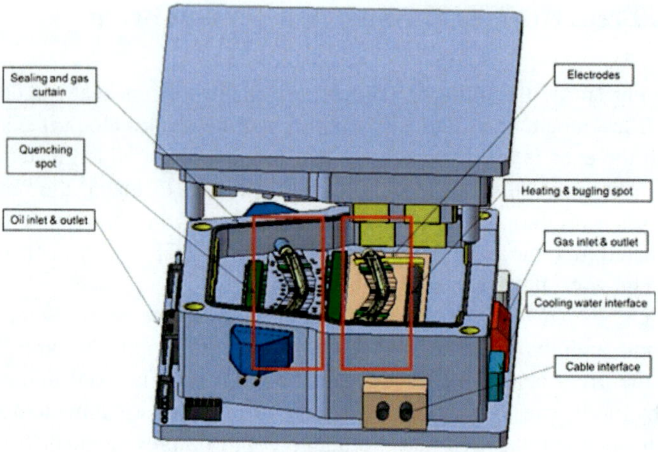

Fig. 15.3 Die tools of hot gas forming

on a 15,000 kN CNC hydraulic press. Finally, the hot gas forming step is the most complicated process, which is also conducted on a 15,000 KN CNC hydraulic press. The apparatus is shown in Fig. 15.3. The forming cavity is sealed by the upper and lower dies with sealing and gas curtains, so that a Nitrogen gas environment can be obtained. In this case, oxidation of heated steel tube could be prevented.

Two forming stations are assembled in the die cavity. One is the resistance heating & bugling spot. On this station, each end of the die cavity has a semicircular copper electrode so that the tube can firmly grasp the tube and heating efficient is improved. Considering that the heating rate is highly influenced by the resistance of the tube, a low voltage and high current power supply is employed with the 12 V/10000A. The pre-formed tube can be rapidly heated up to 900 °C in 30 s, as shown in Fig. 15.4. It is notable that the highest temperature of the electrode is about 200 °C so there is not a risk of melting.

After heating, then die closes in several seconds and the tube is bugling up in the same station with the punch and die less in 1 s. The gas up to 70 MPa pressure is filled into the pre-formed tube and the initial internal pressure of tube is not controlled during bugling. The parameters of the gas system are listed in Table 15.2.

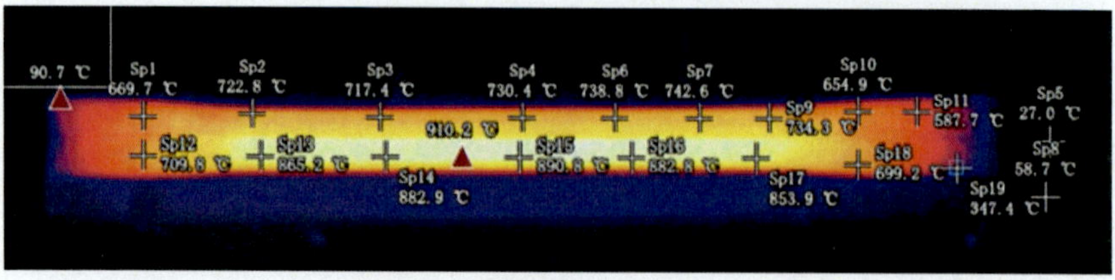

Fig. 15.4 Resistance heating the pre-formed tube

Table 15.2 Gas system parameters

Maximum pressure	Maximum volume	Minimum open distance
70 MPa	8 L	450 mm

After the bugling, an in-die quenching is conducted in the same spot for 20 s. In this step, the high-pressure gas supplement keeps. The cooling water system is design as shown in Fig. 15.5. The diameter of the water ducts is 10 mm and the displacement from the ducts to the surface is 8 mm. The water velocity of the inlet is about 3 m/s, which makes it 25 L/s. Thus, A 120 kW screw water-cooled chiller is employed.

Then the die opens, and the formed part is translated to the other spot. In this spot, the tube with a residual temperature is cooling to the room temperature in a second process routine to further prevent oxidation of the tube. The entire process time line is shown in Fig. 15.6.

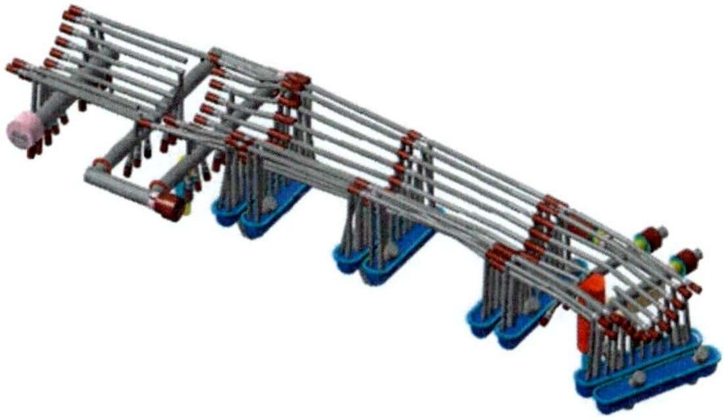

Fig. 15.5 Cooling water system

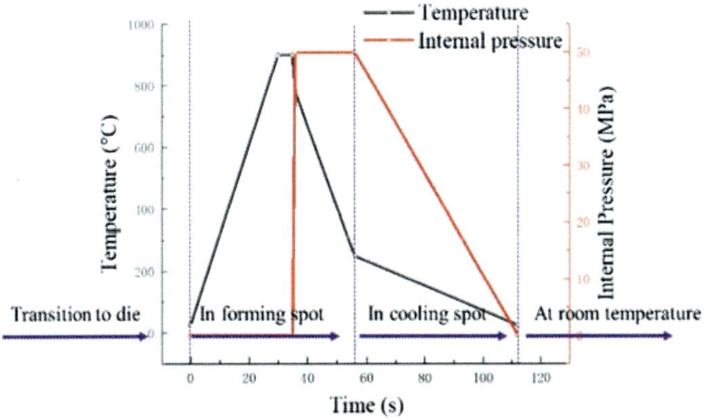

Fig. 15.6 Sequence of the entire process

15.3 Material Description

Hot rolled BR1500 press harden steel is employed in the study. Its mechanical properties are significantly influenced by the temperature and strain rate. It would undergo a phase transformation to austinite during heating process. The material has a Young's modulus about 210 GPa at room temperature and decreases along with the temperature increasing, dropping down to 120 GPa at 1000 °C. On the other hand, the flow stress of the material is elevated by higher strain rate [15]. Determined flow stress of the austenite between 600 and 900 °C at different strain rates are shown in Fig. 15.7. Higher strain rate leads to a high flow stress.

Within this temperature window the BR1500 exhibits an excellent formability as a consequence of the reduced flow stress at elevated temperatures. The in-die

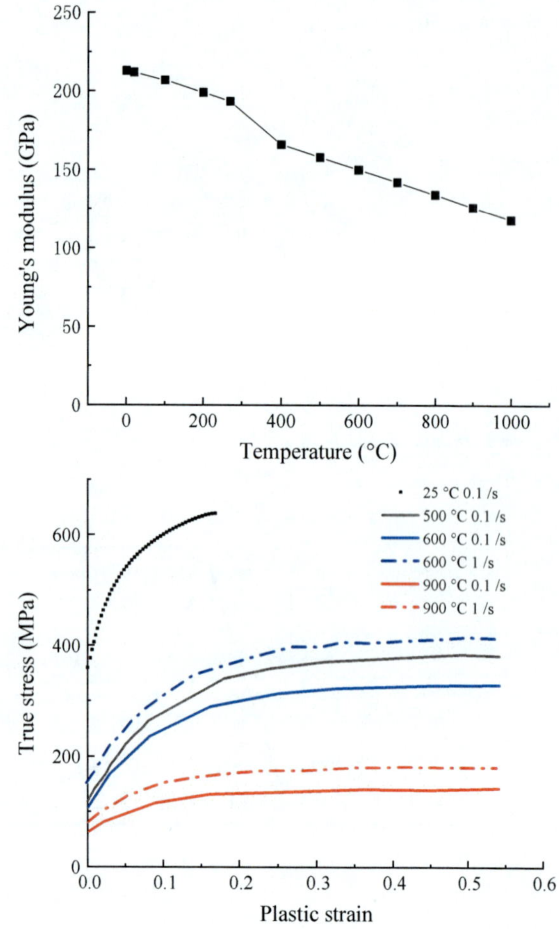

Fig. 15.7 Material properties of BR1500 of different temperatures and strain rates

quenching results in a phase transformation from austenite to martensite and significant increase of the final strength of the part. A cooling rate of at least 27 K/s is essential for preventing bainitic phase transformation according to the continuous time temperature transformation (TTT) diagram. To achieve a full martensitic, an increase of the minimum cooling rate required for martensite transformation up to 35 K/s is necessary [16]. Thus, components with homogeneous mechanical properties at a final high tensile strength level of above 1500 MPa and a sufficient remaining ductility of 6–8% uniform elongation as well.

15.4 Simulation of Indirect HGMF

The simulations are performed using LS-DYNA R10.0 finite element codes, which has several features that are useful to FE modeling hot forming [17, 18]:

Heat conduction, convection and radiation transfer of tube blank before tool to part contact.

Contact conductance between tools and parts during forming, which is a pressure dependent;

Phase transform and temperature dependency are considered in UHSS material model.

The middle layer of tube is modeled as Belytschko-Tsay shell elements with the size of 5 mm. The dies are modeled as rigid bodies. The thickness of the tube is considered in the simulation while that of the dies are not considered. In three cold forming processes, MAT_PIECEWISE_LINEAR PLASITICY is used for the tube with flow curve at 1 /s. The friction coefficients between the tube and dies is set to be 0.15.

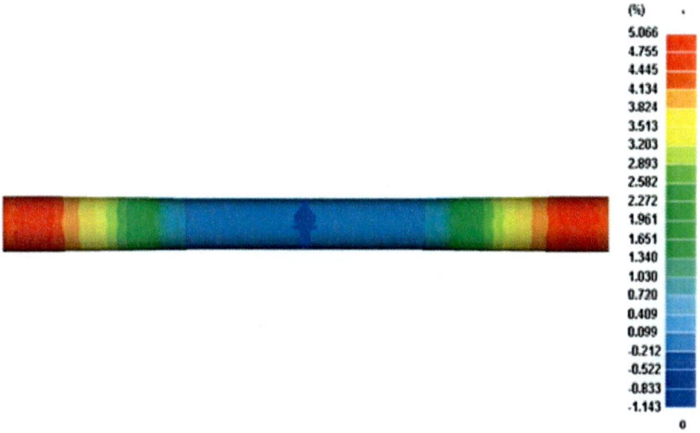

Fig. 15.8 Thinning contour after end expansion

Figure 15.8 shows the thinning contour of the thickness thinning after end expansion. A 5% thinning is observed at the both ends while a slight thicken is found in the middle region. Besides, the thinning contour after press bending is also shown in Fig. 15.9. In this step, the thickness change concentrates in the middle region. The inner side of the bending direction is thickened further while the wall thickness of the out side reduces about 3%. No obvious risk of crack nor wrinkle is found in the bending step.

In addition, the thickness distribution of the pre-forming is shown in Fig. 15.10. It can be seen that the most thinning area has a thickness more than 3.2 mm, which is about 8% thinning. And most thickened areas locate on the top of the V-shape in the inner side of the bending, where the material flowing is hindered by the die surface.

The cold forming simulation result is transferred to the hot gas forming simulation with a dynain file, which includes positions of the nodes, the stress and strain information of the elements, as well as the thickness.

Fig. 15.9 Current thinning contour after press bending

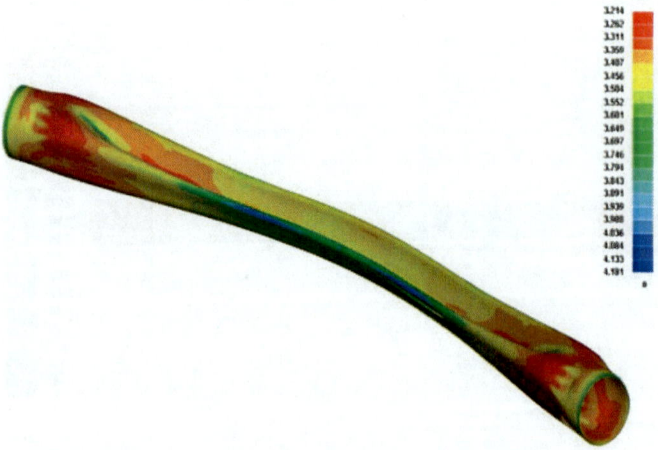

Fig. 15.10 Thickness distribution after cold forming process

A thermo-mechanical coupled numerical model is developed involving material properties and thermo physical parameters in detail. MAT_UHS_STEEL is employed in simulation with the flow curves at 1 /s since the bugling is a rapid process. The material model developed by P. Akerstrom is suited for the hot forming process that the phase transformation happens [15]. A fully austenitized phase is assumed to be obtained when the initial temperature of the blank is higher than 900 \degcIt is notable to be mention that the initial stress and strain state from the previous simulation should be removed in case of error in calculation. The phase distribution during cooling is calculated by solving the rate equation for each phase transition as Eq. (15.1):

$$\dot{X}_k = g_k(G, C, T_k, Q_k) f_k(X_k) \tag{15.1}$$

where g_k is a function dependent on the grain number G, the chemical composition C, the temperature T and the activation Q. Besides, the function*f* is dependent on the actual phase $X_t = x_k/x_{eq}$ as Eq. (15.2):

$$f_k(X_k) = X_k^{0.4(X_k-1)}(1 - X_k)^{0.4X_k},\ k = 2, 3, 4 \tag{15.2}$$

where $k = 2,3,4$ present the ferrite, pearlite and bainite respectively. The true amount of martensite as defined in Eq. (15.3) is modelled by using the true amount of the austenite left after the bainite phase:

$$x_5 = x_1\left(1 - e^{-\alpha(MS-T)}\right) \tag{15.3}$$

where x_1 is the true amount of the austenite, \alpha is a material constant and MS is the martensite start temperature. It is notable to be mention that the initial stress and strain state from the previous simulation should be removed in case of errors in calculating MS. In addition, the material model can provide the further information in history variables such as phase amounts, yield strength.

The thermal properties of the tube are defined using MAT_THERMAL_ISOTROPIC_TD_LC keywords, which allows isotropic thermal properties that are temperature dependent specified by load curves to be defined, which is shown in Fig. 15.11.

The thermal contact between blank and tools are implemented using CONTACT_FORMING_SURFACE_TO_SURFACE_THERMAL_FRICTION keyword, which can define the mechanical friction coefficients and thermal contact conductance as functions of temperature and pressure, as shown in Fig. 15.12. The process flow is illustrated as:

Heating of the blank to 900 °C;

Gravity loading applied for tube dropping into the die cavity for 3 s;

Die closing for 5 s;

Bulging time for 0.5 s with the pressure increases up to 50 MPa, and

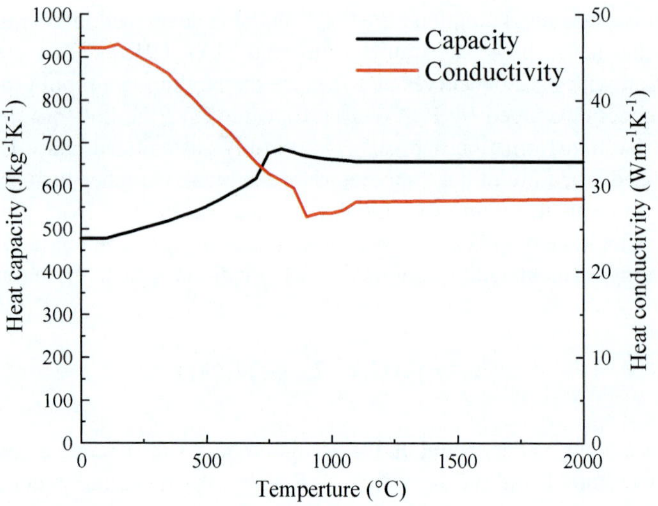

Fig. 15.11 Heat capacity and conductivity versus temperature

Quench time 20 s.

Figure 15.13 shows the thickness distribution in the blank with a range from 4.169 mm (dark blue) to 3.169 mm (red) after bugling. The cross-section views can also be found and they demonstrate the fitness of the part to the die is quite satisfying.

The material phase fractions distribution calculated in the simulation after in-die quenching is shown in Fig. 15.14, as well as the residual temperature. It can be seen that the edge on the top of the V-shape has the higher residual temperature. When most of the part has been quenched down to 100 °C, it is still above 200 °C. It indicates that the cooling rate on this region may not reach the martensite transforming rate. It is agreed by the phase distribution. Although most of the part is martensite phase, but bainite transformation occurs on the edge region. The result implies that the geometry of the die has caused a problem for cooling and it should be improved.

15.5 Conclusions

A numerical simulation the studies on indirect hot metal gas forming process for a V-shape torsion beam are performed using BR1500 press-harden steel. An integrated manufacturing system is used. Based on the result, the guidance to the design of the part and die will be implemented as well. Practical Hot Metal Gas Forming tests will be conducted soon and future studies would focus on the cyclical production processes.

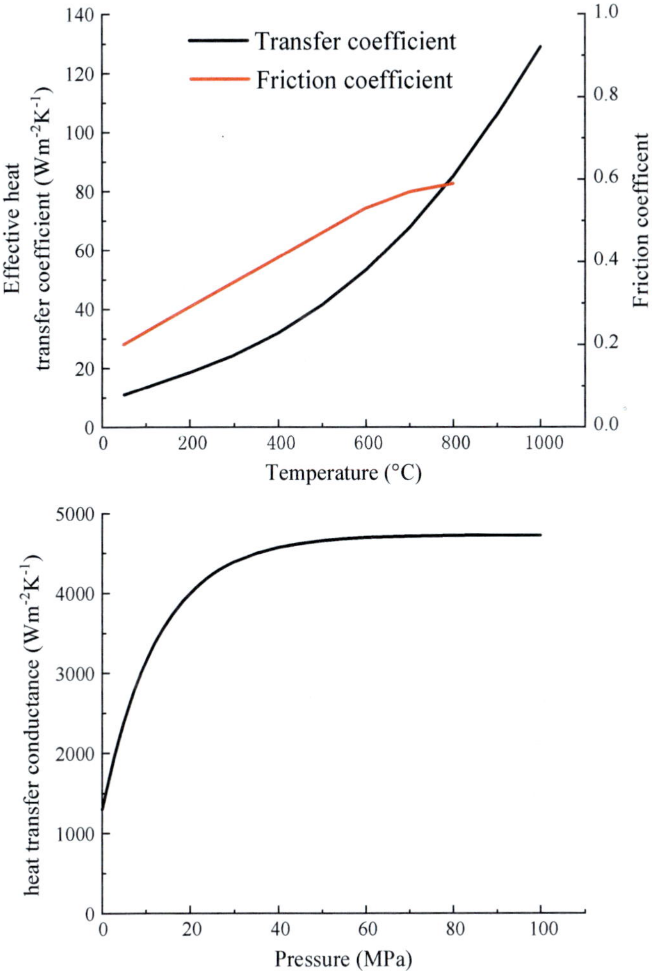

Fig. 15.12 Contact interface parameters

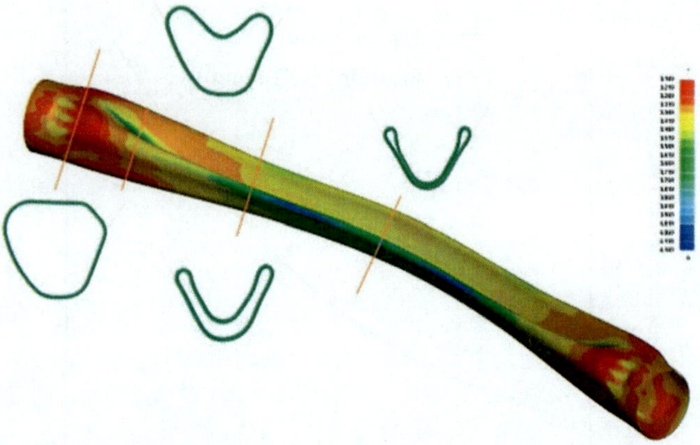

Fig. 15.13 Thickness distribution and cross-section fitness of the part after bugling process

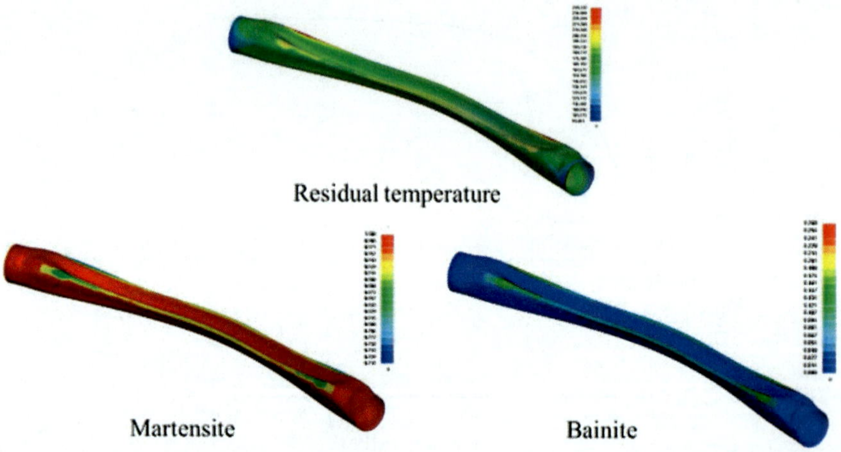

Fig. 15.14 Residual temperature and material phase fractions

Acknowledgements 1. High-end CNC Machine Tools and Basic Manufacturing Equipment (No. 2019ZX04002030).

2. National Key Research and Development Program of China (2017YFB0304404).

References

1. Merklein, M., Lechler, J.: Investigation of the thermo-mechanical properties of hot stamping steels. J. Mater. Process. Technol. **177**(1–3), 452–455 (2006)

2. Neugebauer, R., Schieck, F., Werner, M.: Tube press hardening for lightweight design. In: ASME 2011 International Manufacturing Science and Engineering Conference, pp. 495–502. MSEC, Cornvallis, Oregon (2011).
3. Mori, K., Maki, S., Tanaka, Y.: Warm and hot stamping of ultra high tensile strength steel sheets using resistance heating. CIRP Ann. **54**(1), 209–212 (2005)
4. Merklein, M., Wieland, M., Lechner, M., Bruschi, S., Ghiotti, A.: Hot stamping of boron steel sheets with tailored properties: a review. J. Mater. Process. Technol. **228**, 11–24 (2016)
5. Karbasian, H., Tekkaya, A.E.: A review on hot stamping. J. Mater. Process. Technol. **210**(15), 2103–2118 (2010)
6. Vadillo, L., Santos, M., Gutierrez, M., Pérez, I., Gonzalez, B., Uthaisangsuk, V.: Simulation and experimental results of the hot metal gas forming technology for high strength steel and stainless steel tubes forming. AIP Conf. Proc. **908**(1), 1199–1204 (2007)
7. Drossel, W., Pierschel, N., Paul, A., Katzfuß, K., Demuth, R.: Determination of the active medium temperature in media based press hardening processes. J. Manuf. Sci. Eng. **136**(2), 021013 (2014)
8. Paul, A., Strano, M.: The influence of process variables on the gas forming and press hardening of steel tubes. J. Mater. Process. Technol. **228**, 160–169 (2016)
9. Paul, A., Reuther, F., Neumann, S., Albert, A., Landgrebe, D.: Process simulation and experimental validation of Hot Metal Gas Forming with new press hardening steels. In: 36th IDDRG Conference - Materials Modelling and Testing for Sheet Metal Forming, Vol. 896, pp. 012051. IOP Publishing Ltd, Munich, Germany (2017).
10. Yi, H.K., Pavlina, E.J., Tyne, C.J.V., Moon, Y.H.: Application of a combined heating system for the warm hydroforming of lightweight alloy tubes. J. Mater. Process. Technol. **203**(1–3), 532–536 (2008)
11. Maeno, T., Mori, K., Fujimoto, K.: Development of the hot gas bulging process for aluminium alloy tube using resistance heating. Key Eng. Mater. **410–411**, 315–323 (2009)
12. Maeno, T., Mori, K., Adachi, K.: Gas forming of ultra-high strength steel hollow part using air filled into sealed tube and resistance heating. J. Mater. Process. Technol. **214**(1), 97–105 (2014)
13. Maeno, T., Mori, K., Fujimoto, K.: Hot gas bulging of sealed aluminium alloy tube using resistance heating. Manufacturing Review **1**, 5 (2014)
14. Zhang, W., Han, C., Yuan, S.: Hot forming processing simulation of torsion beam of high strength steel 22MnB5. Mater. Sci. Technol. **22**, 16–22 (2014). (in Chineses)
15. Åkerström, P., Oldenburg, M.: Austenite decomposition during press hardening of a boron steel—Computer simulation and test. J. Mater. Process. Technol. **174**(1–3), 399–406 (2006)
16. Merklein, M., Lechler, J.: Determination of material and process characteristics for hot stamping processes of quenchenable ultra high strength steels with respect to a FE-based process design. SAE Int. J. Mater. Manuf. **1**(1), 411–426 (2008)
17. Arthur, S.: inite element modeling of hot stamping. Steel Res. Int. **80**(9), 658–664 (2009)
18. Shapiro, A.B.: Using LS-Dyna for hot stamping. In: 7th European LS-DYNA Users Conference, Salzburg, Austria (2009).

Chapter 16
Modeling and Parameter Identification for Active Lubricated Hydrostatic Bearing-Rotor System

Shuo Zhan, Wei Pan, Yixin Zhang, and Changhou Lu

Abstract Active lubricated hydrostatic bearing-rotor system can achieve large bearing capacity, high motion accuracy, high anti vibration ability and long service life. Models from first principles are often complex and inaccurate caused by the discrepancy between the ideal hypothesis and the actual situation. And the multiple-input multiple-output (MIMO) characteristics of the system further increase the complexity of modeling, but the system requires low complexity dynamic model over its entire operating range. In order to solve the problem, this paper presents a modeling and identification method. Firstly, a segmented linear model is established in a small displacement of the shaft, and the parameters are solved by the prediction error method. The parameter variation law of the system is obtained by fitting identified system parameters. Then, the dynamic equation of the system can be obtained. Finally, the identified model is validated by impulse signals. The experimental results demonstrate the validity and accuracy of the identified model.

Keywords Active lubricated bearings · Grey-box modelling · Linear parameter-varying system · System identification

16.1 Introduction

Hydrostatic bearings have wide applications in precision and ultra-precision machinery. Although systems composed of hydrostatic bearings and other functional units have the self-adjusting abilities, they are usually "passive". It means that when the external load changes, the equilibrium position of the rotor changes, resulting in vortex trajectory and loss of mechanical accuracy. The traditional passive hydrostatic bearing-rotor system is relatively mature and has low potential for significant improvements [1, 2]. The active lubricated bearing-rotor system, which combines

S. Zhan · W. Pan (✉) · Y. Zhang · C. Lu
Key Laboratory of High-Efficiency and Clean Mechanical Manufacture of MOE, National Demonstration Center for Experimental Mechanical Engineering Education, School of Mechanical Engineering, Shandong University, Jinan 250061, Shandong Province, China
e-mail: panw@sdu.edu.cn

J. Xu and K. M. Pandey (eds.), *Mechanical Engineering and Materials*, Mechanisms and Machine Science 100,
https://doi.org/10.1007/978-3-030-68303-0_16

tribology, control techniques and informatics, has significant advantages in the following aspects: controlling the lateral vibration of flexible rotating shafts; modifying bearing dynamic characteristics, as stiffness and damping properties; increasing the rotational speed ranges by improving damping and eliminating instability problems [3]. However, modeling, control and other related problems of the active control bearing-rotor system, especially of the active lubricated oil bearing-rotor system, have not been well solved.

In recent years, a variety of active lubricated bearings have been put forward. These papers mainly include two types of bearings: gas bearing and oil bearing. Mizumoto [4, 5] has carried out further research on the active lubricated bearings, and designed a high-speed gas bearing. Moros [6] put forward a gas bearing with active force generated by piezoelectric actuator installed on the back of the bearing sleeve. Li and Liu [7, 8] present an active lubricated system based on hydrostatic oil bearing and restrictor. In this system, the restrictor can control the rotor trajectory by changing the flow rate into the oil chamber of the hydrostatic bearing. Nicoletti [9] researched the active lubricated oil bearings with tilting-pad, and designed a non-linear controller for this system.

The above research explored the application potential of the active lubricated bearing. At present, there are still some problems in controlling this kind of bearing: (a) the dynamic equation derived by theory is too complex to be used in active control of the system; (b) there are many unknown factors in the system, which are difficult to describe by mathematical model. Based on the above factors, parameter identification of active lubricated bearing has been studied. Thiesen [10] carried out modeling and parameter identification of the gas bearing, and designed a controller to enhance the stability of its system. Matta [11] uses the impact hammer to identify the model parameters of the pneumatic bearing. Chuang [12] applies the state feedback control to the system, and then identify the parameters of the system model. These papers indicate the identification and control method of the lubricated bearings. In addition, active magnetic bearings (AMB) also have potential applications in this system, so the identification research of such kind of bearing is also of great significance. In [13], a predictor-based subspace identification method was proposed to identify the AMB system, and the obtained model was used to design robust controllers. In [14], a frequency-based method was proposed for identification of the transfer function matrix model of a non-rotating shaft supported by AMBs. The method consisted of steps identifying the submodels separately and finally combining them together. Wang [15] proposed a weighted instrumental variable (WIV) method to identify the damping ratio of the AMB system. The experimental results show that this method can be more accurate than the traditional methods.

In summary, modeling and parameter problems of active gas bearings and AMBs have been studied, but there is little literature on active hydrostatic oil bearings. The hydrostatic oil bearing is different from the other two, in terms of large bearing capacity and self-stability. Therefore, based on the active lubricated bearing-rotor system proposed by Liu and Li [7, 8], the equipment in this paper adopts the hydrostatic oil bearing and the piezoelectric membrane restrictor to form an active control system. Firstly, the system is established as a segmented linear model, and the parameters are solved by the error minimization criterion. After obtaining the segmented parameters, the parameter variation law of the system is obtained by fitting multiple segmented parameters. Finally, the identification results demonstrate the validity and accuracy of the proposed method.

16.2 Principle and Modeling of Active Lubricated Bearing-Rotor System

16.2.1 Principle of Active Bearing Rotor-System

The structure diagram of the system is shown in Figs. 16.1 and 16.2. The device uses piezoelectric actuator (PZA) and the membrane restrictor to form the control unit to control the system. The upper and left oil chambers of hydrostatic bearing are throttled by piezoelectric membrane restrictor, and the lower and right oil chambers are throttled by orifice restrictors. The basic principle of the system is to modulate the amplitude and phase of the equilibrium position of the rotor, so as to achieve the expected equilibrium point position or the expected equilibrium point trajectory. As a result, this system changes the membrane position of the piezoelectric membrane restrictor by changing the input voltage of the PZA, so as to change the throttling clearance. The change of clearance can change the oil chamber pressure of the four

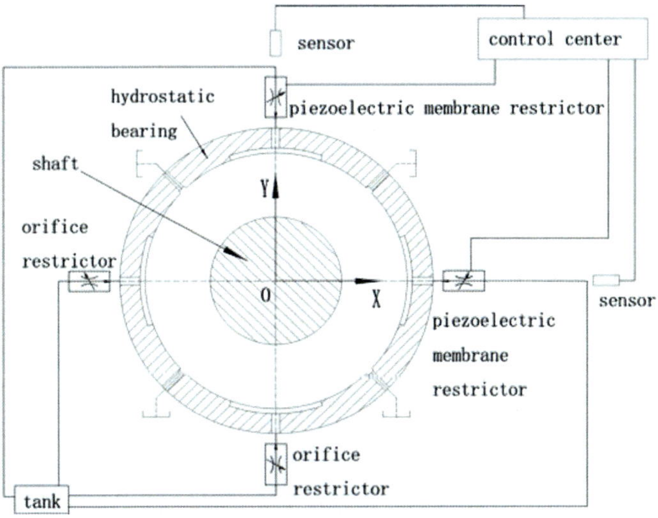

Fig. 16.1 The structure of active lubricated bearing-rotor system

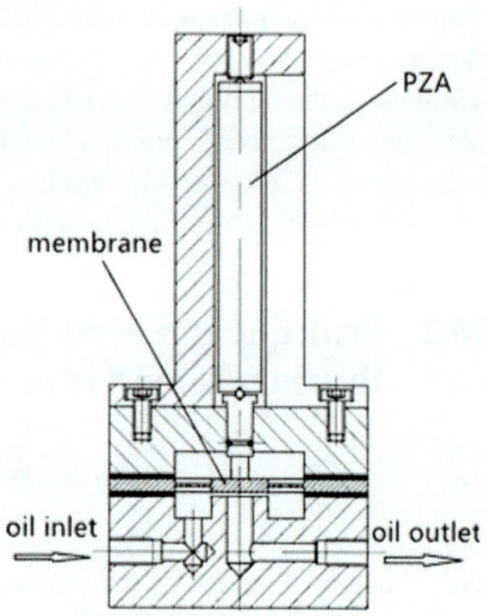

Fig. 16.2 The structure of piezoelectric membrane restrictor

hydrostatic oil chambers, and then realize the control of the orbit of the shaft center through the pressure difference between the oil chambers.

16.2.2 Modeling of Active Lubricated Bearing-Rotor System

The structure of the system can be known from the Fig. 16.1, so the whole system is modeled as a grey-box model. In this system, the parameters are not constant. These parameters change with the shaft position, so it is a nonlinear system. In order to identify these parameters, the system is regarded as a linear parameter-invariant system in a small displacement of the shaft. And the whole system can be a linear parameter-varying system. In each small displacement, the force equation of the shaft can be equivalent to a 2 DOF coupled mass-spring-damper system model, so the dynamic equation of the system can be rewritten as Eq. (16.1).

$$M\ddot{d}(t) + c\dot{d}(t) + kd(t) = f(t). \tag{16.1}$$

where M is the equivalent mass matrix, c is the equivalent damping matrix, k is the equivalent stiffness matrix, d(t) is the displacement of the shaft. In this system, the change of external force applied to the shaft is mainly realized by the change of the elongation of PZA. It is further known that the change of external force should be related to the change of input voltage. In each small displacement, the system is regarded as a linear system, it can be considered that the external force has a linear relationship with the input voltage, so the external force can be written as $f(t) = b \times u(t)$. u(t) is the input voltage. Therefore, the dynamic equation of the system can be rewritten as following Eq. (16.2):

$$M\ddot{d}(t) + c\dot{d}(t) + kd(t) = bu(t). \tag{16.2}$$

This grey-box modelling is eased by reformulation of dynamic equation to state space form, where u(t) is the input of the system and d(t) is the output of the system. The model inevitably gives rise to modelling errors, which are include as process noise v(t), measurement noise is modelled as an additive signal w(t). This combination of process noise and measurement noise can be reduced to the equivalent noise term e(t) entering both the state and the output equation. Therefore, model errors and measurement noise are included as a signal e(t) entering both through an input gain E and the measurements directly as given in Eq. (16.3):

$$\begin{aligned} \dot{x}_0(t) &= Ax_0(t) + Bu(t) + Ee(t) \\ d(t) &= Cx_0(t) + e(t). \end{aligned} \tag{16.3}$$

where the system matrix, input gain matrix, output matrix and state vector are defined in Eq. (16.4):

$$A = \begin{bmatrix} 0 & I \\ K & D \end{bmatrix}, B = \begin{bmatrix} 0 \\ B \end{bmatrix}, E = \begin{bmatrix} 0 \\ E \end{bmatrix}, C = \begin{bmatrix} I & 0 \end{bmatrix}, x_0 = \begin{bmatrix} d_x, d_y, \dot{d}_x, \dot{d}_y \end{bmatrix}^T \tag{16.4}$$

and $K = -M^{-1}k, D = -M^{-1}c, B = M^{-1}b$. The parameters in Eqs. (16.3) and (16.4) are identified by recasting the problem to a model parameterized in $\theta = \{K, D, B, E, x_0\}$ as $\Phi(\theta)$. Each matrix K, D, B, E has four elements. e.g. $K = \begin{bmatrix} k_{xx} & k_{xy} \\ k_{yx} & k_{yy} \end{bmatrix}$.

16.3 Parameters Identification of Active Lubricated Bearing-Rotor System

16.3.1 Description of Experiments

The experimental equipment setup is shown in Fig. 16.3. Physical dimensions of the equipment are shown in Table 16.1. In the experiment, the oil supply pressure $p = 1.1$ MPa and the rotational speed $\Omega = 0$ rad/s, all measurements are sampled with period $t_0 = 0.2$ ms.

In the experiment, the input signal is PRBS. Because of the incompressibility of the oil, the stiffness and damping of the system are not constant. These parameters will change with the position of the shaft. Therefore, in this experiment, the initial position of the shaft can be changed by changing the valley value of the PRBS, and

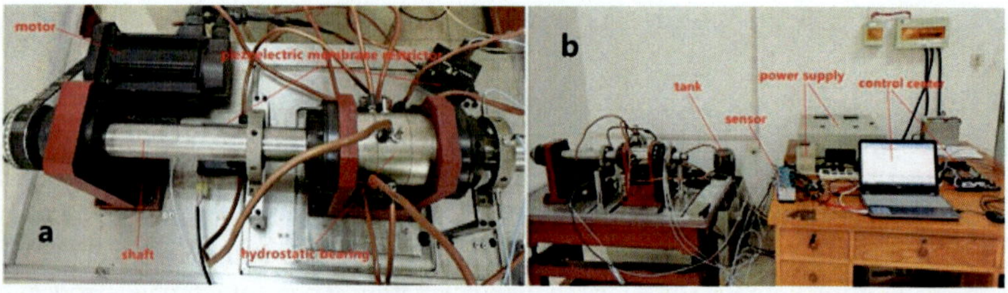

Fig. 16.3 The experimental equipment: **a** hydrostatic bearing and piezoelectric membrane restrictor; **b** control system

Table 16.1 Experimental equipment physical dimensions

Dimension	Bearing length [mm]	Bearing diameter [mm]	Shaft length [mm]	Shaft mass [kg]	Orifice diameter [mm]
value	48	100	894.8	11.7	10

the identified parameters are the average parameters of the response displacement. Some pre-experiments are needed to determine the parameters of the signal before the experiment. When the input voltage is too small, the signal noise will have a great impact on the collected data, and the cross-coupling response of the system is not obvious, so the input voltage amplitude in horizontal and vertical directions is 40/30 V. For the signals collected for identification, a moving average filter of size 50 is used to reduce measurement noise from the shaft movement measurements.

16.3.2 Identification of System Parameters

The optimal system model from each data set is chosen as the one associated with the minimum cost of a prediction error cost function $W(\theta) = \sum \varepsilon^2(t)$, where the prediction error $\varepsilon(t)$ is defined as the difference between the one step ahead measured and the predicted output $\varepsilon = d(t) - \hat{d}(t)$. The minimum is sought using the prediction error method (PEM) [16].

After calculating the error cost function, it needs to be solved numerically. At first, the parameter range of $\{k_{xx}, k_{yy}, c_{xx}, c_{yy}, b_{xx}, b_{yy}\}$ can be estimated through the step response in horizontal and vertical directions. Then all the parameters are searched and solved by particle swarm optimization (PSO) [17] algorithm. When the number of iterations reaches 30, the iteration is terminated, and the system parameter can be obtained.

16.4 Analysis for the Identified Parameters of Model

16.4.1 Results of Parameters Identification

After the system parameters are identified, the identified models are verified by using different data collected under the same conditions. Figure 16.4 shows an example of measured shaft movement and their predicted. The horizontal fit is 93.97%, and the vertical fit is 89.53% in this example. Figure 16.5 shows the validation fit percentage of the identified model. The horizontal fit mean value is 91.35% and its standard

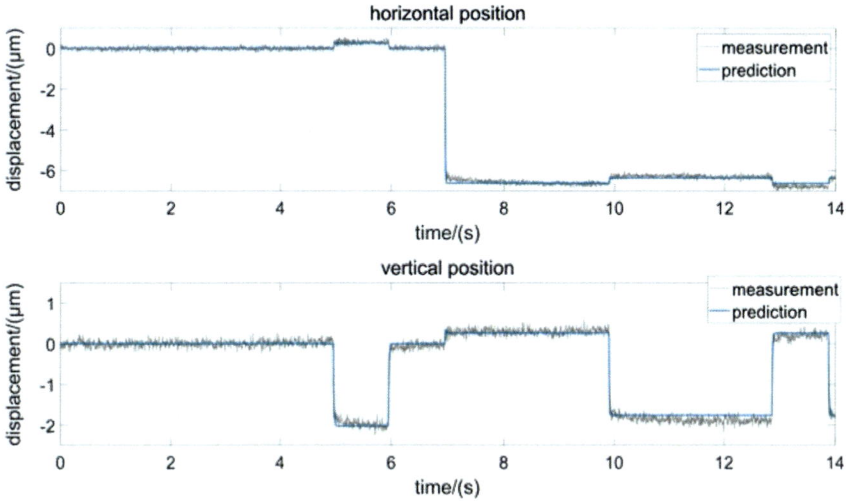

Fig. 16.4 Example of identification from data collected at x = 2.6 μm and y = 0.75 μm

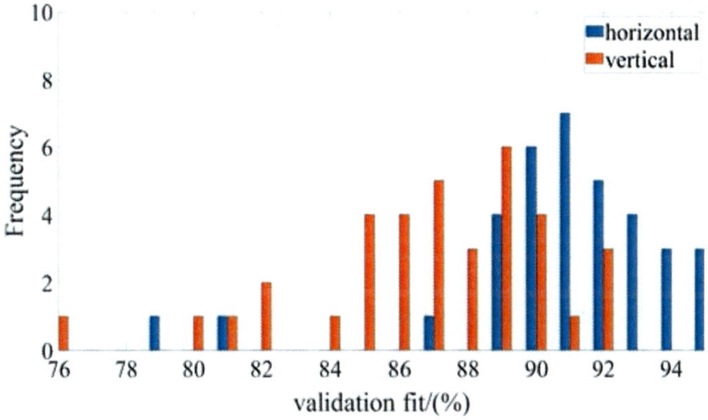

Fig. 16.5 Fit percentages in validation of the identified models on other data

deviation is 3.261%, and the vertical fit mean value is 87.44% and its standard deviation is 3.629%. These are high fit percentage indicating the models can well describe the system. In addition, the infinity norm of the residuals is small, indicating good significance of the results.

16.4.2 Parameters Variation Law

In this system, its stiffness and damping will change with shaft positions. However, this variation law is difficult to identify directly, so we will identify the equivalent stiffness and damping instead. The stiffness and damping are regarded as the average value of the model parameter in this section, and then the average parameter values of all the sections are fitted to obtain the variation law of the parameters. These bearing-rotor model parameters are expected to depend continuously on the horizontal position and vertical position. Each key parameter is therefore approximated onto a polynomial surface. Thereby the identified stiffness, damping matrix coefficients are modelled as Eq. (16.5):

$$\begin{aligned} k(x, y) &= k_0 + k_1 x + k_2 y + k_3 xy + k_4 y^2 + k_5 xy^2 + k_6 y^3 \\ c(x, y) &= c_0 + c_1 x + c_2 y + c_3 xy + c_4 y^2 + c_5 xy^2 + c_6 y^3 \end{aligned} \tag{16.5}$$

This describes a surface in space. The parameters of the polynomial models are fitted using MATLAB toolbox and the curved surfaces are listed in Figs. 16.6. and

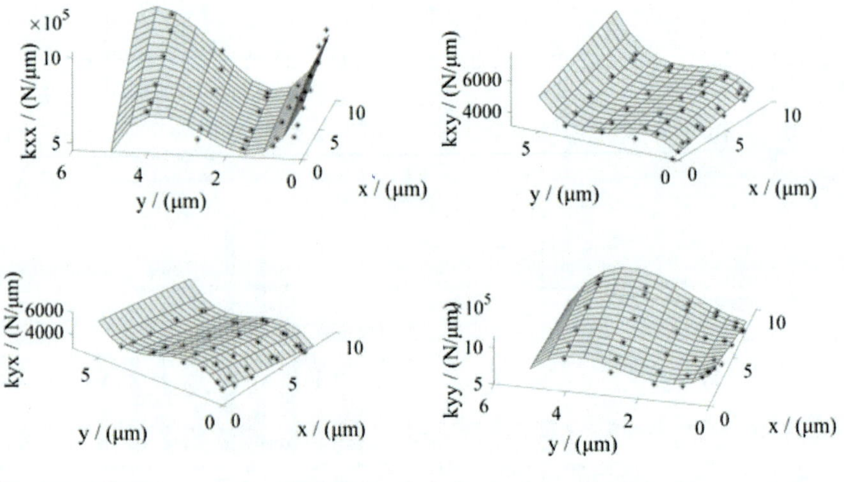

Fig. 16.6 The variation law of identified coefficients, identified stiffness coefficients as function of the horizontal and vertical displacement

16.7. Tables 16.2 and 16.3 show the estimated coefficients with polynomial approximations. The variation of each key parameter's estimate in the same condition is small, indicating a good consistency across different data sets.

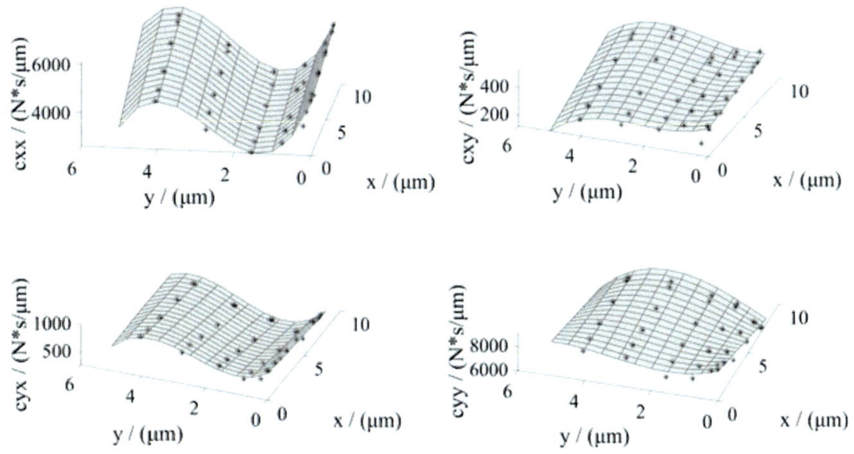

Fig. 16.7 The variation law of identified coefficients, identified damping coefficients as function of the horizontal and vertical displacement

Table 16.2 Coefficients of fitted polynomials of the form Eq. (16.5) for the key parameters (stiffness)

Stiffness	k_0	k_1	k_2	k_3	k_4	k_5	k_6
k_{xx}	7.52e5	2.31e4	−4.78e5	−1.08e4	2.34e5	3.89e3	−3.00e4
k_{xy}	3.64e3	1.42e2	1.44e3	−1.47e1	−8.23e3	9.19e0	1.21e2
k_{yx}	4.17e3	−1.63e2	2.59e3	−4.92e1	−1.33e3	8.80e0	1.77e2
k_{yy}	9.13e5	−4.09e4	−2.74e5	5.25e4	9.81e4	−9.75e3	−831e3

Table 16.3 Coefficients of fitted polynomials of the form Eq. (16.5) for the key parameters (damping)

Damping	c_0	c_1	c_2	c_3	c_4	c_5	c_6
c_{xx}	4.74e3	1.36e2	−1.53e3	−1.01e1	1.74e3	6.20e0	−2.17e2
c_{xy}	3.18e2	1.17e1	−9.43e2	7.96e0	4.95e2	−7.66e-1	−7.69e0
c_{yx}	7.48e2	−1.95e1	−3.65e2	−2.59e1	2.43e2	5.96e0	−3.44e1
c_{yy}	8.42e3	−2.54e2	−1.07e3	3.05e2	4.87e2	−6.03e1	−5.15e1

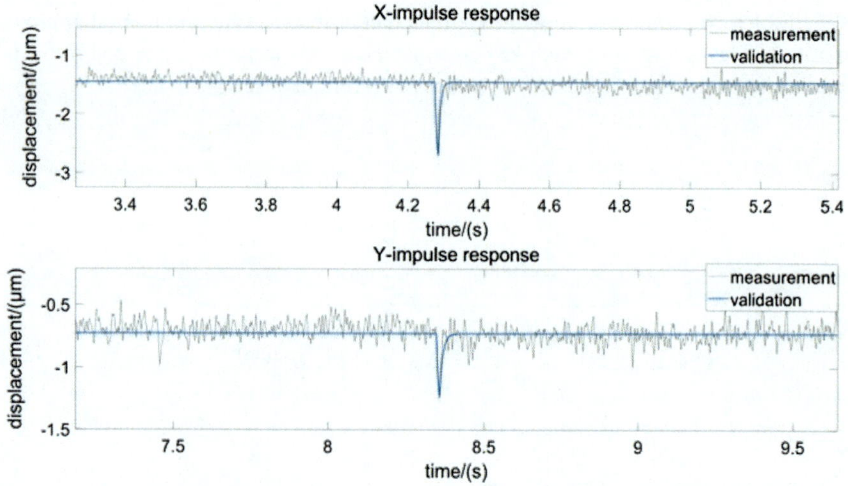

Fig. 16.8 Response of system for shaft position at x = 1.15 μm and y = 0.75 μm when the system is input the impulse signal of the horizontal and vertical directions respectively

16.4.3 Model Validation

After getting the variation law of the system parameters, the system can be verified by an impulse signal. Therefore, an impulse signal is input at the horizontal and vertical positions of the shaft at 1.15 μm and 0.45 μm respectively. The response of the system is shown in Fig. 16.8. According to the response of the system, the fit percentages of the horizontal and vertical directions are 84.13% and 78.41% respectively. The results show that the identified model describes the characteristics of the system well.

16.5 Conclusions

In this paper, an active control system mainly composed of hydrostatic oil bearing and piezoelectric membrane restrictor is modeled and identified. The equivalent stiffness and damping in the dynamic equation change with the position of the shaft, which leads to inaccurate modeling of the system and makes the system unable to move according to the expected trajectory. In order to solve this problem, the grey-box modeling technology is used to model the system. The parameters in the segmented model are resolved by the error minimization function. The variation law of the system parameters is realized by fitting the segmented parameters, and then the whole model of the system is obtained. Finally, the system model is verified by impulse signal, and the results show that the system model reflects the characteristics of the system well.

Acknowledgements This research is supported by the National Natural Science Foundation of China (No. 51975337) and Natural Science Foundation of Shandong Province of China (No. ZR2019MEE064).

References

1. Andres, L.S., Lubell, D.: Imbalance response of a test rotor supported on squeeze film dampers. J. Eng. Gas Turbines Power **120**(2), 397–404 (1998)
2. Pietra, L.D., Adiletta, G.: The squeeze film damper over four decades of investigations. part i: characteristics and operating features. Shock and Vibration Digest **34**, 3–27 (2002)
3. Santos, I.F.: Trends in controllable oil film bearings. In: Gupta, K. (eds) IUTAM symposium on emerging trends in rotor dynamics. IUTAM Bookseries, vol. 1011. Springer, Dordrecht (2011)
4. Mizumoto, H., Arii, S., Kami, Y., Goto, K., Yamamoto, T., Kawamoto, M.: Active inherent restrictor for air bearing spindles. Precision Eng. **19**(2–3), 141–147 (1996)
5. Mizumoto, H., Arii, S., Yabuta, Y., Tazoe, Y.: Vibration control of a high-speed air bearing spindle using an active aerodynamic bearing. In: Proceedings of 2010 international conference on control, automation and systems (ICCAS 2010), pp. 2261–2264. IEEE. Gyeonggi-do, South Korea (2010)
6. Morosi, S., Santos, I.F.: Active lubrication applied to radial gas journal bearings. Part 1: Modeling. Tribol. Int. **44**(12), 1949–1958 (2011)
7. Liu, Z.C., Pan, W., Lu, C.H., Zhang, Y.T.: Numerical analysis on the static performance of a new piezoelectric membrane restrictor. Ind. Lubric. Tribol. **68**, 521–529 (2016)
8. Li, H.C.: Reasearch on the Performance and Application of the New Single-Sided Membrane Restrictor. Shandong University (2017)
9. Nicoletti, R., Santos, I.F.: Linear and non-linear control techniques applied to actively lubricated journal bearings. J. Sound Vib. **260**(5), 927–947 (2003)
10. Theisen, L.R.S., Niemann, H.H., Santos, I.F., Galeazzi, R., Blanke, M.: Modelling and identification for control of gas bearings. Mech. Syst. Signal Process. **70–71**, 1150–1170 (2016)
11. Matta, P., Arghir, M.: Identification method for rotordynamic coefficients of cylindrical air bearing using an impact hammer. Proc. Inst. Mech. Eng., Part J: J. Eng. Tribol. **226**(30), 199–212 (2012)
12. Chuang, J.C., Chen, C.Y., Tu, J.Y.: Active control of multi-input hydraulic journal bearing system. J. Phys: Conf. Ser. **744**, 012062 (2016)
13. Balini, H.M.N.K., Houtzager, I., Witte, J., Scherer, C.W.: Subspace identification and robust control of an AMB system. In: Proceedings of the 2010 American Control Conference, pp. 2200–2205. IEEE, Baltimore (2010)
14. Sun, Z., He, Y., Zhao, J.J., Shi, Z.G., Zhao, L., Yu, S.Y.: Identification of active magnetic bearing system with a flexible rotor. Mech. Syst. Signal Process. **49**(1–2), 302–316 (2014)
15. Wang, W.M., Li, Q.H., Gao, J.J., Yao, J.F., Allaire, P.: An identification method for damping ratio in rotor systems. Mech. Syst. Signal Process. **68–69**, 536–554 (2016)
16. Ljung, L.: System Identification: Theory for the User, 2nd edn. Prentice Hall PTR, Upper Saddle River (1999)
17. Kennedy, J., Eberhart, R.: Particle swarm optimization. In: Proceedings of ICNN'95 - International Conference on Neural Networks, pp. 1942–1948 vol. 4. IEEE, Perth, WA, Australia (1995)

Chapter 17
The Influence of Pressure Angle of Spur Gears on Bending Stress Considering the Effect of Root Fillet Radius

Layue Zhao, Minggang Du, and Yang Yang

Abstract Tooth breakage from excessive bending stress and surface pitting from excessive contact stress are the two primary fatigue failure modes for gears. Tooth breakage will end the gear life, so gear bending stress has to be accurately calculated for reliable operation. Many efforts have been made to increase gear bending strength including improving gear material property. With increasing demand for high power density gear applications, the need to optimize gears for minimum stress becomes increasingly important. It is imperative to understand the effect of gear parameters on bending stress in the initial stage of designing gears. Gear pressure angle is an important parameter affecting tooth bending stress. Because the critical section occurs in the gear root fillet, the root fillet radius largely affected by tool (hob) tip radius also has a great significance for improving the bending strength. The maximum tool tip radius will vary with the pressure angle changing due to the geometrical relationship of basic rack. This paper investigates the influence of pressure angle of spur gears on bending stress considering the variation of root fillet radius, and provides a recommendation on how to optimize the pressure angle for high bending strength gears. For analysis and validation of results, three methods of predicting bending stress– ISO standard, 3D-TCA (tooth contact analysis) method and Finite Element Analysis are applied and discussed.

Keywords Gear bending stress · Gear pressure angle · Root fillet radius · Gear material property

17.1 Introduction

Gears are commonly used in automotive, mechanical engineering and other industries. Nowadays, high power density gear applications are increasingly in demand, it is critical to design gears with improved load capacity. Gear tooth pitting due

L. Zhao · M. Du (✉) · Y. Yang
Science and Technology on Vehicle Transmission Laboratory, China North Vehicle Research Institute, Beijing, China
e-mail: mgdu@noveri.com.cn

J. Xu and K. M. Pandey (eds.), *Mechanical Engineering and Materials*, Mechanisms and Machine Science 100,
https://doi.org/10.1007/978-3-030-68303-0_17

to excessive surface contact stress and root fillet fractures due to excessive tooth bending stress are two main fatigue failure modes for gears [1]. The accumulation of defects and a high concentration of stresses in micro defects are also dangerous in the event of shocks which occur during the operation of mechanisms. Micropitting on the tooth surface of the gear can cause it to break [2]. Tooth breakage causes a catastrophic failure, so gear bending stress analysis methods must be reliable and understanding the key influence parameters on bending stress is very important to improve bending strength. Pressure angle is one which plays an essential role in determining the bending stress.

A pressure angle 20° was adopted for standard gears according to ISO 53:1998(E) [3]. The standard pressure angle 20° is a compromise value and cannot meet all the needs of the different applications because of the limited load capacity on root fillet. So in many cases, non-standard pressure angles need to be designed in order to improve the gear performance.

Recently, there have been many efforts made to explore the application of non-standard pressure angle gears. Gupta [4] calculated and compared the maximum bending and contact stress for the low dedendum spur gears with different pressure angles using finite element method. Handschuh [5] investigated the effects of high pressure angle gears compared with typical gear designs, the analysis of contact and bending stress had been done on three gears- standard 3.18 module, 28 tooth and 20° pressure angle, 2.12 module, 42 tooth and 25° pressure angle and 1.59 module, 56 tooth and 35° pressure angle. Sankar [6] studied the effects of pressure angle and tip relief on the failure of a helical gear pairs. Dadhanlya [7] presented a study on the effect of pressure angle on bending stress and deformation of asymmetric involute spur gear using FEA. Oda [8] introduced a study on the effect of pressure angle, helix angle and whole tooth depth on the bending strength.

However, a very critical parameter– root fillet radius is defined by tool (hob) tip radius was not concerned in the research. Since the bending critical section occurs in the gear root fillet, root fillet radius has a great influence on gear bending strength. The root strength can be improved by using a circular fillet design or optimized fillet design according to [9–11]. And the maximum tool tip radius will vary with the pressure angle changing due to the geometrical relationship of basic rack [3]. So in this paper, the bending stress of a spur gear pair with identical dedendum and different pressure angles (14.5°, 17.5°, 20°, 22.5° and 25°) and the variation of root fillet radius are studied.

After reviewing the references, gear bending stress in spur gears can be evaluated with four methods namely, standard methods like ISO standards [12] and AGMA standards [13], 3D-TCA method, Finite Element Analysis (FEA) and experimental methods. In this paper three methods of predicting bending stress—ISO standard, 3D-TCA method and Finite Element Analysis are applied to example gear geometry and compared to make sure the results are valid.

17.2 Geometric Specifications of the Studied Gears

In this study, a spur gear pair with different pressure angles (14.5°, 17.5°, 20°, 22.5° and 25°) and the same dedendum ($h_{fp}/m_n = 1.4$) was investigated. A dedendum equal to 1.4 m_n permits the finishing tool to work without interference, while maintaining the maximum fillet radius and is recommended for high-precision gears transmitting high torques. It is also suitable for gears with tooth flanks finished by grinding [3]. The geometry parameters of the spur gears have been summarised in Table 17.1.

Tool tip radius $\rho_{fp}/m_n = 0.39$ is equivalent to a full radius form for the fillet and is the maximum fillet radius when pressure angle equal to 20° ($h_{fp}/m_n = 1.4$) [3]. To ensure consistency of root radius, only investigate the influence of pressure angle on bending stress, the tool tip radius ρ_{fp}/m_n of spur gear with pressure angle equal to 14.5°, 17.5° and 20° are defined as 0.39. When pressure angle is increased to be greater than 20°, the maximum tool tip radius ρ_{fp}/m_n cannot reach 0.39 according to the geometric relationship of basic rack showed in Fig. 17.1.

A basic rack $h_{fp}/m_n = 1.4$ gives a full radius form for the fillet as showed in Fig. 17.1. The centre of $\rho_{fp\text{-}max}$ is on the centre of the rack space. According to the geometric relationship, the maximum tool tip radius can be derived from Eq. (17.1) [3].

$$\rho_{fp}\max = \frac{\left[\frac{(\pi m)}{4} - h_{fp}\tan\alpha_p\right]}{\tan\left[\left(90^\circ - \alpha_p\right)/2\right]} \tag{17.1}$$

where, h_{fp} is the dedendum of basic rack.α_p is the pressure angle of the basic rack.

Table 17.1 Gear parameters for spur gear pair with different pressure angles

Parameters	α = 14.5°		α = 17.5°		α = 20°		α = 22.5°		α = 25°	
	Pinion	Wheel	Pinion	Wheel	Pinion	Wheel	Pinion	Wheel	Pinion	Wheel
Tooth number Z	39	40	39	40	39	40	39	40	39	40
Module m_n [mm]	3	3	3	3	3	3	3	3	3	3
Face width b [mm]	20	20	20	20	20	20	20	20	20	20
Dedendum h_{fp}/m_n	1.4	1.4	1.4	1.4	1.4	1.4	1.4	1.4	1.4	1.4
Addendum h_{ap}/m_n	1.0	1.0	1.0	1.0	1.0	1.0	1.0	1.0	1.0	1.0
Tool tip radius ρ_{fp}/m_n	0.39	0.39	0.39	0.39	0.39	0.39	0.308	0.308	0.208	0.208
Contact ratio ε_α	2.048		1.846		1.711		1.6		1.51	

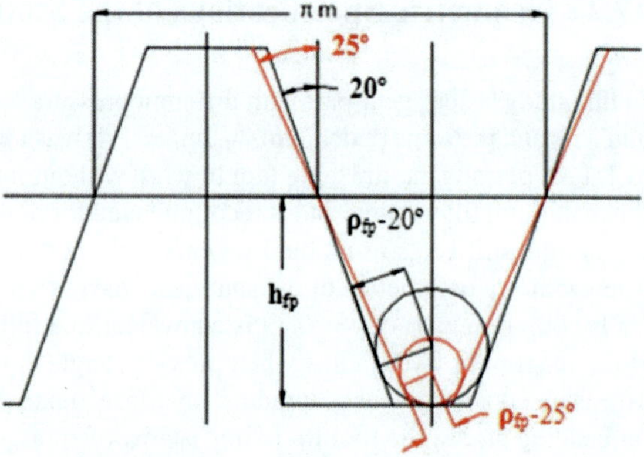

Fig. 17.1 The relationship between pressure angle and maximum tool tip radius ($h_{fp}/m_n = 1.4$)

From the Eq. (17.1), we can see that the maximum tool tip radius ρ_{fp} is correlated to pressure angle and is reduced with pressure angle increasing. The maximum tool tip radius ρ_{fp}/m_n of spur gear with pressure angle equal to 22.5° and 25° showed in Table 17.1 are determined according to Eq. (17.1) equal to 0.308 and 0.208 respectively.

17.3 Methods for Calculating Gear Bending Stress

17.3.1 ISO Standard

ISO standard is a commonly used international standard to determine gear stress. The bending stress equation in ISO 6336-3:2006 [12] based on cantilevered beam bending is given as Eq. (17.2):

$$\sigma_F = K_A K_V K_{F\alpha} K_{F\beta} \frac{F_t}{bm_n} Y_F Y_S Y_\beta Y_B Y_{DT} \tag{17.2}$$

where σ_{F0} is the nominal tooth root stress; Y_F is the form factor, which is defined in Eq. (17.3):

$$Y_F = \frac{\frac{6h_{Fe}}{m_n} cos\alpha_{Fen}}{\left(\frac{S_{Fn}}{m_n}\right)^2 cos\alpha_n} \tag{17.3}$$

Y_S is the stress correction factor. Equations (17.4), (17.5) and (17.6) define Y_S, L and q_s respectively:

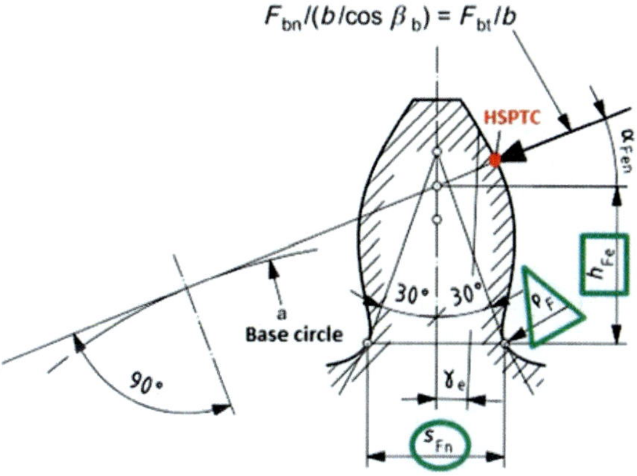

Fig. 17.2 The determination of s_{Fn}, h_{Fe} and ρ_F from [12]

$$Y_s = (1.2 + 0.13L)q_s^{\left[\frac{1}{1.21+\frac{2.3}{L}}\right]} \tag{17.4}$$

$$L = \frac{s_{Fn}}{h_{Fe}} \tag{17.5}$$

$$q_s = \frac{s_{Fn}}{2\rho_F} \tag{17.6}$$

Y_β is the helix angle factor.
Y_B is the rim thickness factor.
Y_{DT} is the deep tooth factor.
$K_A K_V K_{F\beta} K_{F\alpha}$ are the load correction factors.
s_{Fn} is the tooth root chord at the critical section.
h_{Fe} is the bending moment arm for tooth root stress relevant to load application at the outer point of single pair tooth contact.
ρ_F is the tooth root fillet radius at the critical section.
The determination of s_{Fn}, h_{Fe} and ρ_F is showed in Fig. 17.2.

17.3.2 3D-TCA Method

Analytical Tooth Contact Analysis (TCA) method is another method to predict gear tooth root bending stress. A3D-TCA software named GATES [14] (Gear Analysis for Transmission Error and Stress) which was initially developed and tested at Design Unit, Newcastle university is used to calculate gear bending stress in this study. It is an FE based analysis package, using a full 3D FEA stiffness model to estimate the gear stiffness and then using contact analysis to estimate the load distribution, stress and other functional parameters.

The bending stress determined in GATES is similar to ISO 6336 standard, calculating the bending stress at 30 degree tangent position, except the load distribution is calculated from the TCA, stiffness variation across the face width and compressive load is considered.

A critical point for bending stress determined by GATES is extended contact which occurs as the loaded tooth is restored to its original un-deflected state at the end of active profile [15]. If tip relief is not applied, the extended contact will increase the contact ratio thus lowering the actual HPSTC (for spur gear, the bending stress is calculated with load applied at the highest point of single tooth contact (HPSTC)) and then lower the actual bending stress. If too much tip relief is applied, the loading point will increases which will increase the bending stress. Therefore bending stress derived from GATES must be calculated with correct tip relief to make sure the gear loaded at exactly the same position as ISO (HPSTC). In this study, to ensure the results from GATES are valid, the correct tip relief are applied to gears.

17.3.3 Finite Element Analysis

In recent years, finite element analysis is widely used to evaluate gear stress, resulting in an abundance of published research [16, 17]. According to [16], in this study, the finite element analysis of gear bending stress is set as follows:

(1) Three teeth are used to calculate gear stress.
(2) Boundary conditions: fix the three free faces as showed in Fig. 17.3.

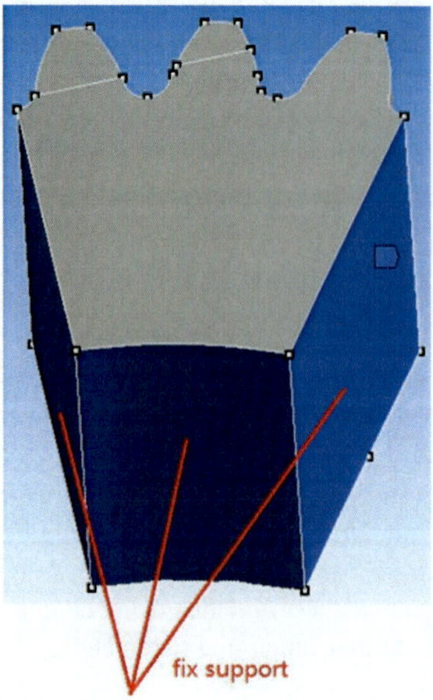

Fig. 17.3 The boundary conditions of bending stress calculation in FEA

(3) Mesh types: automatic mesh method.
(4) Element size is defined in Eqs. (17.7) and (17.8):

$$\mathrm{M_{bending}} = 0.2\rho_\mathrm{F} + 0.15 \text{ For } \rho_\mathrm{F} > 0.5\mathrm{mm} \quad (17.7)$$

$$\mathrm{M_{bending}} = 0.366\rho_\mathrm{F}^{0.614} \text{ For } \rho_\mathrm{F} \le 0.5\mathrm{mm} \quad (17.8)$$

(5) Maximum principal stress is used to represent the gear stress.
(6) Use smaller element sizes to verify the validity of the maximum bending stress to ensure the consistency of the results.

17.4 Results and Discussions

In this paper three independent methods of calculating bending stress—ISO standard, 3D-TCA method and Finite Element Analysis are used to make sure the results are valid. The bending stresses for spur gears with different pressure angle (14.5°, 17.5°, 20°, 22.5° and 25°) from GATES and FEA are showed in Figs. 17.4 and 17.5. The bending stresses from ISO and TCA and FEA are summarized in Table 17.2 and Fig. 17.6. In this study, only bending stresses of the pinion are calculated and compared.

From Table 17.2, it can be seen that:

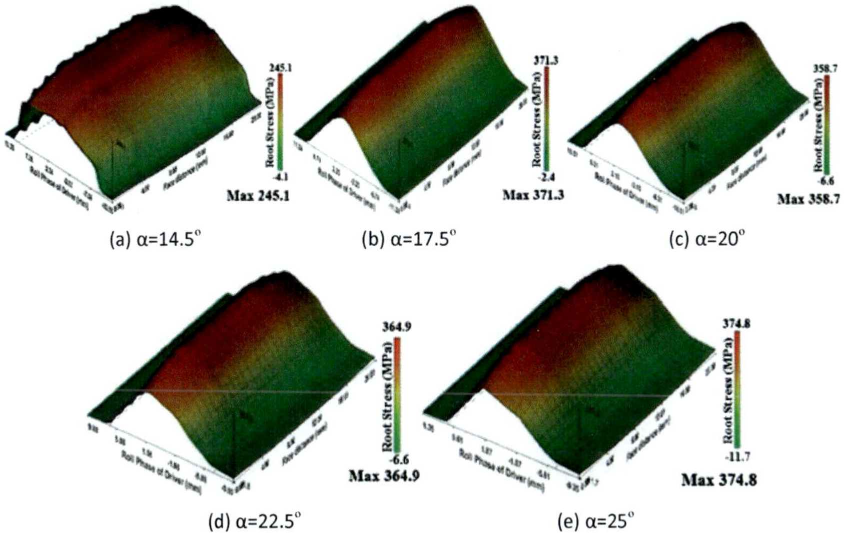

Fig. 17.4 Bending stress from GATES with tip relief for spur gear

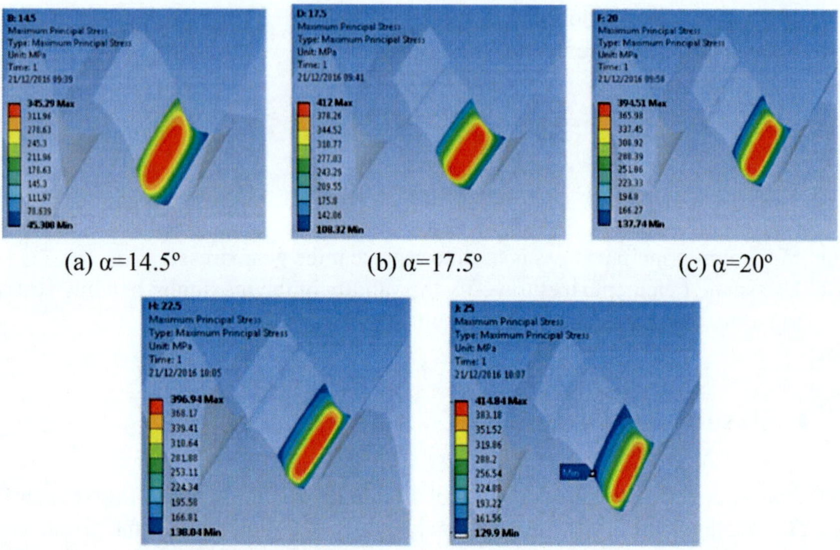

(a) α=14.5° (b) α=17.5° (c) α=20°

(d) α=22.5° (e) α=25°

Fig. 17.5 Bending stress from FEA for spur gear

Table 17.2 Bending stresses (MPa) for gears with different pressure angle from ISO, GATES and FEA

ε_α	α	σ_{F_ISO}	$\sigma_{F\text{-}FEA}$	$\sigma_{F\text{-}GATES\text{-}tip\ relief}$
2.048	14.5	289.7797	345.29	245.1
1.846	17.5	394.376	412	371.3
1.711	20	381.779	394.51	358.7
1.6	22.5	391.368	396.94	364.9
1.51	25	404.729	414.84	374.8

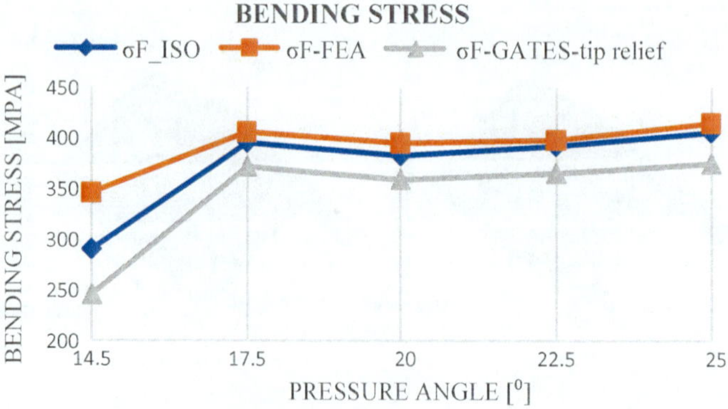

Fig. 17.6 Bending stresses for spur gears with different pressure angle from ISO, GATES and FEA

(1) The variations of bending stresses from ISO, GATES with correct tip relief and FEA with pressure angle changing have the same trends, which can be clearly seen from Fig. 17.6.
(2) The bending stress is minimum when pressure angle equal to 14.5°.
(3) Bending stress is reduced when pressure angle changing from 17.5° to 20°.
(4) While bending stress is increased when pressure angle changing from 20° to 22.5° and 25°.

There are some reasons to explain the results.

(1) The transverse contact ratio ε_α is 2.048 greater than 2.0 when pressure angle equal to 14.5°, which means there are at least 2 teeth sharing load when gear meshing as shown in Fig. 17.7. This results in lower bending stresses, without considering the effects of manufacturing deviations and misalignment.
(2) When pressure angle is increased from 17.5° to 20°, the tooth root chord at the critical section s_{Fn} is increased, the bending moment arm h_{Fe} is also slightly higher, while the root radius ρ_F is almost same. The detailed s_{Fn}, h_{Fe} and ρ_F are showed in Table 17.3, which will lead to a lower form factor Y_F and a slightly higher stress correction factor Y_s, while the decrement of form factor is greater than the increment of stress correction factor, so bending stress decreases when pressure angle changes from 17.5° to 20°.
(3) When pressure angle is increased from 20° to 22.5° and 25°, the tooth root chord at the critical section s_{Fn} and bending moment arm h_{Fe} are also increased, and the root radius ρ_F is largely reduced due to the significant declining of tool tip radius

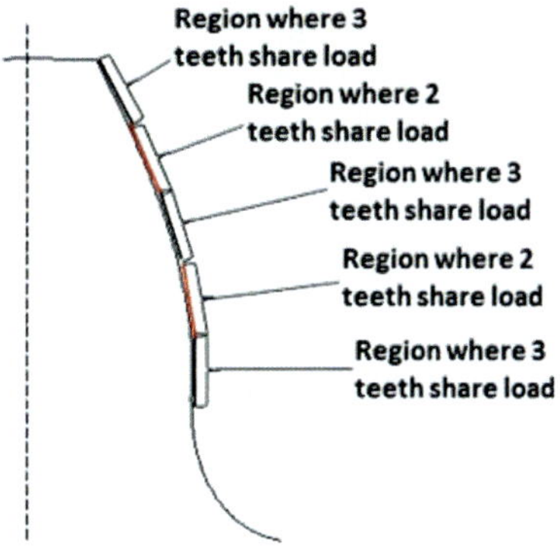

Fig. 17.7 Number of teeth sharing load condition when $2 < \varepsilon_\alpha < 3$ for spur gears

Table 17.3 The details of bending stress key factors in the calculation from ISO

α [°]	S_{Fn}/m_n	h_{Fe}/m_n	ρ_{fp}/m_n	ρ_F/m_n	q_s	Y_F	Y_S
17.5	2.053	1.047	0.39	0.585	1.755	1.503	1.842
20	2.158	1.097	0.39	0.583	1.85	1.421	1.886
22.5	2.273	1.183	0.308	0.528	2.153	1.378	1.994
25	2.405	1.28	0.208	0.463	2.599	1.329	2.138

ρ_{fp}/m_n from 0.39 to 0.308 and 0.208. The calculated S_{Fn}, h_{Fe} and ρ_F are showed in Table 17.3. The large reduction root radius ρ_F leads to significant increase in notch parameter q_s, and dramatic increases stress correction factor Y_s which is greater than the form factor Y_F reduction. So bending stress increases when pressure angle changes from 20° to 22.5° and 25°.

17.5 Conclusions

From the analysis presented in this study, the following conclusions can be drawn:

(1) The variations of bending stresses from ISO, GATES with correct tip relief and FEA with pressure angle changing have the same trends. The results in this study are valid.
(2) For basic rack $h_{fp}/m_n = 1.4$, according to the geometric relationship, the maximum tool tip radius (ρ_{fp}/m_n) of spur gear with pressure angle equal to 20°, 22.5° and 25° are equal to 0.39, 0.308 and 0.208 respectively. It can greatly affect the bending stress, so we must consider its influence when we investigate the effect of pressure angle on bending stress.
(3) Smaller pressure angle such as 14.5° in this study, can result in high contact ratio spur gears ($\varepsilon_\alpha > 2$), at least two teeth sharing the load when gears are in mesh, so it can get lower bending stress. But we must make sure this is maintained when considering manufacturing deviations, elastic deflections and misalignment of shaft and micro modifications, the gears always have transverse contact ratio greater than 2.0, otherwise, the benefit of lower bending stress will disappear.
(4) For the studied gears with $h_{fp}/m_n = 1.4$, the calculated bending stress reduced when pressure angle changing from 17.5° to 20° ($1 < \varepsilon_\alpha < 2$). And the bending stress increases when pressure angle changes from 20° to 22.5° and 25° ($1 < \varepsilon_\alpha < 2$).

References

1. ISO 10825:1995, Gears - Wear and damage to gear teeth –Terminology. 1st edn. UK (1995)
2. Panin, S.V., Moiseenko, D.D., Maksimov, P.V., Vlasov, I.V., Byakov, A.V., Maruschak, P.O., Berto, F., Schmauder, S., Vinogradov, A.: Influence of energy dissipation at the interphase boundaries on impact fracture behaviour of a plain carbon steel. Theoret. Appl. Fract. Mech. **97**, 478–499 (2018)
3. ISO 53:1998, Cylindrical gears for general and heavy engineering- Standard basic rack tooth profile. 2nd edn. UK (1998)
4. Gupta, K., Chatterjee, S.: Effect of pressure angle of spur gears on bending and contact stresses: a comparative study using finite element software. Int. J. Adv. Res. Sci. Eng. **4**, 517–526 (2015)
5. Handschuh, R.F., Zakrajsek, A.J.: High-pressure angle gears: comparison to typical gear designs. J. Mech. Des. **133**(11), 114501 (2011)
6. Sankar, S., Kumaresan, M., Nataraj, M.: Effects of pressure angle and tip relief on the life of speed increasing gearbox: a case study. SpringerPlus **3**(1), 746 (2014)

7. Dadhanlya, K.D., Hirpar, K.P., Vyas, K.M.: Effect of pressure angle on bending stress and deformation of asymmetric involute spur gear using FEA. IOSR J. Eng. **2**(4), 616–620 (2012)
8. Oda, S., Koide, T., Tottori, U., Sawa, Y.: Effect of standard pressure angle on the bending fatigue strength of helical gears. Int. J. Fatigue **31**, 802–807 (1988)
9. AGMA 13FTM03. Analysis of gear root forms: a review of designs, standards and manufacturing methods for root forms in cylindrical gears. USA, (2013)
10. Shukla, A.K., Sarda, A.: Optimization of bending stress of Spur Gear using response surface optimization technique. Int. J. Adv. Res. Innov. Ideas Educ. **4**(2), 317–320 (2018)
11. Shukla, A.K., Sarda, A.: A review of bending stress on spur gear using DOE technique. Int. J. Adv. Res. Innov. Ideas Educ. **4**(2), 313–316 (2018)
12. ISO 6336-3:2006, Calculation of load capacity of sour and helical gears: part 3: calculation of tooth bending strength. 2nd edn. UK (2006)
13. ANSI/AGMA 908-B89, Geometry factors for determining the pitting resistance and bending strength of spur, helical and herringbone gear teeth. USA (1989)
14. Dontyne Systems, Gear production suite brochure [EB/OL]. http://www.dontynesystems.com/gear-design-software/. 14 Sept 2011
15. Palmer, D., Fish, M.: Evaluation of methods for calculating effects of tip relief on transmission error, noise and stress in loaded spur gears. Gear Technol. **1**, 56–67 (2012)
16. Lisle, T.: A Finite Element Approach for Validating Gear Contact and Root Bending Stress. Newcastle University, UK (2013)
17. Lisle, T.: Finite element stress analysis for advanced aerospace gears. In: BGA gears 2011 conference, Shrivenham, UK (2011)

Chapter 18
The Effect of Surface Ultrasonic Rolling Processing on the Rolling Contact Fatigue Performance of Valve Train Tappet Material

Zhijian Wang, Mengyang Ma, Chenyang Jiang, Lei Wang, and Yupeng Wang

Abstract To improve the rolling contact fatigue performance (RCFP) of a kind of tappet material GCr15 in valve train system of engine, surface ultrasonic rolling processing (SURP) with different static rolling loads (600, 800, 1000, and 1200 N) was introduced to reinforce its surface. A variety of experimental instruments were applied to study various properties including surface morphology, surface roughness, micro-hardness, surface residual stress and rolling contact fatigue of the samples. Results point out that compared with the untreated samples, the properties of the samples treated by SURP were significantly improved. With the static rolling load of 1000 N, the samples obtain the best surface quality and fatigue life. On the other hand, the samples with 1200 N gain the highest microhardness (842.3 HV0.3, which is 12.8% higher than the untreated samples) and residual compressive stress. With the static rolling load increase, the RCFP of the tappet material shows a trend of first rising and then falling, and obtain the best RCFP with the static load of 1000 N.

Keywords Tappet material · Surface ultrasonic rolling processing · Surface property · Rolling contact fatigue

18.1 Introduction

As the modern automobile industry develops significantly towards high speed and load, the working performance of the core components, engine, is also improved. At the same time, running speed of the valve train and pressure between the contact

Z. Wang · L. Wang · Y. Wang
State Key Laboratory of Engine Reliability, Weifang, Shandong 261061, China

Weichai Power Co. Ltd, Weifang, Shandong 261061, China

M. Ma (✉) · C. Jiang
School of Mechanical and Automotive Engineering, South China University of Technology, Guangzhou, Guangdong 510640, China
e-mail: mmy222210@163.com

J. Xu and K. M. Pandey (eds.), *Mechanical Engineering and Materials*,
Mechanisms and Machine Science 100,
https://doi.org/10.1007/978-3-030-68303-0_18

parts are also increased. This results in an significant increase to the mechanical load, thermal load, friction and wear, vibration and noise of the valve train system [1]. In the valve train system, the tappet and the camshaft consist a contact pair in the form of point or line generally. When the camshaft rotates at high speed, a large alternating stress is generated between the tappet and the camshaft, which lead to the fatigue wear of the tappet easily [2]. As one of the core components in the valve train system, the performance of the tappet has a direct impact on the power performance, reliability and service life of the engine [3].

Contact fatigue failure is a typical failure mode. Its mechanism has been widely used in key components of the machine manufacturing industry, including the automotive industry, to improve its performance and service life. Based on this actual demand of production, the research on the contact fatigue behavior of the tappet and its strengthening mechanism has become an urgent problem to be solved [4]. There are many factors influencing fatigue life, such as lubrication conditions, surface hardness of contact materials, mechanical properties, etc. [5–8]. Surface quality is one of the most significant factors affecting the fatigue behavior of the tappet. Domestic and foreign scholars have carried out extensive research on the surface strengthening process of tappet materials, such as surface ultrasonic rolling processing (SURP), shot peening, surface coating, etc. [9–11]. As a mature surface strengthening process, SURP is widely used to improve the surface properties of metal parts [12, 13]. Zhang Fei, etc. [14] discussed the influence mechanism of various SURP parameters including static load, amplitude, feed rate, etc. on the surface properties of 45 steel, and compared the friction and wear behavior of materials at different SURP parameters. Liu Yu et al. [15] applied SURP to the surface strengthening of 40Cr alloy steel, and measured the elastic modulus, nano hardness and residual stress of the surface layer. It was found that after ultrasonic surface rolling, the elastic modulus and hardness of the 40Cr surface layer significantly improved. The maximum value of elastic modulus and hardness were found to be located on the surface, and increased with the increase of processing times. At the same time, a certain value of residual compressive stress was obtained on the surface. A study by Liu et al. [16] indicates that SURP can refine the surface layer, improve the microhardness and residual stress. These qualities result in the significant improvement of the alloy's (Ti-6Al-4 V) fatigue properties.

GCr15 alloy steel has good mechanical properties and is the most widely applied as high chromium bearing steel. After quenching and tempering, it has high and uniform surface hardness, good wear resistance and contact fatigue properties. The components made of GCr15 alloy steel are usually used in high load, multi cycle contact and sliding friction environment. The main failure modes are contact fatigue, friction wear and corrosion. The surface hardness of GCr15 after heat treatment is slightly higher than that of medium carbon steel used for camshaft, which ensures good matching with cam material. Due to the traditional tappet without special surface treatment, the fatigue resistance is insufficient under high strength alternating stress, which greatly affects the service life of tappet and even the whole engine. Therefore, it is necessary to further strengthen the surface of the tappet to meet the increasing demand of automobile industry. In this paper, the influence mechanism of SURP on

Microstructure and fatigue behavior of valve train tappet material will be explored, and the crack propagation and fatigue damage form of strengthening surface will be discussed. Finally, the optimal SURP process parameters for tappet material will be given.

18.2 Organization of the Experiment

The tappet material used in the experiment is GCr15 alloy steel, and its composition is shown in Table 18.1. The raw material is an extruded rod-shaped material with diameter of φ50, and SURP is applied on the samples with φ45 × 6 mm, which is line-cut from the rod-shaped raw material. The disc-shaped samples were heat treated in a furnace with a heating temperature of 860 °C for 0.5 h, and then remove it for quenching with oil as the medium. After quenching, keep the sample at 160 °C for 4 h, and then take it out and cool it to normal temperature. The heat treatment scheme is shown in Fig. 18.1. After heat treatment, the surface hardness of the material is 62.2 ± 1.21 HRC, the tensile strength is 1903 ± 3.2 MPa, and the yield strength is 1391 ± 2.5 MPa.

After the heat treatment, the samples were finely ground to remove the surface decarburization layer, and then cleaned with alcohol vibration and blow-dried, finally they can be SURP treated and used for subsequent rolling contact fatigue test. The samples were processed on the self-made ultrasonic rolling platform under the condition of oil lubrication. The equipment mainly consists of lathe and ultrasonic vibration equipment. The processing diagram is shown in Fig. 18.2.

In the process of processing, the static rolling pressure is perpendicular to the surface of the workpiece, and is transferred from the rolling ball head to the surface of the sample by controlling the air pressure; the adjustable frequency range of the

Table 18.1 Chemical composition of camshaft material (wt%)

Material	C	Si	Mn	Cr	Ni	Mo	V	Cu	P	S
GCr15	1.099	0.234	0.423	1.496	0.055	0.011	0.002	0.089	0.016	0.020

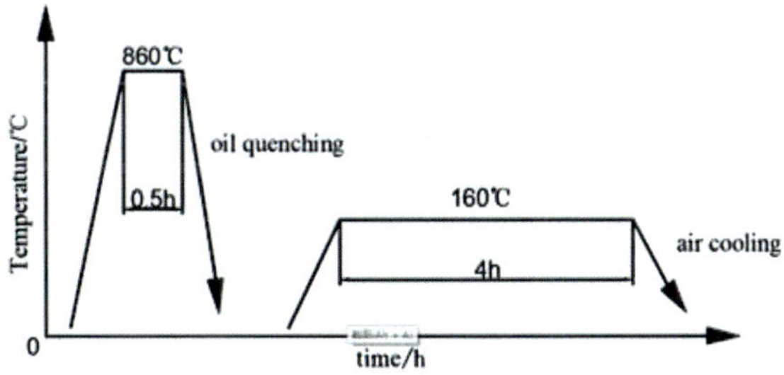

Fig. 18.1 Heat treatment scheme of samples

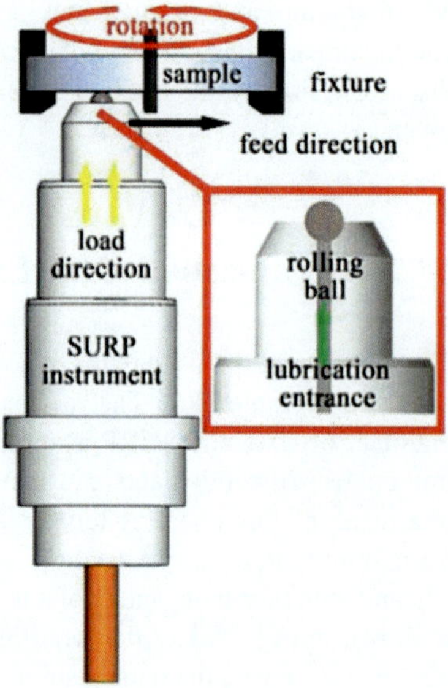

Fig. 18.2 Schematic diagram of SURP treatment

Table 18.2 SURP parameters

Material	Static pressure [N]	Processing times	Vibration amplitude [μm]	Vibration frequency [Hz]	Spindle speed [r/min]	Feed rate [mm/r]
GCr15	600/800/1000/1200	1/3/5	8	27000	45	0.047

ultrasonic vibration equipment is between 20 ~ 30 kHz, and the amplitude range is between 5 ~ 15 μm. In this experiment, the SURP parameters are shown in Table 18.2.

After SURP treated, the surface roughness of the samples was measured with MARSURF-M300C roughness tester. The SCTMC-HV50 Vickers microhardness tester was used to test the hardness distribution of the cross-section of the samples. The test load was 300 g with holding 15 s. The rolling surface and cross section of the sample were corroded with 4% nitric acid alcohol solution, and the microstructure was observed under LEICA-M165C optical microscope. The residual compressive stress on the surface of the sample was detected by PROTO-LXRD X-ray stress analyzer. A QUANTA 200 environmental scanning electron microscope was used to observe the rolling surface and fatigue damage morphology of the sample.

Schematic sketch of the rolling contact fatigue tester is shown in Fig. 18.3. The samples to be tested were φ45 × 6 mm discs, which were fixed on the fixture at the bottom of the lubricating oil container, and the treated surface of samples were in contact with the thrust ball bearing. The bearing ball material is GCr15 steel, the hardness is 60 HRC, the number is 14, the diameter is 5.58 mm and evenly distributed. The spindle speed of the tester is 1600 r/min, and the spindle pressure is set as 8000 N, which is evenly distributed on each bearing ball and perpendicular to the surface of

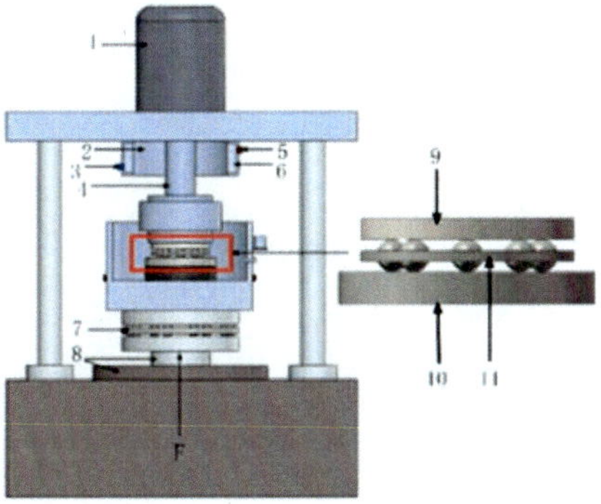

1. Spindle motor 2.Spindle bearing, 3.Coolant inlet 4.Spindle
5. Coolant outlet 6. Cooling chamber 7. Cylinder bearing
8. Hydraulic loading system 9. Sample 10. Sample 11. Rolling bearing

Fig. 18.3 Schematic sketch of the rolling contact fatigue tester

samples. The maximum Hertzian contact stress applied by each bearing ball is 5.7 GPa; the lubricating oil is Mobil 0 W-40CF; the lubricating oil temperature is 80 ± 1 °C, which is consistent with the actual service condition of the tappet. During the test, parameters such as spindle load, spindle rotation speed, vibration signal, test time, etc. are all monitored. The bearing ball is in rolling contact with the treated surface of samples under the spindle pressure. When the surface of sample is damaged by pitting, peeling, etc., the vibration signal values increases sharply; when it exceeds the preset value, the tester stops. And the rotation number of the spindle is recorded as the value for judging the fatigue life of the sample. Each set of rolling parameters takes 9 samples for fatigue test.

18.3 Results and Discussion

18.3.1 Surface Morphology and Roughness

The SEM images of surface morphology of the samples at different SURP static pressure are shown in Fig. 18.4. It can be seen that compared to the untreated area, the surface quality of the treated area is significantly improved. When the static pressure is 600 N, as shown in Fig. 18.4a, there are still many cutting scratches in the rolling area, which are obvious (dents with the same machining direction). When the static pressure is increased to 800 N, the scratch size is significantly reduced compared to the 600 N sample. This observation is presented in Fig. 18.4b. Although there are still machining marks in the processing area of the sample (Fig. 18.4b), the number of scratches is significantly reduced compared with that of the 600 N sample (Fig. 18.4a). Increasing the static pressure to 1000 N, the surface morphology of the

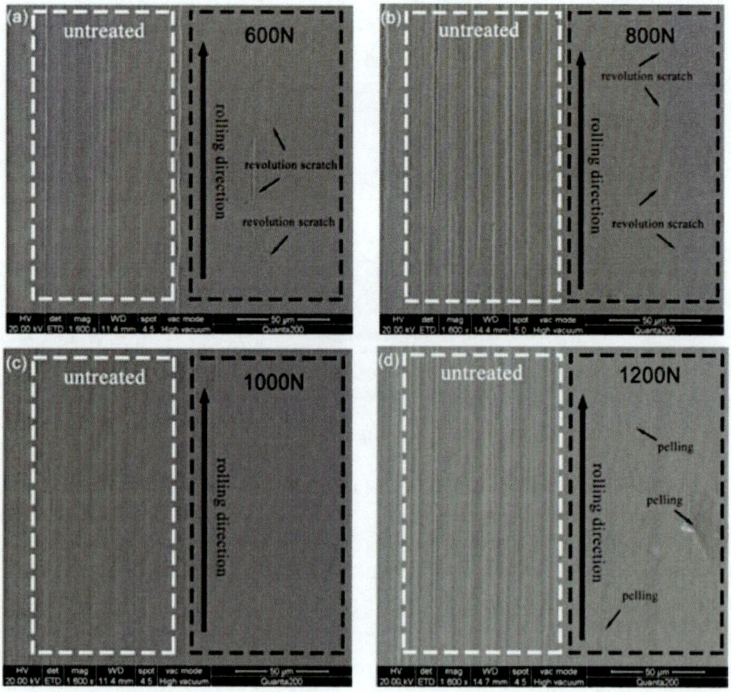

(a) 600 N; (b) 800 N; (c) 1000 N; (d) 1200 N

Fig. 18.4 SEM images of surface morphology of the samples at different static rolling loads

sample is shown in Fig. 18.4c, there are almost no defects in the processing area, and the whole tends to be flat. However, when the static pressure continues to increase to 1200 N, as shown in Fig. 18.4d, a small number of cracks and obvious spalling phenomenon appear on the surface of the processing area.

Figure 18.5 shows the surface roughness of the sample at different static loads. After SURP treatment, the surface roughness of the sample is obviously improved. The results of the surface roughness test are consistent with the morphology analysis of the sample surface. For the tappet material, the plastic deformation caused by 1000 N static pressure makes the sample obtain the best surface quality, and the surface roughness of the sample is the lowest, which is 0.067 μm. The plastic deformation caused by 600 and 800 N can not completely remove the machining marks and other defects on the sample surface. When the static pressure exceeds 1000 N, it will cause damage to the sample surface leading to the decline of surface quality and the increase of roughness. Ren Xuechong et al. [17] explored the influence of SURP process on the surface state of axle steel and found that roughness seriously affected the fatigue performance of the material.

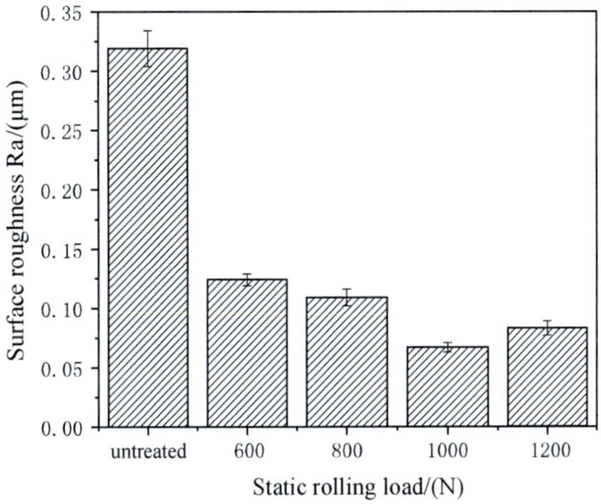

Fig. 18.5 Surface roughness of the samples at different static rolling loads

18.3.2 Microstructure and Microhardness

The mechanism of SURP is the combination of Static Rolling and ultrasonic mechanical high-frequency vibration, and the corresponding severe plastic deformation occurs on the surface of the material where it is impacted [18]. With the continuous linkage load acting on the surface of materials at different positions, the microstructure units inside the surface continuously deform, which makes the grains become fine and even reach the nanometer level [19]. Figure 18.6 shows the surface microstructure of the sample under untreated and four groups of different static pressure, mainly including tempered martensite, retained austenite and granular carbide. Compared to the untreated sample (Fig. 18.6a), it can be found that the surface layer microstructure of the treated sample is significantly refined. It can be observed from Fig. 18.6b–e that the size of microstructure decreases with the increase of static pressure. This is due to the high frequency impact of the rolling ball head on the surface of the sample during the rolling process, dislocation and slip of the grain structure occur, resulting in a certain plastic deformation, thus refining the sample structure [20]. With the increase of SURP static pressure, the plastic deformation of the surface layer of the sample increases, the dislocation and slip caused by microstructure become more severe, and the degree of grain breakage will also increase, and the refining effect on the material is more significant. The strengthening effect of SURP is closely related to the refinement degree of microstructure, because the refined microstructure can significantly improve the surface hardness and strength of the material. When the static pressure increases from 600 N to 1200 N, the microstructure of the sample is refined in turn, which indicates that the static pressure of 1200 N or above can make the microstructure reach the most refined state. This finding is consistent with the research results of Liu Senzhong [21] on the effect of SURP on the surface properties of GCr15 materials.

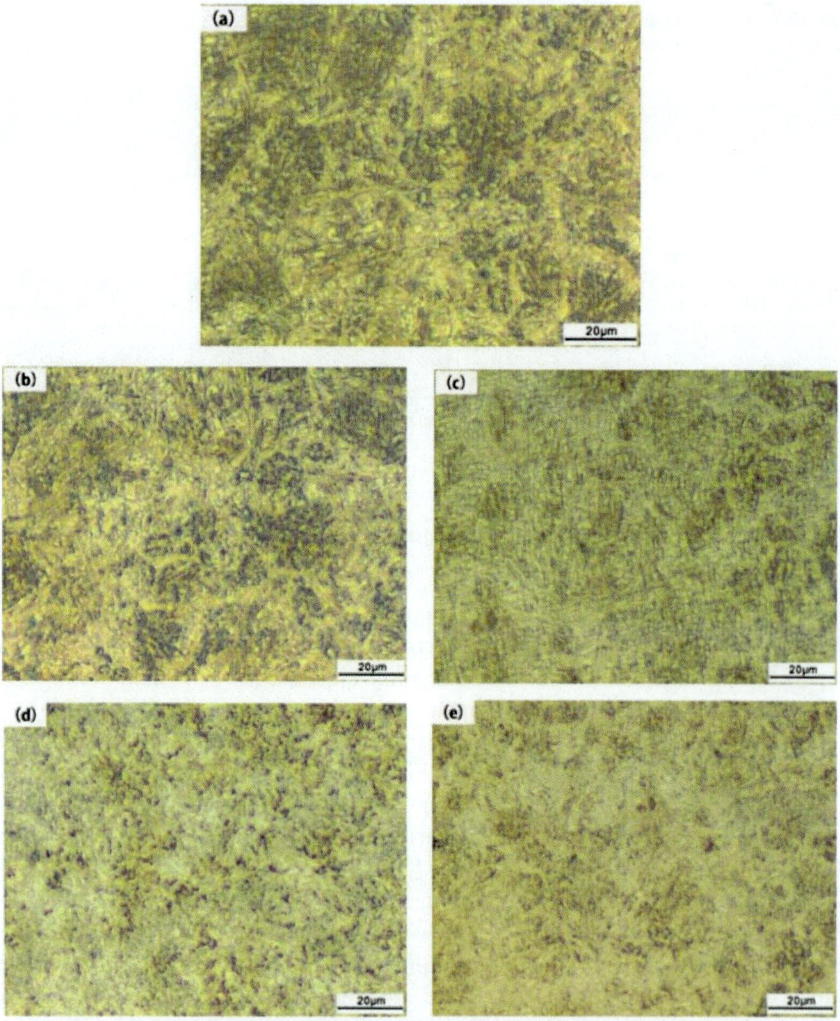

(a) untreated; (b) 600 N; (c) 800 N; (d) 1000 N;(e) 1200 N

Fig. 18.6 Optical microstructure of samples at different static rolling loads

In Fig. 18.7, it can be seen that the phase of the main diffraction peak of all the curves is α-Fe, and the 2θ is about 44.5°. In the process of increasing static load from 0 to 1000 N, the main diffraction peaks of the sample shows an obvious broadening trend, and the peaks increase successively, indicating that as the static load increases, the refined extent of the surface microstructure of the sample increases and accompanied by the transformation of γ-Fe (retained austenite) to α-Fe (tempered martensite) [22]. This analytical method has been used in the study of cemented carbide WC-Co ultrafine powder by Cao Lihong et al. [23]. In addition, when the

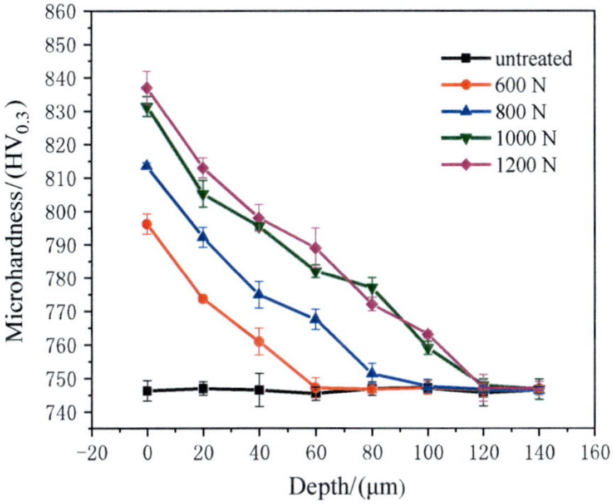

Fig. 18.7 X-ray diffractometer (XRD) patterns of samples at different rolling static loads

static load is increased from 1000 N to 1200 N, there is no significant change in the XRD main diffraction peak, which means that there is no obvious difference between the extent of microstructure refinement and martensite content of the 1000 N and 1200 N samples.

The observation of the microstructure on the surface of the above samples is consistent with the XRD test results, indicating that the ultrasonic rolling static load can significantly affect the microstructure of the material surface, and increasing the static load within an appropriate range can notably refine the grains. Ma et al. [24] pointed out that grain strengthening can effectively improve the properties of steel, including strength, hardness, and fatigue resistance. In this study, the static load of 1200 N or above can refine the material grains in the largest degree.

Figure 18.8 shows the microhardness distribution curve of the samples section at different static loads. The microhardness of the untreated sample is 746.5 $HV_{0.3}$, and

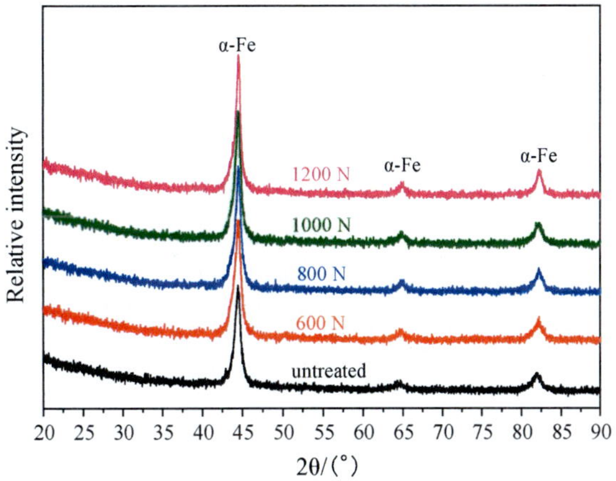

Fig. 18.8 Cross-sectional microhardness distribution of samples at different rolling static loads

the not change along the depth of the section. It can be seen that the microhardness of the treated samples surface has improved to various degrees, and the maximum microhardness of each group of samples is on the surface. Among them, the highest value is 842.3 $HV_{0.3}$ at 1200 N, which is 12.8% higher than that of the untreated sample. By comparing the surface microstructure of the samples at different static loads (Fig. 18.6), it can be found that the value at 1200 N is higher than that of the other group, since the surface of the sample can obtain a relatively finer microstructure, the effect on the material refinement is the most significant at the this static load. At 1000 N static load, SURP has a slighter effect on the surface microstructure compared with 1200 N, and the surface hardness is 831.4 $HV_{0.3}$. When the static load is 800 and 600 N, although SURP has a certain refinement effect on the surface microstructure of the samples, it is relatively limited. Compared with the 1200 N sample, the surface microstructures of the above two groups of samples are coarser, the surface hardness is 813.5 and 796.2 $HV_{0.3}$, respectively. From the analysis above, it is can be known that surface hardness of the material can be improved consistently by increasing the static load in the range of 0–1200 N.

In addition, as the increase of static load within a certain range, the depth of refined layer of the sample is also improved. The depth of the refined layer of the 600, 800 and 1000 N samples are 40, 80 and 100 μm, respectively. When the static load increases to 1200 N, the depth of the refined layer is 100 μm, which is same as the 1000 N sample. From the surface to 60 μm, the microhardness of the 1200 N sample is higher than that of the 1000 N sample at the same depth; however when the depth exceeds 60 μm, the two groups of samples share the same microhardness at the same depth, so it cannot be proved that when the static load exceeds 1000 N, the depth of the refined layer can be improved.

18.3.3 Surface Residual Compressive Stress

Figure 18.9 shows the residual compressive stress on the surface of the sample at different static loads. The surface residual compressive stress of the untreated sample is −191.3 MPa. As the static load increases, the value on the surface of the sample is also improved. The value of the sample at static loads of 600, 800, 1000 and 1200 N is −399.2, −493.6, −610.1 and −674.8 MPa, respectively. Among them, 1200 N sample obtains the most significant growth of the residual compressive stress on the surface. but there is no the marked gap of residual compressive stress between 1000 N and 1200 N samples (only increased by −74.7 MPa). Although the nonuniform plastic deformation in the treated area is improved during this process, a small number of defects and micro-cracks appear on the surface of the 1200 N sample (Fig. 18.4d), resulting in the release of a small amount of residual compressive stress [25].

In the SURP process, the nonuniform plastic deformation occurs on the surface of the sample, which causes the lattice distortion of the surface microstructure, leading to residual compressive stress, and the residual compressive stress on the surface of the material can partially offset the stress on the working surface to improve the wear

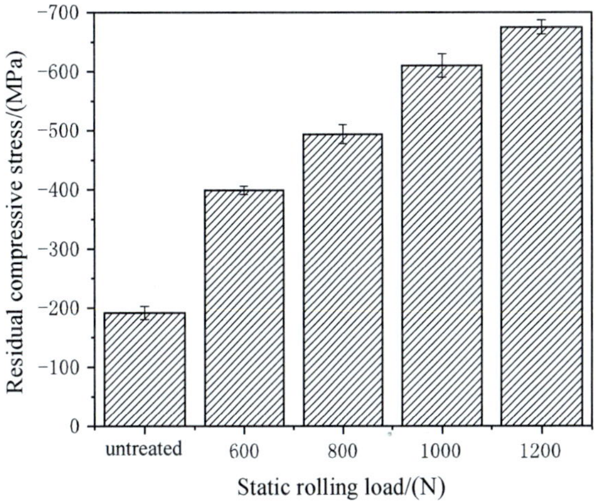

Fig. 18.9 Surface residual compressive stress of samples at different rolling static loads

and fatigue resistance [26–28]. When the static load increases to 1200 N, although the degree of the nonuniform plastic deformation in the treated area is increased, a small amount of defects and microcracks appear on the surface of the sample (Fig. 18.6d), resulting in the release of a small part of the residual compressive stress. Therefore, compared with other contiguous groups of samples, the increase of residual stress is insignificant from 1000 N to 1200 N.

18.3.4 Contact Fatigue Life

The two-parameter Weibull distribution is widely used in fatigue life analysis [29]. The two-parameter Weibull distribution function is defined in Eq. (18.1):

$$F(N) = 1 - \exp\left[-\left(\frac{N}{N_{\alpha}}\right)\right]^{\beta} \quad (18.1)$$

In the equation: *F(N)* is the failure probability; *N* is the contact fatigue life; Na is the characteristic life parameter; β is the shape parameter of the fatigue life.

Table 18.3 shows the number of rolling contact cycles of GCr15 samples at different rolling static loads. From the two-parameter Weibull distribution function, the two-parameter Weibull distribution curve of the fatigue life of the sample can be fitted.

Figure 18.10 is the Weibull distribution curve of the contact fatigue life of GCr15 samples at different rolling static loads fitted by the above data. Compared with the untreated samples, the contact fatigue life of each group of treated samples has been significantly prolonged, and the stability of fatigue performance has also been improved in different degrees. Among them, when the static load is 1000 N, the

Table 18.3 Rolling contact cycle numbers of samples at different static rolling loads

Static pressure [N]	Rolling contact cycle Numbers [10^6]									
Untreated	0.2134	0.5131	0.5832	0.7323	0.9138	1.5379	1.5325	2.3432		6.3218
600	0.6432	0.8521	1.2214	1.5896	2.5846	3.5341	5.4678	6.7457		8.3841
800	0.9645	1.2095	2.0721	3.2385	4.1137	4.5704	6.4519	8.2194		1.2546
1000	3.3758	5.5541	7.6862	8.3258	10.428	12.659	14.3791	17.5694		21.3731
1200	0.7957	1.0253	2.5479	4.1921	4.7138	6.4318	7.1139	13.3648		20.7337

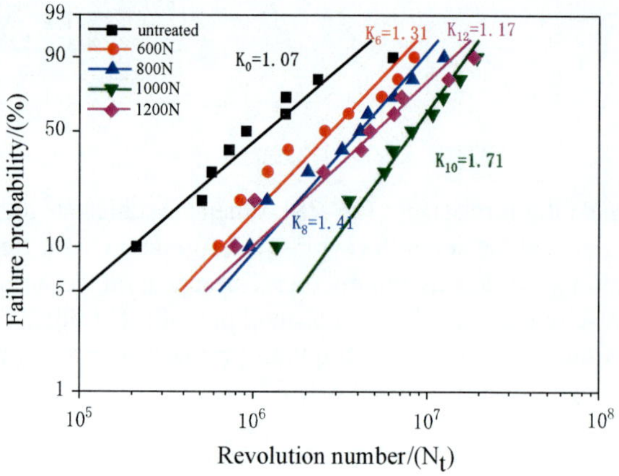

Fig. 18.10 Weibull distribution curves of contact fatigue life of samples at different rolling static loads

contact fatigue life of this group of samples is the longest, and the curve convergence is the highest with the K10 value of 1.71. Secondly, the curves of the 800 and 1200 N samples are intersection. The K8 value of the 800 N samples is 1.41, and the K12 value of the 1200 N samples is 1.17, indicating that the 800 N samples process better stability; When the static load is 600 N, the Weibull curve is only to the right of the curve of the untreated samples, indicating that the contact fatigue life of the samples at this static load is higher than that of the untreated samples, but lower than that of other groups.

Based on the Weibull distribution curve of fatigue life of samples at different rolling static loads, three classical fatigue life values of each group of samples are obtained, including rated life (L10), median life (L50), and characteristic life (L63.2) as shown in Table 18.4. The rated life (L10), median life (L50) and characteristic life (L63.2) of the untreated sample are only 0.2134×10^6, 0.9138×10^6 and 1.5249×10^6 revolutions, while at the static load of 1000 N, the tappet material obtain the longest contact fatigue life, and the three classic lives are 1.3758×10^6, 8.1428×10^6, and 11.2821×10^6 revolutions respectively, which are 6.4, 8.9 and 7.4 times of the corresponding fatigue life of the untreated samples. the static load of 800 N and 1200 N have a similar effect on the fatigue life of the samples. The median life

Table 18.4 Typical values of the contact fatigue life of samples at different static rolling loads

Static rolling load/(N)	Rated life (L10) [$\times 10^6$]	Median life (L50) [$\times 10^6$]	Characteristic life (L63.2) [$\times 10^6$]
Untreated	0.2134	0.9138	1.5249
600	0.6432	2.5846	4.1529
800	0.9645	4.1137	5.1707
1000	1.3758	8.1428	11.2821
1200	0.7957	4.7138	6.6501

(L50) and characteristic life (L63.2) of the 1200 N sample are higher than 800 N, but the rated life (L10) is lower. It can be seen from Table 18.3 that there are lower and higher values in the fatigue life of 1200 N samples, and the data is relatively discrete. This is because cracks and defects on the contact surface of samples at 1200 N bring about the unstable fatigue life. When there are more cracks on the sample surface, the fatigue life is lower and more unstable. Combining the previous analysis of the surface properties of the tappet samples, the samples at 1200 N possesses the highest hardness and residual compressive stress, but the surface quality is poor (more defects on the treated surface and higher roughness). When the above properties comprehensively perform on the contact fatigue performance of the samples, there is a notable gap between the contact fatigue life of samples in this group. On the other hand, the surface quality of the 1000 N samples is relatively higher, and the surface hardness and residual compressive stress are only slightly lower than that of 1200 N sample. In this case, this group of samples possess the best performance in the contact fatigue test, which means that the surface quality has a significant influence on the contact fatigue performance of the material. This finding is consistent with Lee et al. [30] in the study of the effect of ultrasonic nanometer bearing steel surface microcracks on contact fatigue.

18.3.5 Contact Fatigue Damage

The surface damage morphology of the samples after the contact fatigue text is shown in Fig. 18.11. By observing these images, the failure mechanism of the material contact fatigue at different rolling static loads can be analyzed. It can be illustrated that the main mechanism of the contact fatigue failure of the untreated sample in Fig. 18.11a is delamination. Obvious delamination occurred at the edge and bottom of the spalling pit, indicating that the initial micro-cracks initiated on the subsurface. After the expansion and connection of these micro-cracks on the subsurface, them expanded to the surface and finally led to the delamination failure [31].

Figure 18.11b, c are the fatigue damage morphologies of the 600 N and 800 N samples, respectively. The failure modes of the two groups of samples are typical

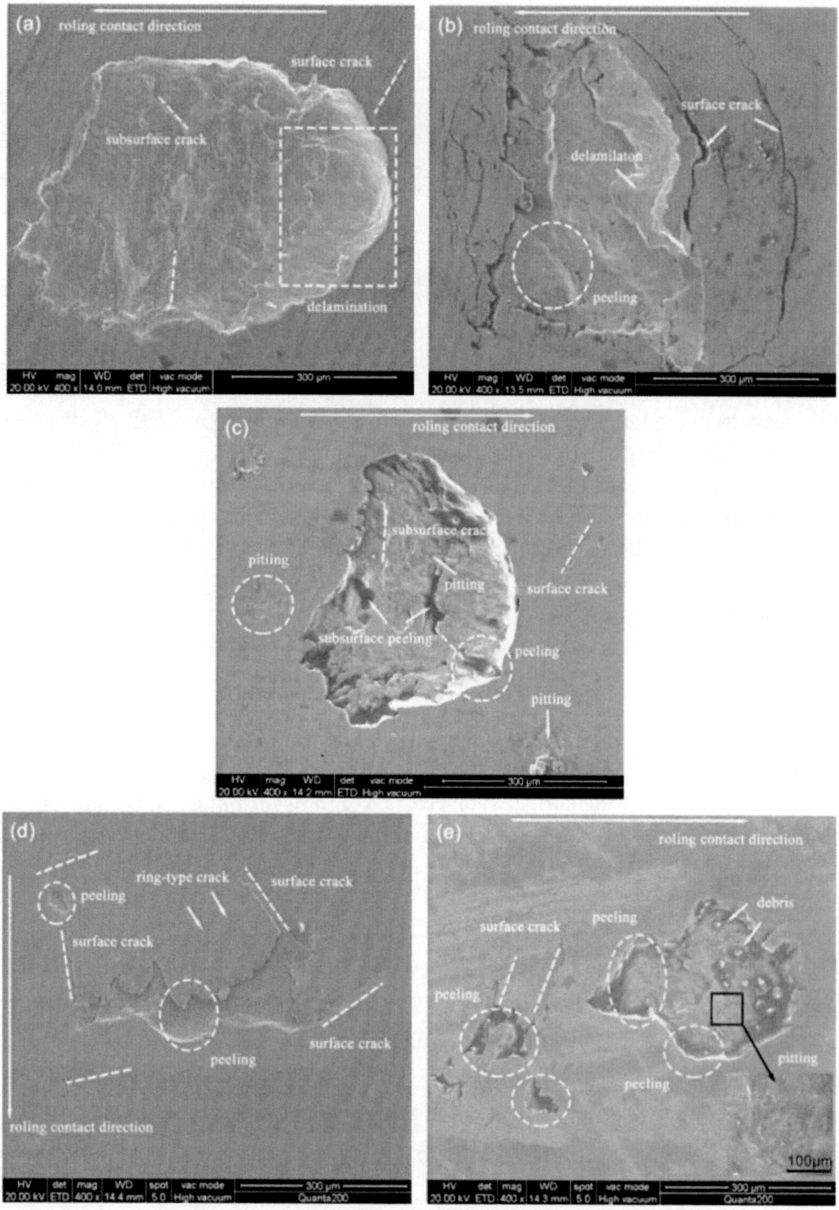

(a) untreated; (b) 600 N; (c) 800 N; (d) 1000 N; (e) 1200 N

Fig. 18.11 Surface rolling contact fatigue failure morphology of samples at different static rolling loads spalling. Because there were cracks in the rolling contact area of them, the lubricant was pressed into the existing cracks during the high-speed contact between the bearing balls and the samples. With the impact of cyclic shear stress caused by high-pressure lubricating oil, the material at both ends of the cracks fractured, causing the crack to further expanded, and finally caused spalling failure [32]. It can be seen that delamination occurs at the edge of the spalling pit of the 600 N sample, and there are pretty obvious cracks distributed around the spalling pit, and the area of the spalling pit is smaller than that of the untreated sample. Because there are still revolution scratches and pores on the surface of the 600 N sample, it is easy to cause stress concentration on the defects on the surface of the sample during the contact process, which promotes the initiation and extension of cracks on the contact surface [33]. This is because during the process, the particles that has fallen off got into between the ball and subsurface of the sample forming three-body abrasive grains system, and continuously contacted with subsurface of the sample, causing scratches and pitting corrosion. This damage mechanism has been verified by Zhang et al. [34] when they studied the contact fatigue of 17Cr2Ni2MoVNb alloy steel. Therefore, it can be inferred that the involvement of more abrasive particles due to surface defects and higher surface roughness can affect the contact fatigue resistance of the material.

Secondly, a large number of studies show that the maximum residual compressive stress occurs below the SURP [35–37]. There is no residual compressive stress in untreated specimens. Therefore, the surface of untreated specimens will produce contact fatigue cracks. After 600 N SURP, the surface residual compressive stress is less than the subsurface residual compressive stress, and there are still rotary scratches and pores on the surface of 600 N sample. In the contact process, it is easier to produce stress concentration on the surface defects of the sample, so the 600 N sample is more likely to produce cracks on the surface. Garnham J.E. [38] finds that under the combined action of residual stress and surface defects, it is easier to crack on the specimen surface. This reason also applies to 800 N samples.

The fatigue failure morphology of the 1000 N sample is shown in Fig. 18.11d. The failure mode of this sample is fatigue spalling. The spalling pit area is smaller and shallower than other samples. The subsurface is smoother. There are fewer cracks at the bottom of the spalling pit but more cracks around the spalling pit, indicating that the location of the crack initiation and propagation were mainly on the contact surface. As we can see, the spalling pit area and the depth of 1000 N sample is smallest and shallowest. According to the previous analysis of the surface properties of the material, we know that the 1000 N sample has smooth surface and high hardness, which can obtain excellent anti fatigue crack initiation and propagation ability. It is easy to draw the conclusion that the 1000 N static load has the most significant effect on the material fatigue process.

Figure 18.11e is the fatigue morphology of the 1200 N sample. The damage morphology is similar to that of the 800 N sample. The failure mechanism is also fatigue spalling, and there are more pitting at the bottom of the spalling pit. Compared with the 800 N sample, the 1200 N sample has a smaller spalling pit area. This is because the surface quality of the 1200 N sample is better than that of the 800 N sample, and the surface of the 800 N sample has more initial cracks and defects,

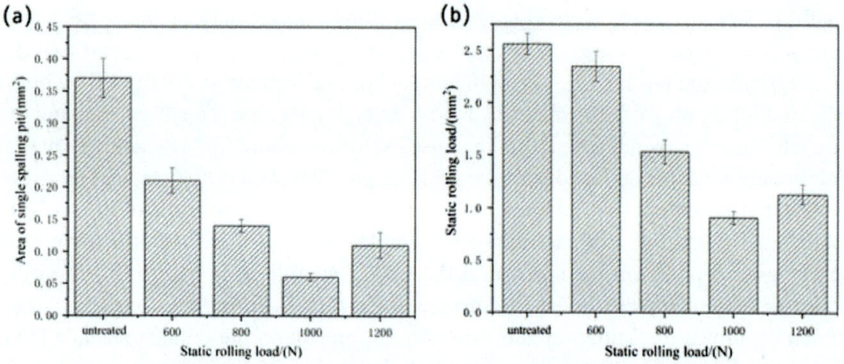

(a)Average area of single spalling pit; (b)Total spalling area

Fig. 18.12 Spalling area of samples at different static loads

therefore, when subjected to the same cyclic stress, the fatigue crack development speed of the 800 N sample is faster than that of the 1200 N sample. And because the surface structure of the 1200 N sample is more refined, it has a certain hindrance to the propagation and interconnection of the cracks. Therefore, the fatigue cracks of 800 N samples are more likely to be connected in a large area during the development process, and finally a larger area of spalling pits are formed. Although the damage morphology is similar and the failure mechanism is the same, under a static load of 1200 N, SURP inhibits the initiation and development of material fatigue cracks better than 800 N.

Figure 18.12 shows the area of spalling pits on the surface of samples at different static loads after rolling contact fatigue test, and Fig. 18.12a shows the average area of single spalling pits. The average areas of spalling pits of untreated, 600 N, 800 N, 1000 N and 1200 N samples is 0.37 mm^2, 0.22 mm^2, 0.14 mm^2, 0.09 mm^2 and 0.11 mm^2 respectively. It can be seen from the figure that the single spalling pit of untreated sample is the largest, while that of 1000 N sample is the smallest, and the size of the spalling pit is only one fourth of that of untreated sample. It can be proved that the sample has the highest resistance property of fatigue crack growth at 1000 N static load, which can explain the test result that the contact fatigue life of 1000 N sample is the longest. Figure 18.12b shows the total area of all spalling pits. From the figure, it can be illustrated that the total spalling area of the sample at 1000 N static pressure is also the smallest, which indicates that the contact fatigue property of tappet material can be most significantly improved at 1000 N static pressure.

Based on the analysis above, it can be found that the main failure mode of the samples after SURP treatment is spalling, which shows that the effect of SURP to refine the microstructure improves the properties of the surface of the samples. Thereby strengthening the ability of the material surface to resist cyclic shear stress and inhibit the delamination. As the static load increases, the area and depth of the spalling pits are decreased, which shows that the strengthening effect of SURP on the surface of the sample is also more significant as the static load increases. The increase of static load not only gradually improves the surface quality of the sample and reduces the phenomenon of stress concentration on the contact surface, but also the ability of the surface material to inhibit the initiation and development of fatigue cracks can be increased accordingly. Therefore, increasing static load in a suitable range can significantly enhance the fatigue resistance of the tappet material. At the same time, grain refinement can improve the surface hardness of the material and increase the fatigue life of rolling contact fatigue [39–41]. Obvious grain refinement can be observed from Fig. 18.6. Compared with the untreated sample (Fig. 18.6a), the surface microstructure of the treated sample is obviously refined. It can be seen from Fig. 18.6b–e that the size of microstructure decreases with the increase of static pressure. Therefore, the fatigue life of the specimens is gradually improved. Fine grain can inhibit the crack growth and reduce the fatigue crack growth rate. The slower the fatigue crack growth rate, the higher the fatigue life. This indicates that the higher the fatigue resistance of the material [42, 43]. Therefore, the specimen has the highest resistance to fatigue crack growth under 1000 N static load. Compared with the 1000 N sample, although the 1200 N sample has higher surface hardness and residual compressive stress, there are many cracks and peelings on the surface. Through the comprehensive analysis of fatigue life and damage morphology, it is found that the anti-fatigue performance of the 1000 N sample is better than that of the 1200 N sample, which shows that the quality of the contact surface has a remarkable influence on the fatigue performance of the contact pair. Therefore, excessive static load may result in defects on the surface of the sample, which has a negative effect on the anti-fatigue performance and accelerate the process of fatigue failure.

18.4 Conclusion

This paper explored the mechanism of SURP on the surface quality, surface microstructure, microhardness and residual compressive stress of the tappet material GCr15 in an engine valve train at different static loads, and analyzed how the above properties affect fatigue performance of the material, and then discussed the effect of SURP on the fatigue behavior of this material. According to the results of this study, the following conclusions can be drawn:

(1) SURP treatment can effectively improve the surface quality of this tappet material. When the static load is less than 1000 N, the surface quality can be iproved with the increase of the static load; when the static load exceeds 1000 N, due to fatigue spalling, the surface quality of the sample decreases instead.
(2) SURP treatment can effectively refine the surface microstructure of the tappet material and improve the residual stress. The refinement of the surface microstructure and the residual compressive stress show an upward trend with the increase of static load.
(3) Compared with the untreated samples, the contact fatigue life of the SURP samples are significantly improved, and the fatigue performance are more stable.

Among the four different static loads, the material obtains the longest contact fatigue life and the most stable fatigue performance at 1000 N. Its rated life (L10), median life (L50) and characteristic life (L63.2) are 6.4, 8.9 and 7.4 times respectively of the corresponding life of the untreated samples.

(4) SURP treatment significantly affects the contact fatigue failure mechanism of tappet material. Except the untreated sample the main failure mode is delamination, the main failure mode of the other treated samples is spalling. At the static load of 1000 N, the spalling pit area of the sample is significantly smaller than that of other samples, and the development of fatigue spalling is also relatively slower, indicating that at this parameter, SURP treatment has the most significant inhibitory effect on the initiation and development of fatigue cracks in the tappet material.

References

1. Zhang, G.C.: Dynamic Analysis of Diesel Engine Valve Train and Optimization Design of Cam Profile. Tianjin University (2009)
2. Titolo, A., Tappet system for internal combustion engines having shafts with variable-profile cams. US Patent 4693214 (1987)
3. Zhou, X., et al.: Failure analysis of CNG engine cam surface contact fatigue. In: National Failure Analysis Conference, shanghai (2009)
4. Jiang, Y.B., Duan, Z.H., Wu, R.Z.: Analysis of contact stress between cam and tappet in automobile engine. Mag. Equip. Mach. **01**, 9–13 (2019)
5. Zhang, Y., Qu, S., Lai, F., Qin, H., Huang, L., Li, X.: Effect of quenching temperature on microstructure and rolling contact fatigue behavior of 17Cr2Ni2MoVNb steel. Metals **8**(9), 735 (2018)
6. He, C.G.: Research on Friction Fatigue Damage Mechanism and Microstructure Evolution Behavior of Wheel Materials. Southwest Jiaotong University. pp. 157 (2018)
7. Cheng, X., Jiang, Z., Wei, D., Hao, L., Wu, H., Xia, W., Zhang, X., Luo, S., Jiang, L.: Effects of surface preparation on tribological behaviour of a ferritic stainless steel in hot rolling. Wear **376–377**, 1804–1813 (2017)
8. Tuo, W.H.: Research on laser welding technology and fatigue damage mechanism of advanced high-strength steel for vehicles. Shanghai University of Engineering Science. pp. 85 (2016)
9. Wang, Z.: Research on the microstructure and properties of laser selective melting of Ti6Al4 V alloy and its surface ultrasonic rolling processing. South China University of Technology. pp. 144 (2019)
10. Li, G., Qu, S., Xie, M., Ren, Z., Li, X.: Effect of multi-pass ultrasonic surface rolling on the mechanical and fatigue properties of HIP Ti-6Al-4 V alloy. Materials **10**(2), 133 (2017)
11. Torres, M.A.S., Voorwald, H.J.C.: An evaluation of shot peening, residual stress and stress relaxation on the fatigue life of AISI 4340 steel. Int. J. Fatigue **24**(8), 877–886 (2002)
12. Zhang, F., Shangguan, X.C.: Effect of surface ultrasonic rolling processing on fatigue properties of AISI304 austenite stainless steel. Hot Working Technol. **046**(16), 136–140 (2017)
13. Hu, J.J.: Effect of Ultrasonic Rolling on Surface Integrity and Fatigue Performance of 60Si2CrVAT Spring Steel. Guizhou University. pp. 73 (2017)
14. Zhang, F., Zhao, Y.C.: Research on the improvement of surface layer properties and fatigue properties of 45# steel by ultrasonic surface rolling. Surf. Technol. **09**, 185–190 (2017). (in Chinese)

15. Liu, Y., Wang, L., Wang, D., Wang, Y.: Mechanical properties of 40Cr surface layer after ultrasonic surface rolling processing. J. Tianjin Univ. (Science and Technology) **7**, 656–661 (2012)
16. Liu, C., Liu, D., Zhang, X., He, G., Xu, X., Ao, N., Ma, A., Liu, D.: On the influence of ultrasonic surface rolling process on surface integrity and fatigue performance of Ti-6Al-4 V alloy. Surf. Coat. Technol. **370**, 24–34 (2019)
17. Ren, X.C., Chen, L.Q., Liu, X.G., Xiang, B., Lin, G.B.: The effect of surface ultrasonic rolling treatment on fatigue properties of high-speed train axle steels. J. Mater. Eng. **43**(12), 1–5 (2015). (in Chinese)
18. Wang, T., Wang, D., Shen, Y., Gong, B., Deng, C.: Effect of ultrasonic surface rolling processing parameters on 40Cr surface roughness. J. Tianjin Univ. (Science and Technology) **42**(2), 168–172 (2009)
19. Xu, S.K.: Study on the surface grain refinement of bearing ring ultrasonic rolling processing. Henan University of Science and Technology (2019)
20. Li, F.Q., Zhao, B.: Effect of ultrasonic machining rolling force on surface characteristics of titanium alloy. Surf. Technol. **10**, 34–40 (2019). (in Chinese)
21. Liu, S.Z.: The effect of ultrasonic rolling on the surface properties of GCr15 material. Nanchang University. pp. 56 (2019)
22. Mangonon, P.L., Thomas, G.: Martensite phases in 340 stainless steel. Metallurg. Mater. Trans. B. **1**(6), 1577–1586 (1970)
23. Cao, L., et al.: Preparation of ultrafine cemented carbide WC-Co powder. J. Chin. Ceram. Soc. **05**, 123–127 (1996)
24. Ma, K., Wen, H., Hu, T., Topping, T., Isheim, D., Seidman, D.N., Lavernia, E.J., Schoenung, J.M.: Mechanical behavior and strengthening mechanisms in ultrafine grain precipitation-strengthened aluminum alloy. Acta Mater. **62**, 141–155 (2014)
25. Gong, S.X., Meguid, S.A.: On the effect of the release of residual stresses due to near-tip microcracking. Int. J. Fract. **52**, 257–274 (1991)
26. Hassani-Gangaraj, S.M., Moridi, A., Guagliano, M., Ghidini, A.: Nitriding duration reduction without sacrificing mechanical characteristics and fatigue behavior: The beneficial effect of surface nano-crystallization by prior severe shot peening. Mater. Des. **55**, 492–498 (2014)
27. Roland, T., Retraint, D., Lu., K., Lu, J.: Fatigue life improvement through surface nanostructuring of stainless steel by means of surface mechanical attrition treatment. Scripta Materialia **54**(11), 1949–1954 (2006)
28. Kubiak, K., Fouvry, S., Marechal, A.M., Vernet, J.M.: Behaviour of shot peening combined with WC–Co HVOF coating under complex fretting wear and fretting fatigue loading conditions. Surf. Coat. Technol. **201**(7), 4323–4328 (2006)
29. Liu, K.G., Yan, C.: The influence of structural fatigue test life data distribution type on its service life. J. Jilin Univ. (Engineering Science Edition) **41**(02), 419–423 (2011)
30. Lee, C.S., Cho, I.S., Pyoun, Y.S., Park, I.G.: Study of inner micro cracks on rolling contact fatigue of bearing steels using ultrasonic nano-crystalline surface modification. Key Eng. Mater. **462–463**, 979–984 (2011)
31. Zhang, X.C., Xu, B.S., Xuan, F.Z., Wang, Z.D., Tu, S.T.: Failure mode and fatigue mechanism of laser-remelted plasma-sprayed Ni alloy coatings in rolling contact. Surf. Coat. Technol. **205**(10), 3119–3127 (2011)
32. Wang, J.W., Li, G.L., Wang, H.D., Xu, B.S., Kang, J.J.: Rolling contact fatigue failure mechanism of nitrocarburizing 45 steel. Trans. Mater. Heat Treatm. **10**, 124–129 (2012)
33. Qin, H.D., Zhang, Y.L., Liu, H.P., Qu, S.G., Lai, F.Q.: The effect of shot peening on the rolling contact fatigue properties of 25CrNi2MoV steel. Surf. Technol. **49**(5), 222–229 + 236 (2020)
34. Zhang, Y.L., Lai, F.Q., Qu, S.G., Liu, H.P., Jia, D.S., Du, S.F.: Effect of ultrasonic surface rolling on microstructure and rolling contact fatigue behavior of 17Cr2Ni2MoVNb steel. Surf. Coat. Technol. **366**, 321–330 (2019)
35. Zhang, M., Liu, Z., Deng, J., Yang, M., Zhang, T.: Optimum design of compressive residual stress field caused by ultrasonic surface rolling with a mathematical model. Appl. Math. Model. **76**, 800–831 (2019)

36. Wang, F., Men, X., Liu, Y., Fu, X.: Experiment and simulation study on influence of ultrasonic rolling parameters on residual stress of Ti-6Al-4 V alloy. Simul. Model. Pract. Theory **104**, 102121 (2020)
37. Wang, H., Song, G., Tang, G.: Evolution of surface mechanical properties and microstructure of Ti 6Al 4V alloy induced by electropulsing-assisted ultrasonic surface rolling process. J. Alloy. Compd. **681**, 146–156 (2016)
38. Garnham, J.E., Davis, C.L.: Very early stage rolling contact fatigue crack growth in pearlitic rail steels. Wear **271**(1–2), 100–112 (2011)
39. Ooki, C.: Improving Rolling Contact Fatigue Life of Bearing Steels through Grain Refinement. SAE Technical Papers (1), (2004)
40. Baughman, R.A.: Effect of hardness, surface finish and grain size on rolling contact fatigue life of M50 bearing steel. J. Basic Eng. **82**(2), 287–294 (1960)
41. Zhang, J., Wu, H., Wang, Y., Zhou, J.: Study on crack growth and fatigue life in TC6 titanium alloy by integrated phase proportion and grain size modeling. Eng. Fract. Mech. **237**, 107235 (2020)
42. Cruse, T.A., Meyers, G.J., Wilson, R.B.: Fatigue growth of surface cracks. Astm Special Technical Publication (1977)
43. Chen, S., Cui, Z.: Surface crack growth behaviour under tensile cyclic loading. J. Int. J. Fatigue **10**(1), 43–47 (1988)

Chapter 19
Electro-Mechanical Co-Simulation of a 6-DOF Parallel Robot

Fengchao Liang, Shuang Tan, Long Chen, and Zhe Lin

Abstract Parallel robot has the advantages of high precision, high stiffness, fast speed and strong bearing capacity. When applied in 3D printing, it can avoid the defects caused by the vertical direction layer by layer accumulation additive manufacturing mode, and the 3D printing equipment with 6-DOF parallel robot can print 3D entities with complex shape and smooth surface without support. In order to increase the design efficiency and verify the correctness of the mechanical and control system of a 6-DOF parallel robot used in a 3D printer, an electro-mechanical co-simulation method of the parallel robot was realized based on the electro-mechanical model created using interface technology between ADAMS and MATLAB. First, kinematics mathematical model was built based on theory of inverse kinematics of parallel robots. Then, simulation of kinematics was performed after the virtual prototype of the 6-DOF robot was created in ADAMS, and the simulation results were consistent with the theoretical calculation. Then, the model of electro-mechanical system of the 6-DOF parallel robot was established by embedding the ADAMS virtual prototype into the control model of MATLAB. Finally, step and sinusoidal function responses of the 6-DOF parallel robot pose were studied respectively. The simulation results show that the 6-DOF parallel robot has a high steady accuracy and good dynamic performance, and the simulation parameters provide the necessary design basis for the development of the 6-DOF parallel robot.

Keywords Parallel robot · ADAMS · Kinematics · Dynamics · 3D printing · Virtual prototyping

19.1 Introduction

3D printing technology, also known as additive manufacturing technology, is completely different from the traditional material removal processing technology. It generates 3D entities by adding materials layer by layer. Many scholars have studied

F. Liang (✉) · S. Tan · L. Chen · Z. Lin
Beijing Institute of Space Mechanics and Electricity, Beijing 100094, China
e-mail: fc.liang@qq.com

J. Xu and K. M. Pandey (eds.), *Mechanical Engineering and Materials*,
Mechanisms and Machine Science 100,
https://doi.org/10.1007/978-3-030-68303-0_19

3D printing [1–5], and put forward a concept of multi-directional printing. This multi-directional 3D printing can accumulate materials to the formed surface along different directions, and improve the quality of layered manufacturing. However, robots are required to have at least four degrees of freedom. Lee et al. [6] designed a 5-DOF 3D printing robot to realize multi-directional printing. Keating et al. [7] combined 3D printing technology with multi axis milling technology, and realized multi-functional and multi-material processing with 6-DOF robot. However, the above robots mainly use series mechanism, which has large inertia and large error. At present, most 3D printing robots on the market adopt series structure, and the precision and relative complexity of printed products are relatively low [8]. Parallel robot has the advantages of high precision, high stiffness, fast speed and strong bearing capacity. Moreover, when applied in 3D printing, it has the following advantages: Firstly, the position and posture of the end effector (nozzle, laser or UV light source) can be changed by the 6-DOF additive manufacturing equipment, and the stacking direction can be changed according to different curved surfaces instead of a single vertical direction. It can avoid the influence of 'step effect' on the surface quality of smooth surface. Secondly, for the fused deposition modeling additive manufacturing technology, the 6-DOF additive manufacturing equipment can quickly cool and solidify the molten resin extruded from the nozzle through the appropriate cooling system during the printing process, so as to realize unsupported printing. Thirdly, changing the posture of the end effector of the mechanism can improve the printing quality and avoid the printing part interference, which could greatly increase the complexity of the printable surface. For example, Song et al. of the University of Iowa studied the additive manufacturing technology based on Stewart type 6-DOF parallel robot, and applied the Stewart 6-DOF parallel robot to 3D printing [9].

The development of 6-DOF parallel robot is the key technology of 6-DOF 3D printer. Before manufacturing the parallel robot, the mechanical system and control system of parallel robot must be simulated to provide theoretical basis and main parameters for structure optimization, controller design and motion simulation.

The traditional mechanical and electrical system design is to design the mechanical and control system separately. When the physical prototype is tested, the two design results are combined for the first time, which is easy to cause the mismatch between the mechanical and control system, the design modification is large and the design efficiency is low. Therefore, there is an urgent need for a design method that considers both mechanical and control systems from the beginning of design and establishes them in the same simulation model for electro-mechanical co-simulation. Co-simulation is a method in which different simulation programs exchange data at the same step time, and then solve the problem separately. This method can effectively combine the advantages of many kinds of software, realize the design requirements of parallel robot and improve the efficiency and quality of design. ADAMS can exchange control data with MATLAB. In this paper, the electro-mechanical co-simulation of a Stewart 6-DOF parallel robot is realized through the interface of ADAMS/control module and MATLAB.

19.2 Kinematics Analysis of the Parallel Robot

19.2.1 Theoretical Basis

Kinematics analysis is the basis of dynamic analysis, structural optimization design and control strategy research. As shown in Fig. 19.1, the structure of a Stewart 6-DOF parallel robot is composed of upper and lower platforms and six parallel, independent and freely retractable struts [10]. The six struts and the platform are connected by Hooke hinges A_i and spherical hinges B_i. The posture of the moving platform (upper platform) is adjusted by controlling the length of the six struts. In order to quantitatively express the coordinates of each part of the moving platform, the inertial coordinate system (static coordinate system) O-xyz is established at the center of hinges A_i of the static platform, and the coordinate origin is O. The dynamic coordinate system p-xyz is established at the center of hinges B_i, and the coordinate origin is p. The direction of each coordinate system is shown in Fig. 19.1. In this paper, the finite rotation of a rigid body around a certain axis mentioned in Euler's theorem is decomposed into three finite rotations around the coordinate axis of the connected body in a certain order, and the angle of each rotation can be defined as three generalized coordinates to determine the relative position of the rigid body before and after rotation. The order of rotation around the X → Y → Z axis is selected. The Euler angles of rotation are U, V and W, respectively. After three rotations, the rotation transformation matrix shown in Eq. (19.1) can be obtained from the properties of rotation matrix:

$$ {}_{B}^{A}\mathbf{R} = \mathbf{R}(x, U) \cdot \mathbf{R}(y, V) \cdot \mathbf{R}(z, W) = $$

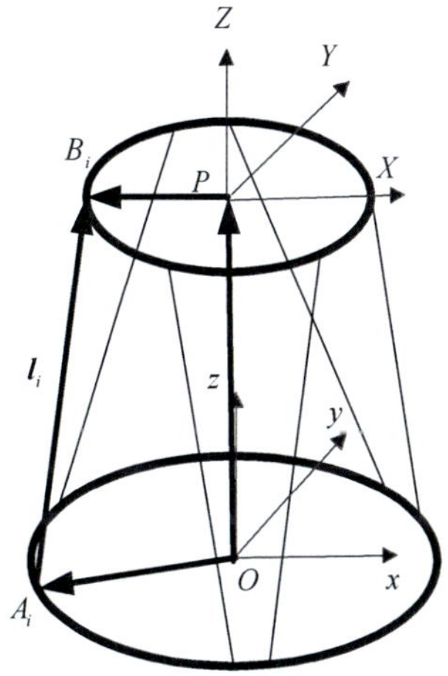

Fig. 19.1 Coordinate system of the 6-DOF parallel robot

$$\begin{bmatrix} cVcW & -sVsW & sV \\ sUsVcW + cUsW & -sUsVsW + cUcW & -cVsU \\ -cUsVcW + sUsW & cUsVsW + sUcW & cUcV \end{bmatrix} \tag{19.1}$$

where $cU = cos\,(U)$, $cV = cos\,(V)$, $cW = cos\,(W)$, $sU = sin\,(U)$, $sV = sin\,(V)$, $sW = sin\,(W)$. When the generalized coordinates of the posture of the moving platform are $q = [X, Y, Z, U, V, W]^{\mathrm{T}}$, the length of each strut is can be defined in Eq. (19.2):

$$l_i = |\mathbf{l}_i| = \sqrt{\mathbf{l}_i^T \cdot \mathbf{l}_i} \tag{19.2}$$

where the vector $\boldsymbol{l}_i$ is the strut vector A_iB_i shown in Fig. 19.1, l_i is the length of the struts, i = 1, 2, …, 6. So far, the inverse kinematics mathematical model of the 6-DOF parallel robot is established.

19.2.2 Kinematics Simulation

In order to study the kinematics simulation of the 6-DOF parallel robot in ADAMS, the virtual prototype model must be established in ADAMS. ADAMS provides the function of parametric modeling, that is, the eigenvalues of the established model are expressed by the design parameters in ADAMS. In this way, the model will automatically change with the modification of design parameters, greatly simplifying the manual modification process, especially conducive to the optimal design of mechanical structure.

In ADAMS/view, the coordinates of the upper hinge points B_i, the lower hinge points A_i and the platform radius are set as variables, and the parametric virtual prototype model as shown in Fig. 19.2 is established. In order to simulate the motion of the parallel robot, it is necessary to add the correct constraints between the components. The static platform is fixed on the earth by fixed pair constraint, and the hinge points

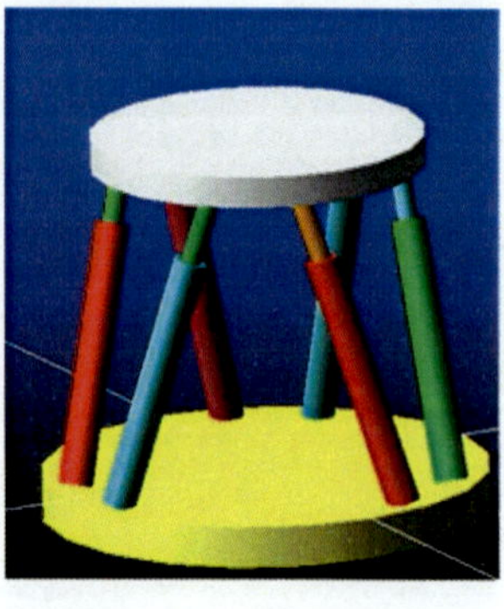

Fig. 19.2 Parametric virtual prototype of the 6-DOF parallel robot

Fig. 19.3 Inverse kinematics solution in ADAMS

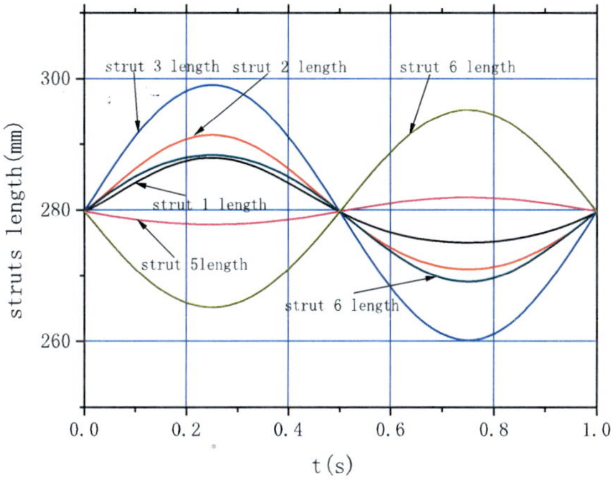

A_i is constrained by T-type hinges, so that there are only two rotational degrees of freedom (DOF) between a strut and the static platform. The hinge points B_i is constrained by spherical hinges, so that there are only three rotational DOF between a strut and the moving platform. The moving pair constraint is used between the moving part and the static part of the strut, so that there is only 1 translational DOF to realize the expansion and contraction. In this way, the kinematics simulation of the 6-DOF parallel robot can be carried out based on the virtual prototype.

The general point motion excitation of 6 DOF is applied at the center of the moving platform at the same time, so that the moving platform can move along the X, Y, Z axis and rotate along the X, Y, Z axis, so as to simulate the general motion of the moving platform in actual work. After the motion simulation, the length variation curve of each strut as shown in Fig. 19.3 is obtained, which is the inverse kinematics solution. Then according to the Eqs. (19.1) and (19.2), the real-time length of each strut is calculated by MATLAB programming. By comparison, the error between the real-time length curve calculated by MATLAB and ADAMS is in the order of 10^{-5} mm, which shows that at the same time, the expansion and contraction of the two models are consistent. The theoretical model and ADAMS virtual prototype are correct. This virtual prototype can be used for electro-mechanical co-simulation.

19.2.3 Electro-Mechanical Co-Simulation

The co-simulation between ADAMS and MATLAB is actually that ADAMS provides parameter interface and MATLAB provides control scheme. The software can automatically establish and solve the dynamic equation. In the process of solving, the data exchange between ADAMS and MATLAB is carried out every time interval.

19.2.4 Establishment of Co-Simulation System Model

Parallel robot is a MIMO, highly nonlinear and strong coupling system, and its control problem is always a difficult problem. Parallel robot relies on reasonable and practical control strategy to give full play to its performance potential. The control strategy will directly determine the final performance index of the system. In practice, the traditional PID control algorithm is still widely used. The most commonly used control strategy of 6-DOF Parallel robot is PID control. In this control method, PID controllers control the posture of the moving platform by controlling six struts to complete their respective actions.

In order to carry out the co-simulation of ADAMS and MATLAB and realize the parameter transfer between mechanical system and control system, it is necessary to use ADAMS/control interface module. After the input and output variables required by the co-simulation system are established in ADAMS, the 6-DOF parallel robot system in ADAMS is imported into MATLAB as a subsystem of MATLAB/Simulink, so as to realize the co-simulation of mechanical and control system.

Invoke the ADAMS/control module in MATLAB and enter the command *adams sys*, MATLAB will generate a mechanical subsystem module of the 6-DOF parallel robot, as shown in Fig. 19.4. It can be seen that there are six input variables *InputDispDrive_Strut(1-6)* at the left end of the module, which are used to receive the control instructions of MATLAB controller. There are 12 output variables at the right end, variables named *Output_Strut(1-6)Length* are the feedback value of struts' length, variables named *MovPlat_(X, Y, Z, U, V, W)* are the posture feedback of the moving platform.

The electro-mechanical co-simulation system model established in MATLAB is shown in Fig. 19.5. Reverse solver is the inverse kinematics calculation module of the 6-DOF parallel robot, and *adams_sub* is the mechanical system of the parallel

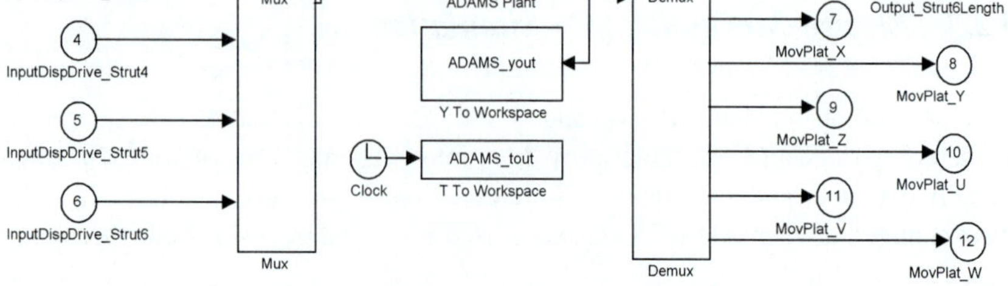

Fig. 19.4 Mechanical subsystem of the co-simulation model

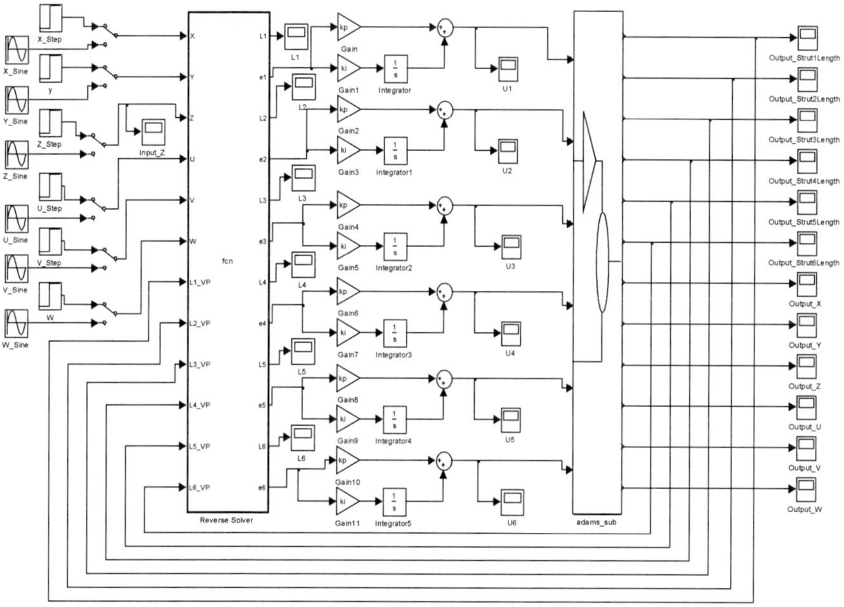

Fig. 19.5 Electro-mechanical co-simulation system model of the 6-DOF parallel robot

robot. After the posture trajectory $q = [X, Y, Z, U, V, W]^T$ of the moving platform are input, the length L_i of each strut and the error between the struts' length instruction and the actual struts length are calculated by reverse solver, $e_i = L_i - L_i\_VP$, and e_i is transformed into the displacement command of each strut by PI controller. Input variables *InputDispDrive_Strut(i)* are responsible for driving multi axis motion, so as to drive the moving platform to move. The actual length of the struts, *Output_Strut(i) Length*, are fed back to reverse solver in real time, and the struts' length error e_i is obtained by comparing and calculating the module. Where e_i is the real-time error of the struts length, L_i is the target struts length obtained by inverse kinematics solution, L_i_VP is the actual length of each strut of the 6-DOF parallel robot, $i = 1, 2, \ldots, 6$. In this way, through the precise closed-loop control of the length of each strut, the precise control of the posture of the moving platform can be realized.

19.2.5 Electro-Mechanical Co-Simulation Experiment

The function of the 6-DOF parallel robot is to accurately adjust the posture of an end effector of a 3D printer. In order to verify the posture adjustment performance of the parallel robot, the steady-state accuracy of the co-simulation system is tested by the step trajectory signal, and the dynamic performance is studied by the sinusoidal trajectory signal.

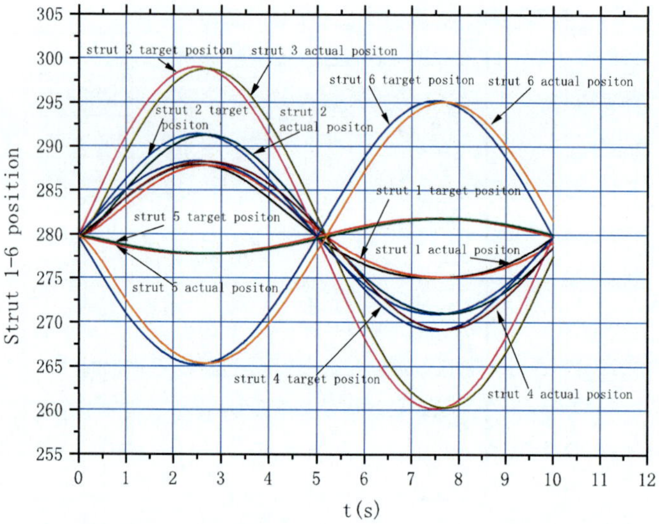

Fig. 19.6 Tracking curves of six struts under the excitation of sinusoidal signal

Under the excitation of step signal $q = [5, 5, 5, 5, 5]^T$, the time from zero position to stable position of struts 1-6 is 3.12 s, 3.29 s, 3.53 s, 3.46 s, 3.08 s and 3.43 s, respectively, reaching steady state rapidly, and the steady-state value is equal to the command value, and there is no steady-state error. In the simulation animation of ADAMS, we can see that the parallel robot moves rapidly from the initial posture to the target posture, and keeps the posture unchanged stably. The simulation results shows that struts 1 ~ 6 can drive the moving platform moving from the initial position to the target position in a short time, and stabilize at the target position given by the command. When the input target trajectories are *5*sin(2*pi*0.1*t)*, the response curve of each strut is shown in Fig. 19.6. It can be seen in Fig. 19.6 that each strut can accurately track the given sine command, and the time delay is about 0.2 s, and each strut has good dynamic characteristics. The electro-mechanical co-simulation results show that the 6-DOF parallel robot has good steady-state and dynamic performance.

It should be pointed out that, in practical engineering, the displacement resolution of each strut is limited, and the ideal continuous motion of simulation is discrete in practical engineering, which will cause the accuracy of the platform to be limited by the resolution of the strut, and the various degrees of freedom of the platform may be coupled. Therefore, in the design of engineering application, it is also necessary to carry out the inverse solution simulation of the displacement resolution of the strut according to the requirements of the displacement resolution of the platform to obtain the required strut resolution, and then carry out the forward solution according to the resolution of the strut to check the displacement resolution of the platform.

19.3 Conclusion

The 6-DOF parallel robot is a core component of 6-DOF 3D printer. For further improving the quality of 3D printing, the electro-mechanical co-simulation of the Stewart 6-DOF parallel robot is realized through the interface of ADAMS/control module and MATLAB. The steady-state accuracy of the 6-DOF parallel robot is simulated by the position and attitude step trajectory signal, and the dynamic performance is simulated by the sinusoidal trajectory signal. The simulation results verified that the control strategy can realize the precise control of the six struts, so as to realize the fast and accurate control of the position and attitude of the moving platform. Which also show that the 6-DOF parallel robot has a high steady accuracy and good dynamic performance, and the simulation parameters provide the necessary design basis for the development of the 6-DOF parallel robot.

References

1. Son, T.A., Minh, P.S., Thanh, T.D.: Effect of 3D printing parameters on the tensile strength of products. Key Eng. Mater. **863**, 103–108 (2020)
2. Huynh, H.N., Bui, T.H., Thai, T.T.H., Nguyen, H.T.: Research on the effect of technical attributes on the tensile strength of FDM products. Key Eng. Mater. **86**, 33–50 (2020)
3. Li, X.D., Zhao, F.: 3D printing technology impact on development of industrial design. Key Eng. Mater. **693**, 1901–1904 (2016)
4. Glasschroeder, J., Prager, E., Zaeh, M.F.: Powder-bed-based 3D-printing of function integrated parts. Rapid Prototyp. J. **21**(2), 207–215 (2015)
5. Chen, Y., Zhou, C., Lao, J.: A layerless additive manufacturing process based on CNC accumulation. Rapid Prototyp. J. **17**(3), 218–227 (2011)
6. Lee, W.C., Wei, C.C., Chung, S.C.: Development of a hybrid rapid prototyping system using low-cost fused deposition modeling and five-axis machining. J. Mater. Process. Technol. **214**(11), 2366–2374 (2014)
7. Keating, S., Oxman, N.: Compound fabrication: A multi-functional robotic platform for digital design and fabrication. Robot. Comput.-Integr. Manuf. **29**(6), 439–448 (2013)
8. Günther, D., Heymel, B., Franz Günther, J., Ederer, I.: Continuous 3D printing for additive manufacturing. Rapid Prototyp. J. **20**(4), 320–327 (2014)
9. Song, X., Pan, Y., Chen, Y.: Development of a low-cost parallel kinematic machine for multidirectional additive manufacturing. J. Manuf. Sci. Eng. **137**(2), 297–310 (2015)
10. Stewart, D.: A Platform with six degrees of freedom. Proc. Int. Mech. Eng. **180**(1), 371–386 (1965)

Chapter 20
Multi-Objective Optimization of Automotive Front Rail Based on Surrogate Model and NSGA-II

Yiwei He, Wangdong Xu, and Fanruo Gu

Abstract With the rapid development of automotive industry, more and more attention has been paid to the lightweight and safety design. Crashworthiness optimization is an essential part in automotive design. In this study, a non-dominated sorting genetic algorithm II (NSGA-II) based on Kriging model is proposed to optimize the structure of automotive frontal rail to meet the requirements of crashworthiness and lightweight. The material of frontal rail is mild steel, which performs well in strength. Kriging surrogate model is employed to replace traditional finite element model, which will reduce much computational time and improve the efficiency. Then NSGA-II is applied to solve the multi-objective optimization problem. The results illustrate that the Pareto optimal front obtained by NSGA-II exhibits good performance on convergence and diversity. And then the optimal scheme for design is selected, the accuracy is proved to be high. Compared to baseline model, the optimized automotive frontal rail shows significant improvement on crashworthiness and achieves 7.5% weight reduction.

Keywords NSGA-II · Kriging model · Lightweight · Pareto optimal front

20.1 Introduction

As is known, crashworthiness optimization problems are always important for automotive design. And plenty of researchers focus on thin-wall structure optimization. Ma et al. had many researches of topology [1]. And there are many topics, such

Y. He (✉)
Mechanical Engineering, Rensselaer Polytechnic Institute, Troy, NY 12180, USA
e-mail: YiweiHEEE@163.com

W. Xu
Mathematics and Computer Science, University of California-San Diego, La Jolla, San Diego, CA 92122, USA

F. Gu
Faculty of Arts and Sciences, College of William & Mary, Williamsburg, VA 23187, USA

J. Xu and K. M. Pandey (eds.), *Mechanical Engineering and Materials*,
Mechanisms and Machine Science 100,
https://doi.org/10.1007/978-3-030-68303-0_20

as material sensitive analysis, shape design and so on [2–5]. Sensitivity analysis of automobile lightweight design was also a hot topic recently. Zuo et al. [6] made stress analysis and optimization of rectangular tubes. And Chen et al. [7] proposed a method of lightweight design under static and dynamic stiffness demands.

Material structure optimization is a vital approach for automobile lightweight design. Topology optimization method [8] is widely used in cellular material. And surrogate model has been an effective way to replace the material structure model in FE software [9], which has been proved to reduce optimization time while guarantee high accuracy.

Among such metamodeling methods, the Kriging method is one of the most accurate and efficient methods even for highly nonlinear performance functions. The Kriging method generates a surrogate model using a mean structure and a Gaussian stochastic process, and the desired prediction value becomes different depending on the type of the mean structure and correlation function, and parameters of the stochastic process. Therefore, in this paper, Kriging surrogate model and NSGA-II are employed to solve crashworthiness and lightweight optimization problem.

20.2 Finite Element Model

The FE model of mini-bus's body-in-white (BIW) established in this study is constructed for simulation of frontal impact test against deformable barrier with 40% overlapping in Hypermesh software. Much computational time is needed to simulate a car collision since the entire vehicle model consisted of many parts. And a reduced FE model is developed to decrease the computational time [10]. The B-pillar front part of BIW is detailed constructed in FE model while other parts are replaced by mass points, and the reduced FE model is shown in Fig. 20.1. Referring to the regular in the C-NCAP [11], the front passenger seat can be replaced by 77 kg weight

Fig. 20.1 Reduced FE model of mini-bus's BIW

Table 20.1 Material property of mild steel

Density	Young's modulus	Poisson's ratio	Name In LS-DYNA
7800 kg/m^3	206 GPa	0.3	material #24

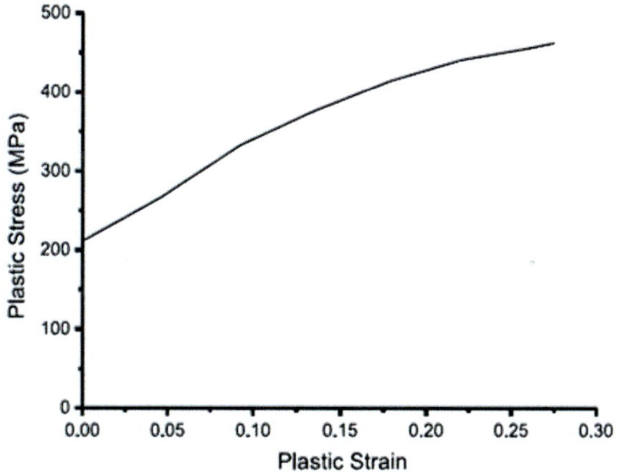

Fig. 20.2 Strain hardening curve of mild steel

mass points.The reduced FE model contains 428,263 elements and 451,820 nodes while the percent of triangular elements less than 5%. According to the standards of frontal impact test against deformable barrier with 40% overlapping in C-NCAP, the speed of impact is set as 64 km/h while the simulation time is set as 100 ms. LS-DYNA is utilized for collision simulation. The contact force in a collision is necessary to be considered in modeling [12], therefore the contact force is detailed construction in FE model.

The structure considered in this study is a front rail that consists of several thin-walled parts, and the material is mild steel, then the properties of material are listed as Table 20.1. And this material has a piecewise linear elastic-plastic behavior with strain hardening. Figure 20.2 illustrates the relationship between the plastic stress and plastic strain of mild steel [13].

The process of optimization design on lightweight and crashworthiness can be draw as Fig. 20.3.

20.3 Definition of Multi-objective Optimization Problem

20.3.1 Optimization Objectives and Constrains

The target of optimization design is to achieve lightweight and improve crashworthiness. In this paper, the maximum acceleration of B-pillar a_{max} and mass of front rail m are considered as objectives while the rearward intrusion of A-pillar D_1,

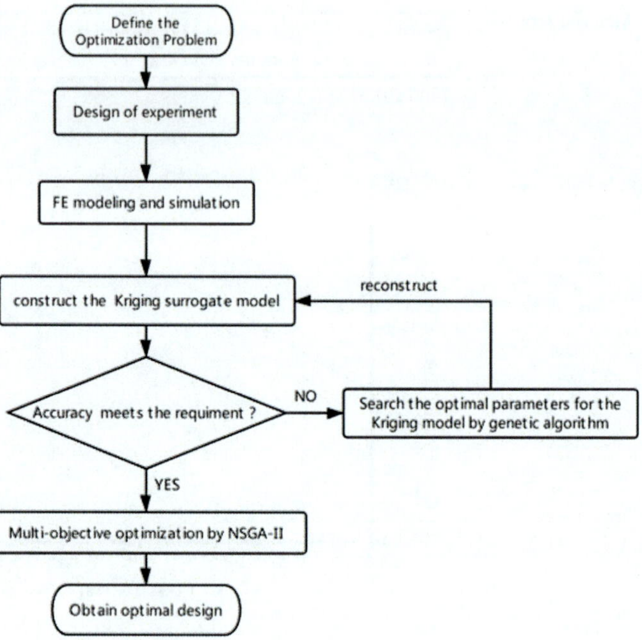

Fig. 20.3 The process of optimization design in this study

the steering column D_2 and the deformation of front rail D_3 are considered as the constrains.

20.3.2 Design Variables

The principal parts of crumple zone on the automotive frontal rail are selected as design parameters, and the explosive view of thin-walled parts in front rail for optimization design is shown in Fig. 20.4.

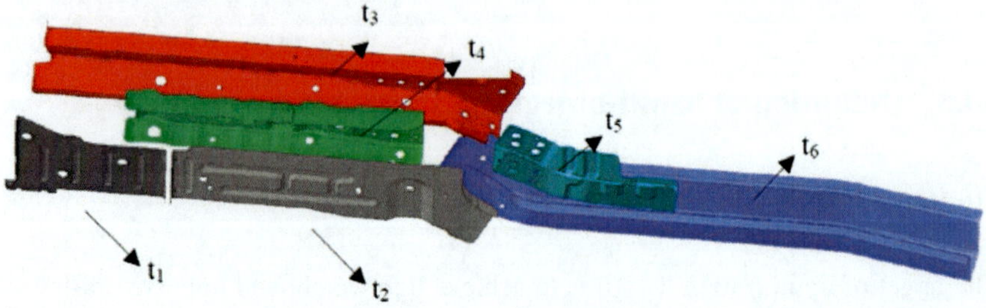

Fig. 20.4 The explosive view of design parts in front rail for optimization design

Table 20.2 The ranges for design variables

Design variable	Initial value [mm]	Lower bound [mm]	Upper bound [mm]
t_1	1.72	0.70	3.20
t_2	1.53		
t_3	1.39		
t_4	2.31		
t_5	2.12		
t_6	0.82		

Then the thickness of six components t_1, t_2, t_3, t_4, t_5 and t_6 are set as the design variables. Considering the actual manufacturing of the thin-walled parts, the thickness should be rounded up to 2 decimal places in this study. The ranges for design variables are listed in Table 20.2.

20.4 Construction of Kriging Surrogate Model

Design of experiment (DoE) is the first step to build a surrogate model. In order to determine a set of training samples, an Optimum Latin Hypercube Sampling is employed to generate 200 sample points in this study.

The Kriging model uses the training samples to construct the surrogate model for predicting the output. It consists of approximate model and a stochastic process as shown in Eq. (20.1):

$$y(x) = f(x)^T \beta + z(x) \tag{20.1}$$

where β is the regression parameters,f(x) is the column vector of basic functions;z(x) denotes the stochastic parameter.

The DoE data are seen as results of a stochastic process, which is described by using a set of random vectors $\mathrm{Y(x)} = \left[\mathrm{Y}\left(\mathrm{x}^{(1)}\right), \cdots, \mathrm{Y}\left(\mathrm{x}^{(\mathrm{n})}\right)\right]$, with mean 1μ(1 is an $\mathrm{n} \times 1$ column vector of ones). The random variables are correlated with each other using the basis function expression as shown in Eq. (20.2):

$$\mathrm{cor}\left(\mathrm{Y}\left(x^i\right), \mathrm{Y}\left(x^l\right)\right) = \exp(-\sum_{j=1}^{k} \theta_j \left|x_j^{(i)} - x_j^{(l)}\right|^p) \tag{20.2}$$

where the θ vector represents the width of each basis function, the parameters p controls the smoothness of the approximation in the proximity of the given sample points, and k is the dimensionality of the problem.

Then an $\mathrm{n} \times \mathrm{n}$ correlation matrix of all the observed data can be conducted as Eq. (20.3):

$$\text{cov}\begin{pmatrix} cov\left(Y^{(1)}, Y^{(1)}\right) & \cdots & cov\left(Y^{(1)}, Y^{(n)}\right) \\ \vdots & \ddots & \vdots \\ cov\left(Y^{(n)}, Y^{(1)}\right) & \cdots & cov\left(Y^{(n)}, Y^{(n)}\right) \end{pmatrix} \tag{20.3}$$

These correlations depend on the absolute distance between the sample points $\left|x_j^{(i)} - x_j^{(l)}\right|$ and the parameters θ and p.

The likelihood function is defined as Eq. (20.4):

$$L\left(Y^{(1)}, \cdots, Y^{(n)}|\mu, \sigma\right) = \frac{1}{\left(2\pi\sigma^2\right)^{\frac{n}{2}}} \exp\left(-\frac{\sum\left(Y^{(i)} - \mu\right)^2}{2\sigma^2}\right) \tag{20.4}$$

which can be expressed in terms of the sample data as Eq. (20.5)

$$L = \frac{1}{\left(2\pi\sigma^2\right)^{\frac{n}{2}}|\Psi|^{1/2}} \exp\left[-\frac{(y - 1\mu)^T\Psi^{-1}(y - 1\mu)}{2\sigma^2}\right] \tag{20.5}$$

where Ψ is the correlation matrix between the random variables. Then we take the natural logarithm to simplify the likelihood function as Eq. (20.6)

$$\ln(L) = -\frac{n}{2}\ln(2\pi) - \frac{n}{2}\ln\left(\sigma^2\right) - \ln|\Psi| - \frac{(y - 1\mu)^T\Psi^{-1}(y - 1\mu)}{2\sigma^2} \tag{20.6}$$

And we can obtain the maximum likelihood estimates (MLE) for μ (Eq. 20.7) and σ^2(Eq. 20.8):

$$\hat{\mu} = \frac{1^T\Psi^{-1}y}{1^T\Psi^{-1}1} \tag{20.7}$$

$$\hat{\sigma}^2 = \frac{(y - 1\mu)^T\Psi^{-1}(y - 1\mu)}{n} \tag{20.8}$$

Then the concentrated ln-likelihood function [14] can be deduced as Eq. (20.9)

$$\ln(L) \approx -\frac{n}{2}\ln\left(\hat{\sigma}^2\right) - \frac{1}{2}\ln|\Psi| \tag{20.9}$$

The value of this function depends on the unknown parameters θ and p. Therefore we are supposed to find values for these parameters which maximize Eqs. (20.3), (20.9) to build the best Kriging model with high accuracy. According to previous study [15], we can fix the parameter at p = 2, which shows the best smooth correlation with a continuous gradient,

To test the error of surrogate model, such metrics as maximum relative error (e_{max}), average relative error (e_{avg}) and R-square (R^2) are employed [16]. And the results of error analysis are shown in Table 20.3.

Table 20.3 Accuracy assessment

Objectives	R^2	e_{max}	e_{avg}
a_{max}	0.97218	0.19162	0.01320
m	0.99986	0.01745	0.00069
D_1	0.99808	0.03888	0.00310
D_2	0.99951	0.02324	0.00169
D_3	0.99892	0.00191	0.00083

It can be concluded from Table 20.3 that the Kriging models show high accuracy, therefore, the surrogate models can be applied in next optimization.

20.5 Multi-objective Optimization by NSGA-II

20.5.1 Mathematical Model

As mentioned above, the maximum acceleration of B-pillar a_{max} and mass of design components m are considered as objectives while the rearward intrusion of A-pillar D_1, the steering column D_3 and the deformation of front rail are determined to be the constrains, the value of D_1 and D_2 are supposed to be less than 300 mm on the demand of C-NCAP and the value of D_3 should be less than 1106.43 mm that is the value of baseline. The aim is to find the optimum design variables t_1, t_2, t_3, t_4, t_5 and t_6 to minimize the values of a_{max} and m. The mathematical models can be expressed as Eq. (20.10):

Find $t_1, t_2, t_3, t_4, t_5, t_6$

$$\begin{cases} Minimize a_{max}(t_1, \cdots, t_6), m(t_1, \cdots, t_6) \\ s.t. D_1, D_2 \ll 300mm \\ D_3 \ll 1106.43mm \\ 0.6mm \ll t_1, \cdots, t_6 \ll 3.2mm \end{cases} \tag{20.10}$$

20.5.2 Application of NSGA-II

NSGA-II (non-dominated sorting genetic algorithm), as a widely used optimization algorithm for multi-objective problem, is employed to find the Pareto optimum front (POF) for the objective function, which is proved to be an effective strategy for multipurpose search [17, 18]. The population size is set as 20, and the crossover probability and mutation probability are 0.9 and 0.1, respectively. The distribution

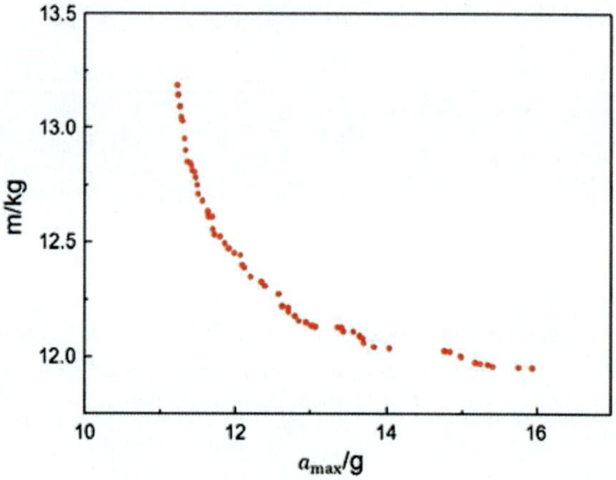

Fig. 20.5 Pareto optimal fronts

index for crossover and mutation are 10 and 20. The fitness function and constrains function can be determined by the Kriging metamodels above.

20.5.3 Result Analysis

The POF of the NSGA-II based on Kriging surrogate model is shown in Fig. 20.5. Refer to Fig. 20.4, the Pareto optimal front of NSGA-II based on the Kriging model performs well in convergence and diversity.

In real engineering optimization design, the lightweight is as important as the crashworthiness. Therefore, one of the optimal points in POF was determined as the optimum solution. Then the optimum solution is compared to baseline model and the accuracy is verified by calculating the objective values in FE model, the results are shown in Table 20.4.

To compare the optimum results in FE models, the deformation of the front rail in frontal impact test against deformable barrier with 40% overlapping in Ls-Dyna software are shown in Fig. 20.6. It is obvious that the optimized front rail has smaller deformation after impact.

20.6 Conclusion

In order to obtain the optimum structure of the parts in front rail, a surrogate-based optimization is developed a frontal rail. The multi-optimization problem is defined to minimum the peak accelerate a_{max} and the mass of frontal rail m with the constraints of the rearward intrusions of A-pillar D_1 and the steering column D_2.To reduce the calculation time, the Kriging surrogate model is proposed to provide high accuracy

Table 20.4 Optimization results

Variable	Baseline model	Optimized model		
t_1	1.1	0.99		
t_2	0.85	0.86		
t_3	0.95	0.98		
t_4	0.75	0.82		
t_5	0.65	0.7		
	/	Predicted	Real	Error
a_{max}	16.79	11.37	11.21	1.43%
m	13.86	12.71	12.82	0.86%
D_1	219.47	217.3	215.24	0.96%
D_2	188.92	112.97	113.55	0.51%
D_3	1106.43	979.81	972.16	0.79%

Fig. 20.6 Deformation of baseline model (left) and optimized model (right)

surrogate model for NSGA-II. Optimized by the proposed approach, the crashworthiness improved significantly and achieve 7.5% weight reduction than that of baseline model.

References

1. Ma, Y., Chen, R., Bai, J., Zuo, W.: Shape optimization of thin-walled cross section for automobile body considering stamping cost, manufacturability and structural stiffness. Int. J. Automot. Technol. **21**, 503–512 (2020)
2. Ma, Y., Wang, X., Zuo, W.: Analytical sensitivity analysis method of cross-sectional shape for thin-walled automobile frame considering global performances. Int. J. Automot. Technol. **21**, 1207–1216 (2020)
3. Lu, S., Ma, H., Xin, L., Zuo, W.: Lightweight design of bus frames from multi-material topology optimization to cross-sectional size optimization. Eng. Optimiz. **51**(6), 961–977 (2019)
4. Bai, J., Li, Y., Zuo, W.: Cross-sectional shape optimization for thin-walled beam crashworthiness with stamping constraints using genetic algorithm. Int. J. Veh. Des. **73**, 76–95 (2017)
5. Zuo, W.: Bi-level optimization for the cross-sectional shape of a thin-walled car body frame with static stiffness and dynamic frequency stiffness constraints. Proc. Inst. Mech. Eng., Part D: J. Automob. Eng. **229**(8), 1046–1059 (2015)

6. Zuo, W., Yu, J., Saitou, K.: Stress sensitivity analysis and optimization of automobile body frame consisting of rectangular tubes. Int. J. Automot. Technol. **17**(5), 843–851 (2016)
7. Chen, W., Zuo, W.: Component sensitivity analysis of conceptual vehicle body for lightweight design under static and dynamic stiffness demands. Int. J. Veh. Des. **66**(2), 107–123 (2014)
8. Qiao, H., Wang, S., Zhao, T., Tang, H.: Topology optimization for lightweight cellular material and structure simultaneously by combining SIMP with BESO. J. Mech. Sci. Technol. **33**, 729–739 (2019)
9. Guo, S., Ang, H., Cai, H.: Construction of an adaptive sampling surrogate model and application in composite material structure optimization. Acta Materiae Compositae Sinica **35**(8), 2084–2094 (2018)
10. Liu, H., Zhang, X., Xiao, L.: Collision simulation analysis of 7075 aluminum alloy car bumper based on LS-DYNA. J. Machine Des. **2**, 18–23 (2011)
11. China Automotive Technology and Research Center, C-NCAP Management Regulation http://www.c-ncap.org/cncap/pjgz (2018)
12. Ma, Y., Chen, X., Zuo, W.: Equivalent static displacements method for contact force optimization. Struct. Multidiscipl. Optimiz. **62**, 323–336 (2020)
13. Technical Transfer Dispatch#6-Body Structure Materials, ULSAB-AVC Consortium (2001)
14. Raponi, E., Bujny, M., Olhofer, M., Aulig, N., Boria, S., Duddeck, F.: Kriging-assisted topology optimization of crash structures. Comput. Methods Appl. Mech. Eng. **348**(1), 730–752 (2019)
15. Forrester, A., Sobester, A., Keane A.: Engineering Design Via Surrogate Modelling: A Practical Guide. John Wiley & Sons, Inc (2008)
16. Tian, L., Mehari, M., Santi, S., Latré, S., Poorter, E., Famaey, J.: Multi-objective surrogate modeling for real-time energy-efficient station grouping in IEEE 802.11ah. Pervasive Mobile Comput. **57**, 33–48 (2019)
17. Deb, K., Pratap, A., Agarwal, S., Meyarivan, T.: A fast and elitist multi-objective genetic algorithm: NSGA-II. IEEE Trans. Evol. Comput. **6**(2), 182–197 (2002)
18. Ardakan, M.A., Rezvan, M.T.: Multi-objective optimization of reliability–redundancy allocation problem with cold-standby strategy using NSGA-II. Reliab. Eng. Syst. Saf. **172**, 225–238 (2018)

Chapter 21
Study on Vibration Characteristics of Metro Bogie with Permanent Magnet Synchronous Direct Drive Motor

Liqiang Yang, Xiaoping Jia, and Li Huang

Abstract In this paper, the vibration characteristics of a metro bogie directly driven by a permanent magnet synchronous motor are introduced. In a field test, by arranging acceleration sensors on the bogie and the car body to obtain the vibration signals of the various parts of the vehicle, the vibration of the motor and axle box during the operation of this type of bogie was acquired and analyzed, and the vibration transmission of the bogie was mastered in detail. Through the time/frequency domain analysis theory, the results demonstrate that the RMS of motor vibration acceleration increases with the increase of vehicle speed, and the main frequency of the motor is identical before and after the motor is removed. In addition, the traction rod seat has 6 characteristic frequencies, and each characteristic frequency exists at both root and end, while the energy amplitude at the end is larger. It can provide a certain technical reference for the application and further research of this bogie.

Keywords Permanent magnet direct drive motor · Bogie's vibration characteristics · Vibration transmission

21.1 Introduction

Compared with three-phase AC asynchronous motor, permanent magnet synchronous motor (PMSM) is more energy-saving, environment-friendly and has better performance [1], it can effectively reduce the volume and weight of the traction drive system, reduce the noise and improve the efficiency of the transmission system

---

L. Yang
Xuzhou Urban Rail Transit Co., Ltd, Xuzhou 221000, China

X. Jia
CRRC Nanjing Puzhen Co., Ltd, Nanjing 211800, China

L. Huang (✉)
State Key Laboratory of Traction Power, Southwest Jiaotong University, Chengdu 610036, China
e-mail: 837120525@qq.com

J. Xu and K. M. Pandey (eds.), *Mechanical Engineering and Materials*, Mechanisms and Machine Science 100,
https://doi.org/10.1007/978-3-030-68303-0_21

[2]. It has become a new development direction for rail transit [3]. Zhang Jiang [4] introduced the development of permanent magnet synchronous motor direct drive technology at home and abroad. By analyzing many advantages of this technology, it showed a broad prospect. The permanent magnet synchronous direct drive motor bogie has always been the research focus of domestic and foreign scholars. Feng Zunwei [5] studied a technical scheme of a flexible bogie directly driven by permanent magnet synchronous motor. By adjusting the structural parameters of the flexible frame beam, a lower Y-direction torsional angle stiffness can be obtained, and it is compared with the traditional B-type metro vehicle. Chu Yongping [6] studied a permanent magnet direct-drive bogie using "cross-plate" flexible beams, which can provide lower torsional stiffness and suitable diamond resistant rigidity. This has changed the characteristics of traditional bogies with primary and secondary suspensions that adapt to track distortion. Zhang Zhihe [7] introduced the structural and technical characteristics of the Locomotive Bogie with the permanent magnet motor and direct drive mode, focusing on the technical features and research contents of the key components using direct drive mode, the analysis shows that the Bogie using this technology has obvious advantages. Lu Zhenggang [8] proposed a new solution for directly driven wheel of disc permanent magnet synchronous motor (PMSM) customized for low floor tramcar, the simulation shows that its dynamic performance is excellent. Dong Shengmin [9] studied the suspension device of the motor of the permanent magnet direct-drive electric locomotive, and analyzed the strength of the motor suspension device using the finite element method, the calculation results meet the design requirements. According to GB/T 5599-2019 "Specification for dynamic performance assessment and testing verification of rolling stock" [10], a certain type of metro vehicle was tested for dynamic performance, and its bogie is shown in Fig. 21.1. Based on the results of the field test, this paper analyzes the dynamic performance of the motor and axle box before and after the motor is removed at different speeds. The vibration acceleration root mean square (RMS) value of the motor and the vibration frequency of the motor and the axle box are used to obtain

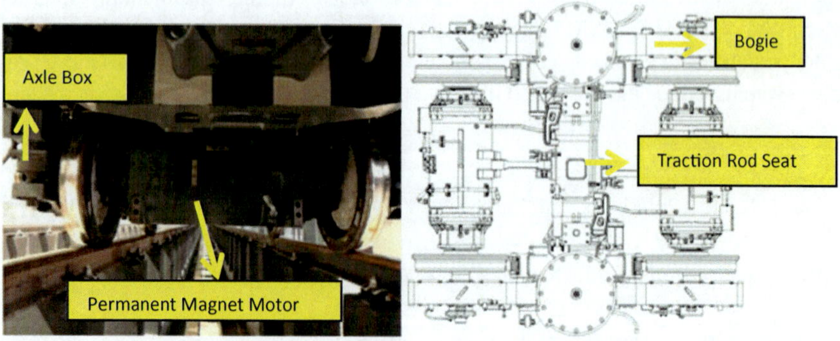

Fig. 21.1 Metro Bogie with permanent magnet synchronous direct drive motor

the vibration characteristics of the motor and the gear box; the vibration transmission characteristics of the axle box, frame, the bottom of car body, the passenger compartment floor and the root-end of traction rod seat are analyzed as well.

21.2 Vibration Characteristics of Motor

Motor vibration has an important impact on the stability of the bogie. Therefore, three-direction acceleration sensors were arranged on the motor of the motor-car bogie. The vibration characteristics of the motor under normal operation and motor removal at different speeds are analyzed and compared, so as to obtain the vibration characteristics of permanent magnet synchronous motor. No matter in normal or removal conditions, the vibration acceleration of the motor is within 5 g on the non-impact section, and the maximum motor acceleration is about 200 g on the impact section of the switch. For the comparison of the acceleration before and after the motor is removed, the RMS value of the acceleration is adopted, and the 1 s window is used, and the moving window is 1 s without overlap. The data of the turnout has been eliminated during the calculation process. The result is shown in Fig. 21.2. As

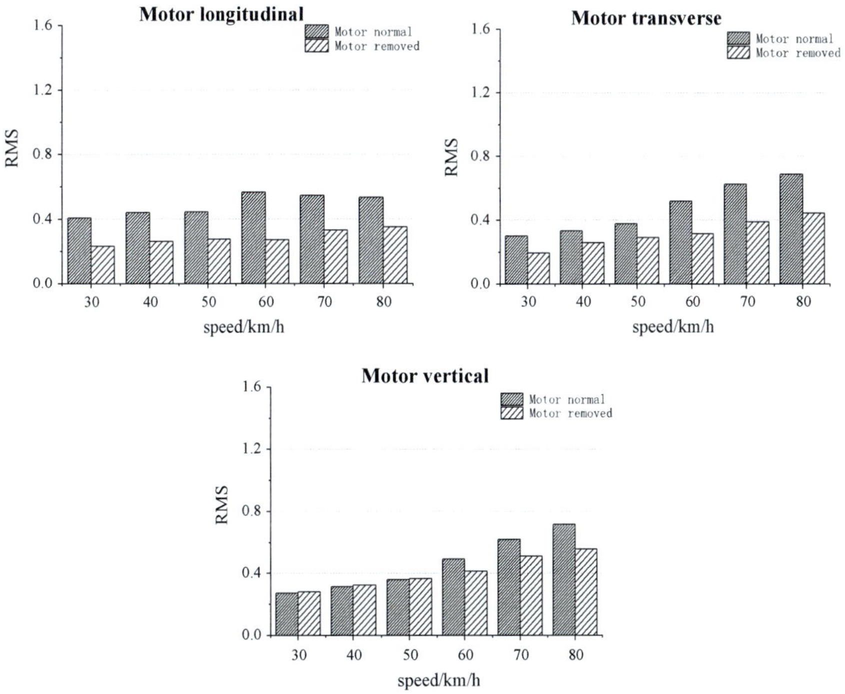

Fig. 21.2 Root mean square value of motor vibration

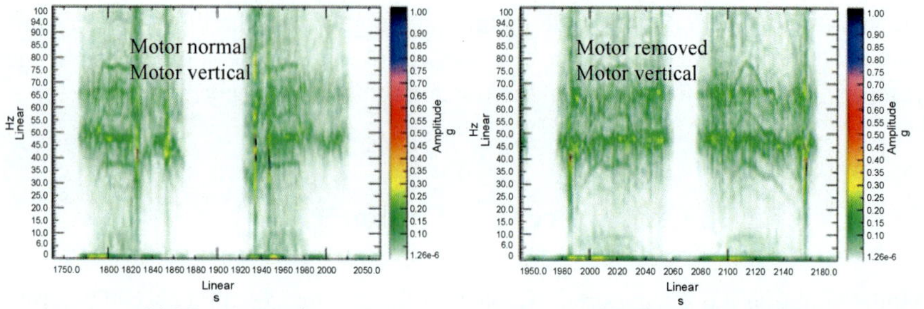

(a) Motor normal (b) Motor removed

Fig. 21.3 Time-frequency diagram of motor vertical vibration

the vehicle speed increases from 30 km/h to 80 km/h, the motor vibration acceleration RMS gradually increases, from low speed to high speed by about 2–4 times. After the motor is removed, the RMS value of the motor vibration acceleration decreased by about 20%.

For the frequency of motor vibration, the frequency within 100 Hz of the vertical time-frequency diagram of the motor is partially amplified, as shown in Fig. 21.3. It can be seen that before and after the removal of the motor, the main frequency of vibration remains unchanged, which is around 45 Hz and 65 Hz, and the main frequency of 60–70 Hz is the P2 force frequency of the motor-car (P2 force is the medium and low frequency force generated by the entire vehicle system and line system by pulse excitation, the frequency is 30 ~ 100 Hz.).

21.3 Analysis of Vibration Characteristics of Axle Box

Three-direction acceleration sensors were arranged at the axle box (as shown in Fig. 21.4), and the vehicle AW0 operating condition data was selected to analyze the vibration characteristics of the bogie axle box. AW0 means no load. Regardless of whether it is a trailer or a motor-car, the motor is normal or removed, the amplitude of the vibration acceleration of the axle box is within 5 g on the normal section, and the axle box acceleration of the switch impact section is up to 46 g. As the speed increases, the vibration amplitudes in the three directions of the axle box continue to increase. After the motor is removed, the RMS of the axle box vibration acceleration increases slightly, but the increase is small, with a maximum of about 20%.

For the vibration frequency of the axle box, the main frequency of the motor is the same before and after the motor is removed, but the energy that related to the main frequency is slightly different, as shown in Fig. 21.5. Partially enlarge the 80 km/h speed level, 0–700 Hz time-frequency diagram, as shown in Fig. 21.6. The axle box has a P2 force frequency of about 60–70 Hz, a natural frequency of about 125 Hz, and a natural frequency band of about 500 Hz. The axle box has a speed-related main

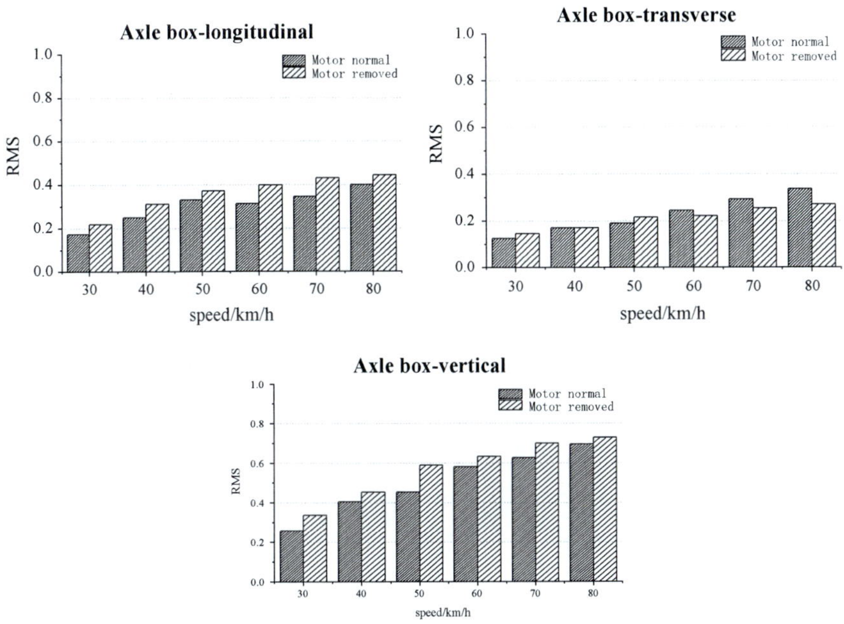

Fig. 21.4 Root mean square value of shaft box vibration before and after motor removal

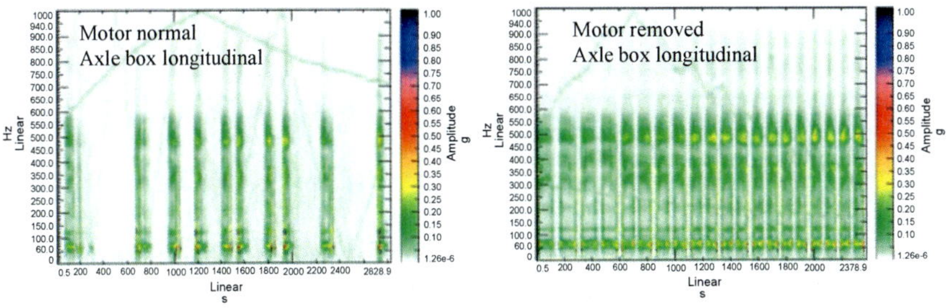

Fig. 21.5 Time-frequency diagram of longitudinal vibration of axle box between motor normal and motor removed

frequency of 150 Hz (speed 80 km/h, wavelength about 0.15 m). The trailer does not have such a main frequency, which may be related to the permanent magnet motor.

21.4 Characteristics of Vibration Transmission

The vibration transmission on the bogie is mainly analyzed in two parts: axle box-end of the frame-bottom of the car body-passenger compartment floor and root of

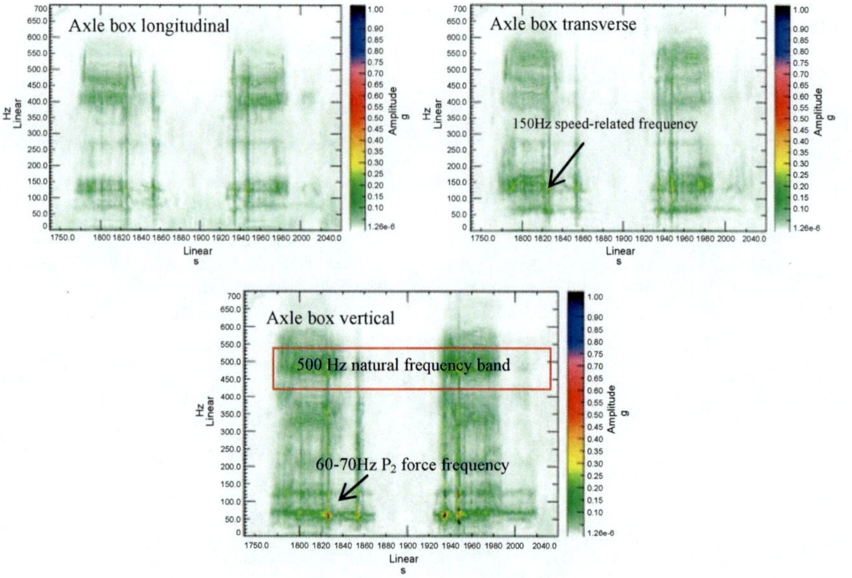

Fig. 21.6 Time-frequency diagram of three-dimensional vibration acceleration of axle box when motor is normal

the traction rod seat-end of the traction rod seat. Vibration transmission analysis is carried out by intercepting vibration data of axle box, frame and car body at 80 km/h under normal motor working conditions of AW0. It can be seen from Figs. 21.7 and 21.8 that the axle box has a P2 force frequency of 60–70 Hz, a natural frequency of 125 Hz, and natural frequency bands of 350 Hz and 500 Hz, of which, the main frequencies of 60–70 Hz and 125 Hz of the axle box are transmitted to the frame. In addition, the frame has natural frequency bands of 400 Hz and 800 Hz. The main frequency of the vehicle body is the suspension modal frequency within 3 Hz, and the vertical mode of the vehicle body is more obvious.

For the traction rod seat, the three-directional vibration acceleration at the root is about 0.5 g, 0.5 g and 1 g, and the three-directional vibration acceleration at the end is about 2 g, 1.5 g and 3.5 g, respectively. The maximum vibration acceleration at the root of the traction rod seat is about 4 g, the maximum vibration acceleration at the end of the traction rod seat is about 19 g. In normal motor and motor removal conditions, the vibration of the root-end of the traction rod seat is amplified, about 1.5-3.2 times. Compare and analyze the vibration magnification of the root-end of the traction rod seat, as shown in Fig. 21.9. In the normal working condition of the motor, 80 km/h speed level, from the RMS point of view, the longitudinal, transverse and vertical vibration magnifications are 2.69 times, 1.81 times and 3.00 times, respectively.

For vibration transmission, from the perspective of vibration frequency, the time-frequency diagram of the root and end of the traction rod seat under the speed of 80 km/h is shown in Fig. 21.10. There are 6 characteristic frequencies in the traction

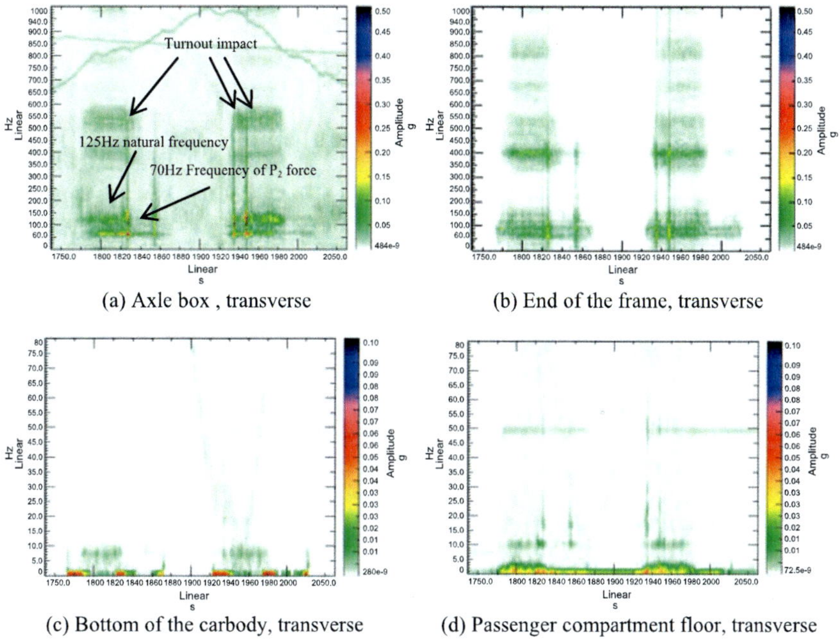

(a) Axle box , transverse

(b) End of the frame, transverse

(c) Bottom of the carbody, transverse

(d) Passenger compartment floor, transverse

Fig. 21.7 Transversal vibration transmission

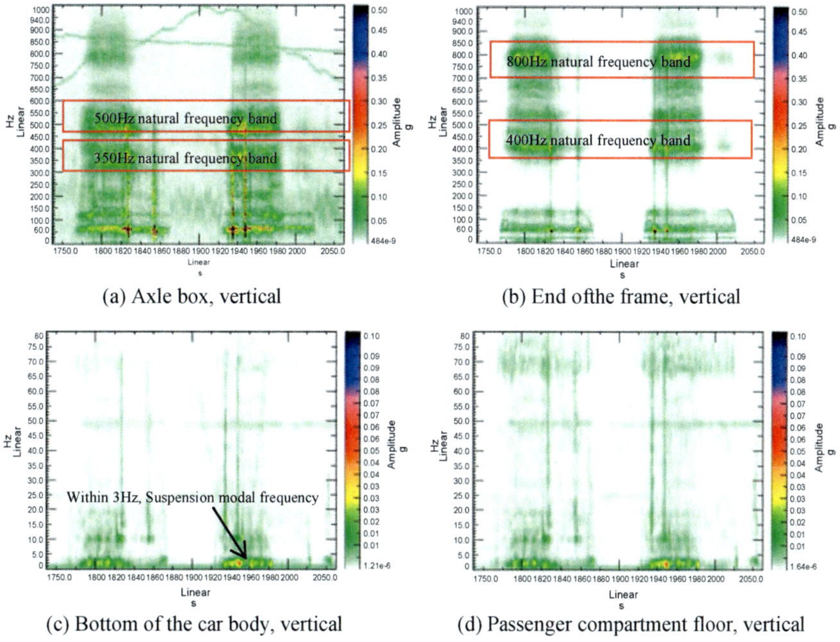

(a) Axle box, vertical

(b) End ofthe frame, vertical

(c) Bottom of the car body, vertical

(d) Passenger compartment floor, vertical

Fig. 21.8 Vertical vibration transmission

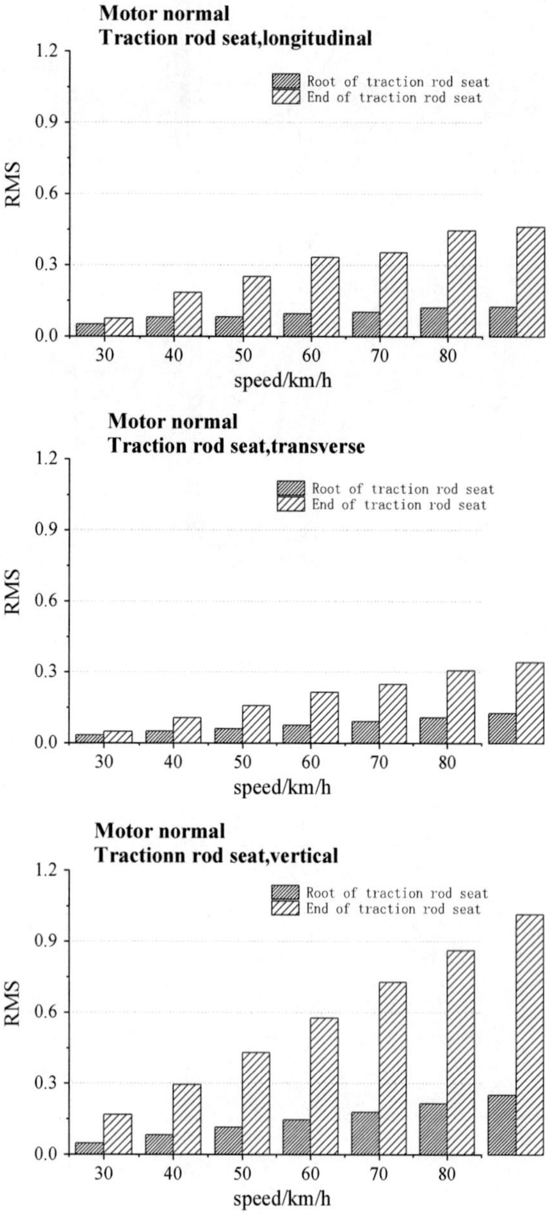

Fig. 21.9 Vibration transmission of traction rod seat (normal motor)

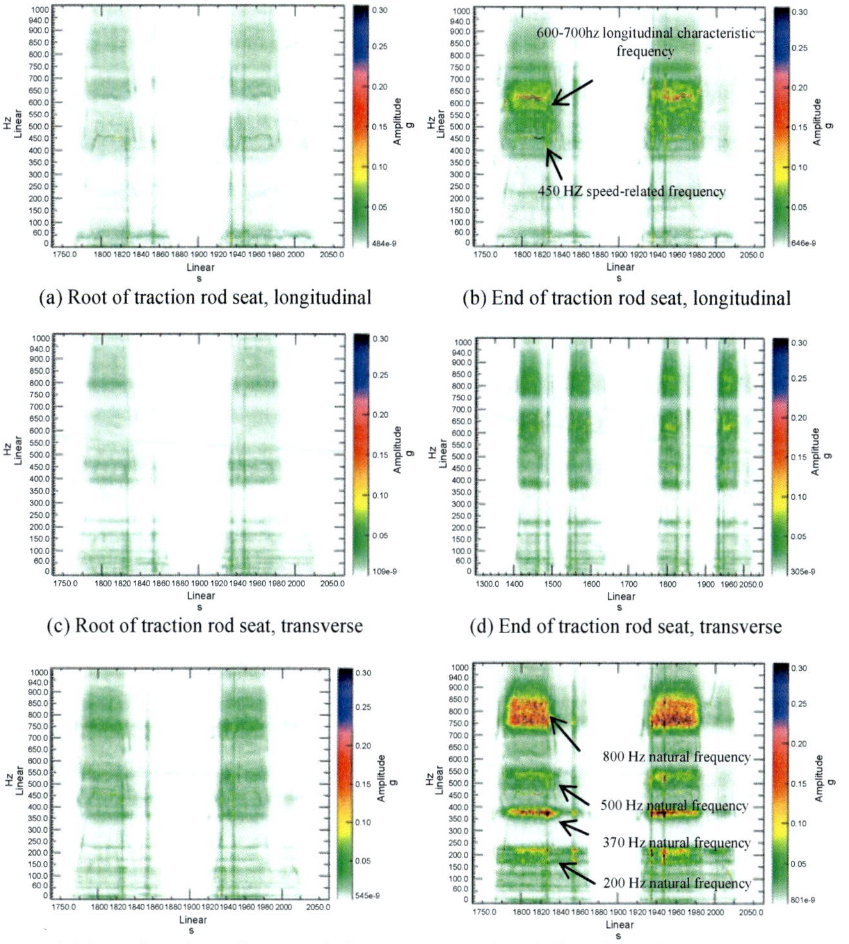

(a) Root of traction rod seat, longitudinal (b) End of traction rod seat, longitudinal

(c) Root of traction rod seat, transverse (d) End of traction rod seat, transverse

(e) Root of traction rod seat, vertical (f) End of traction rod seat, vertical

Fig. 21.10 Time-frequency diagram of traction rod seat root-end (Motor normal)

rod seat, among which, 200 Hz natural frequency, 370 Hz natural frequency, 500 Hz natural frequency and 800 Hz natural frequency are more obvious in the vertical direction of the end; for the speed-related frequency of about 450 Hz, this frequency is related to the torque output of the motor and the frequency disappears after the motor is removed; the 600–700 Hz frequency band is more obvious in the longitudinal direction of the end, both when the motor is normal or removed. Each characteristic frequency exists at both root and end, and the energy amplitude at the end is larger.

21.5 Conclusions

The vibration characteristics of a metro bogie directly driven by a permanent magnet synchronous motor are demonstrated through field test, which can be concluded as follows:

For motor, with the increase of vehicle speed, the RMS of motor vibration acceleration increases gradually, from low speed to high speed by about 2–4 times. The RMS value of vibration acceleration decreased by about 20% after the motor is removed. Before and after the motor is removed, the main frequency of vibration remains unchanged, both at 45 Hz and 65 Hz, of which 60–70 Hz is the P2 force frequency of the motor.

For axle box, as the speed increases, the vibration amplitudes in the three directions of the axle box continue to increase. After the motor is removed, the RMS of the axle box vibration acceleration increases slightly, but the increase is small, with a maximum of about 20%. For the vibration frequency of the axle box, the main frequency of the motor is the same before and after the motor is removed, but the energy that related to the main frequency is slightly different.

For the vibration transmission of the bogie, the 60–70 Hz and 125 Hz main frequency of the axle box are transmitted to the frame. In addition, the frame has 400 Hz and 800 Hz natural frequency bands. The main frequency of the vehicle body is the suspension modal frequency within 3 Hz, and the vertical mode of the vehicle body is more obvious. In both normal motor and motor removal conditions, the vibration of the root-end of the traction rod seat is amplified. The traction rod seat has 6 characteristic frequencies, each characteristic frequency exists at both root and end, and the energy amplitude at the end is larger.

References

1. Yuan, Z.Q., Nie, M.: Development of high-power permanent magnet direct drive passenger locomotive. Electr. Drive Locomot. (01), 40-44 + 50 (2019)
2. Chen, G.S., Xiao, T., Chen, X.H., Gong, X.B., Lu, Z.G., Wang, W.B.: Some key technology problems for independent-wheel and hub direct driving bogie with permanent magnet motor. Electr. Drive Locomot. **05**, 4–7 (2016)
3. Liu, Q.A., Zhang, J., Zhong, W.S.: The influence of balancing link tilt angle on motor motion and primary load of bogies with permanent magnet driven directly. Electr. Drive Locomot. **02**, 37–39 (2015)
4. Zhang, J., Zhong, W.S., Liu, G.K.: General description of the development of the direct-drive technology of permanent magnet synchronous motor of bogie at home and abroad. Railw. Locomot. Car **34**(03), 79–83 (2014)
5. Feng, Z.W., Hu, D.X., Chu, Y.P.: Research on dynamic performance of direct-driven flexible bogie for permanent-magnetic synchronous motor. Railw. Locomot. Car **33**(03), 73–76 (2013)
6. Chu, Y.P., Hu, D.X., Zhou, L.: Design and analysis of a new-type metro bogie with permanent magnet driving motors. Urban Mass Transit 19(06), 17-21 + 27 (2016)
7. Zhang, Z.H., Li, H.X., Li, Q., Zhao, Q.H.: The locomotive bogie with permanent-magnet motor and direct driving pattern. Railw. Locomot. Car **40**(03), 39–42 (2020)

8. Lu, Z.G., Sun, X.J., Zhang, J.: Design and control of disc PMSM directly driven wheel for tramcar. Adv. Mech. Eng. **6** (2014)
9. Dong, S.M., Zhang, G.X., Wang, Y.: Calculation of suspension device and bolt strength of permanent magnet direct drive electric locomotive. Electr. Locomot. Mass Transit Veh. **43**(02), 35–39 (2020)
10. GB/T 5599-2019, Specification for dynamic performance assessment and testing verification of rolling stock. China (2019)

Chapter 22
Influence Study of the Bird Shape on the Birdstrike Response for Flat Blade

Xiamei Zhang and Qiang Zhang

Abstract With the development of the aviation industry, the number of civil and cargo aircrafts is increasing. However, birdstrike is a major threat to the aircrafts. In order to conduct the influence study of the bird shape on the birdstrike response for flat blade, four bird shape models, i.e. cylinder, ellipsoid, sphere and cylinder with hemispherical ends are established. Through the birdstrike simulations of bird with different shapes on the flat blades, the maximum displacement of the flat blade, the peak of the impact contact force and the energy absorption of the flat blade are obtained. The results show that the spherical bird causes the most serious damage to the flat blade, and the bird with the shape of cylinder with hemispherical ends is weaker than spherical bird. Cylindrical bird and ellipsoidal bird are comparable, but both are weaker than the former two. Therefore, a spherical bird should be adopted in the numerical birdstrike study of the engine blade to ensure the safety of the engine.

Keywords Birdstrike · Blade · Bird shape

22.1 Introduction

The number of civil and cargo aircrafts is greatly increasing with the vigorous development of the aviation industry. Accidents related to aviation also occur more frequently because birdstrike is one of the important causes of aircraft failures and accidents. On August 15, 2019, an Airbus A321 of Russia hit a group of seagulls during take-off, resulting in damage to both engines and a huge landing risk. The aircraft may lose its power instantly when the birdstrike happens, which is very substantial danger. For this reason, this paper conducts a numerical simulation study on the birdstrike problem of the engine blades.

Bird body is a key factor in the birdstrike research that affects the impact response. FAR 25.571 requires that the aircraft engine should be controlled to shut down without a fire under the impact of a 1.8 kg bird body, but the shape of bird body is

X. Zhang (✉) · Q. Zhang
Power-Plant Institute, Chinese Flight Establishment, Xi'an 710072, China
e-mail: 710021280@qq.com

J. Xu and K. M. Pandey (eds.), *Mechanical Engineering and Materials*, Mechanisms and Machine Science 100,
https://doi.org/10.1007/978-3-030-68303-0_22

not specified. The national military standard GJB 2464-95 stipulates that the shape of bird body should adopt a cylinder with a length-to-diameter ratio of 2:1. Therefore, some domestic researchers used a cylindrical bird body for birdstrike research. Yongkang and Yulong [1] used the cylindrical bird body to study the bird modelling method under different impact speeds. Guida et al. [2] analysed the anti-birdstrike performance of GLARE plates with different layers using a cylindrical bird body. According to Heimbs's review of birdstrike research [3], the bird shape of cylinders with hemispheres at both ends was most used. Hu et al. [4] used this bird shape to study the response of birdstrike of helicopter, which was consistent with the experiments. In addition, ellipsoidal bird body is also widely used. Diamantakos et al. [5] used ellipsoidal bird body in the airworthiness certification of composite leading edge to establish a set of experimentally validated numerical simulation tools for birdstrike. Georgiadis et al. [6] also used ellipsoid bird body in the airworthiness certification of the Boeing 787 tail structure. On the other hand, the sphere-shaped bird body has few applications. Zammit [7] used the sphere-shaped bird in the birdstrike study of metal blade and composite blade.

Shuhua and Mingbo [8] found that the bird shape is one of the reasons of mismatch between the simulation and test. In their birdstrike study on windshield, a cylindrical bird with hemispheres at both ends was closest to the test. At the same time, there were other research results showing that the cylinder is more in line with the real bird [8, 9]. Different kinds of birds have different density, bone hardness, weight distribution, as well as different impact speeds. Therefore, different studies drew different conclusions. The actual birdstrike situation may be more complicated. There are statistical differences in birds in different natural environments [10], and there are obvious differences in the impact response for different bird parts [11]. Therefore, it is difficult to get a unified conclusion even through experiments. It also makes it less meaningful to find the shape of the bird closest to the experiments.

The purpose of the birdstrike test is to ensure the safety of the aircraft, that is, the aircraft can safely return under the most severe birdstrike damage. Therefore, the aim converts to determine the shape of bird body that causes the most serious damage to the engine. In this paper, four shapes of bird body are considered and their models are established. The response of the four birds impacting the engine blade are investigated by numerical methods, and the impact risks are analyzed to determine the bird shape that can cause the most severe damage at the same mass.

22.2 Birdstrike Model

22.2.1 Bird Modeling

There are four commonly used bird shapes: cylinder, ellipsoid, sphere, and cylinder with hemispheres at both ends, as shown in Fig. 22.1. The lengths of cylinders, ellipsoids, and cylinders with hemispherical ends are all twice their diameters.

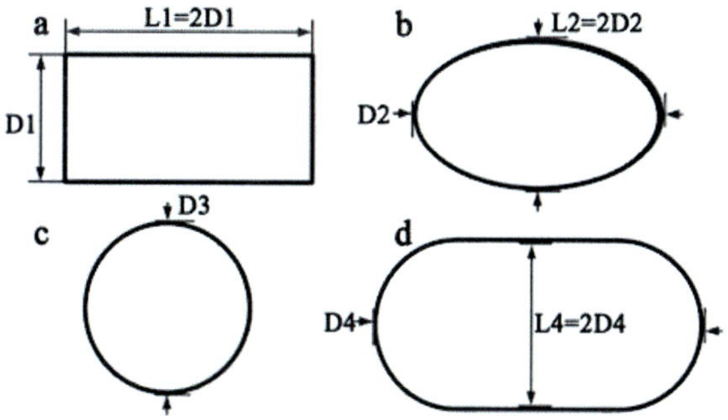

Fig. 22.1 Common bird shape, **a** cylinder shape; **b** ellipsoid shape; **c** spherical shape; **d** cylindrical with hemispheres at both ends

FAR 25.571 stipulates that the weight of the bird impacts on the engine should be 1.8 kg. Therefore, the four bird sizes can be calculated based on their shapes, as seen in Table 22.1. Table 22.1 indicates that the spherical bird has the largest diameter, while the ellipsoidal bird has the largest length.

Four algorithms are mostly used in bird simulation, namely Lagrange algorithm, Euler algorithm, ALE (Arbitrary Lagrange-Euler) algorithm and SPH (Smoothed Particle Hydrodynamics) algorithm. The Lagrange algorithm is simple to calculate, but its grid will be severely distorted under high-speed impact, and therefore the calculation accuracy and efficiency will be significantly reduced. Euler algorithm can ensure a regular grid, but the modeling is complex, and energy and quality leakage will occur, causing inaccuracy. ALE algorithm combines the advantages of Lagrange algorithm and Euler algorithm, but also retains the shortcomings of both. The SPH algorithm is stable in calculation and there is no grid distortion. SPH algorithm can simulate fluid sputtering easily, but it cannot show the stretching effect between particles. Liu et al. [12] used a high-speed camera to study the response of bird at different impact speeds and got conclusions. When the impact speed was about 70 m/s, the bird body became massive fragments after impact, indicating that a solid material model should be used at low speed. When the impact velocity was 120 m/s, the bird body became smaller fragments, indicating that a mixed model of solid and fluid materials should be used for simulation at a medium impact velocity. When the impact velocity was increased to 170 m/s, the bird body almost sputtered in a fluid state, indicating that a fluid material model should be used under high-speed impact. In this paper, the birdstrike speed is 800 km/h, which is the cruising speed of a general transportation aircraft. This speed has far exceeded 170 m/s, so

Table 22.1 Sizes and shapes of the four birds

Bird shape	Diameter [mm]	Length[mm]
Cylinder	106.449	212.899
Ellipsoids	121.854	243.708
Sphere	153.527	153.527
Cylindrical with hemispheres at both ends	113.119	226.239

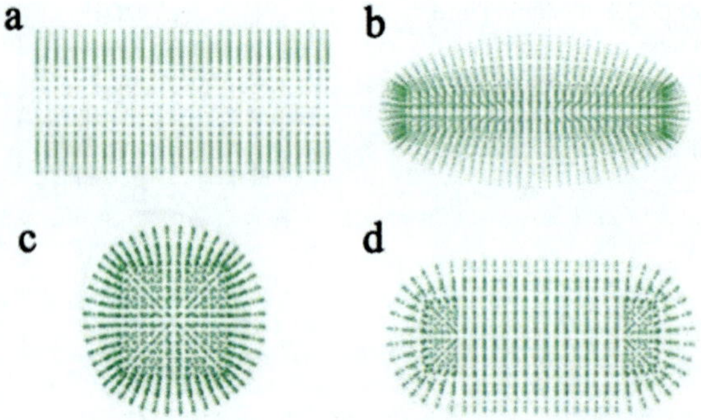

Fig. 22.2 Bird models, **a** cylinder shape; **b** ellipsoid shape; **c** spherical shape; **d** cylindrical with hemispheres at both ends

the fluid material model should be used. SPH uses discrete particles to simulate fluid behavior, which can well simulate the response of a bird's high-speed impact, so it is widely used. The constitutive model of the bird is described by the equation of state. PAMCRASH uses the Murnaghan equation of state [13], as shown as Eq. (22.1).

$$p = p_0 + B((\frac{\rho}{\rho_0})^\gamma - 1) \tag{22.1}$$

where p_0 is the initial pressure of the bird body/Pa; B is the material constant/Pa; γ is the material constant.

The settings applicable to general bird are B = 127 MPa and $\gamma = 7.89$ [14, 15]. Generally, the density of a homogeneous bird is 900–950 kg/m^3, and this paper uses 950 kg/m^3. The four bird models with the sizes shown in Table 22.1 are established using hexahedral elements. The bird mesh grid is about 7 mm, which is determined through mesh sensitivity analysis. Then transform meshes into SPH particles, as shown in Fig. 22.2. The physical properties of the bird body, such as mass, speed, pressure, etc., are reflected by the interrelationship and mechanical behaviours of SPH particles. Since there is no stretching between the particles, it can deform arbitrarily under high-speed impact.

22.2.2 Engine Blade and Boundary Conditions

The research in this paper is not specific to a certain type of engine blade, so a general aluminium flat blade is used to study the general birdstrike response. The size of the flat blade is a square plate of 1000 × 1000 mm and a thickness of 5 mm. It is simulated by hexahedral solid elements. The mesh size is 5 mm. The flat plate is divided into three layers along thickness direction to ensure that the solid element can accurately describe the bending deformation. The flat blade material is Al7075-T651, and the material parameters are shown in Table 22.2.

Table 22.2 Material parameters of Al7075-T651

Density [kg/m^3]	Young's modulus [MPa]	Poisson's ratio	Yield strength [MPa]	Ultimate strength [MPa]	Elongation [%]
2770	71,000	0.33	527	575	11

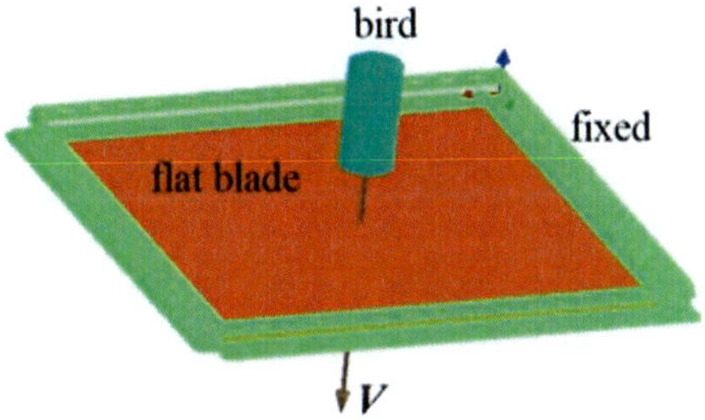

Fig. 22.3 Schematic diagram of the cylindrical bird impacting flat blade model

The boundary elements of the flat blade are restricted by the translational freedom. The bird is hitting the center of the flat at a speed of 223 m/s, and the master–slave contact is set between the bird and the flat blade. The schematic diagram of the birdstrike model is shown in Fig. 22.3 (taking the cylindrical bird hitting a flat blade as an example.

22.3 Dynamic Response and Simulation Results of Birdstrike

The PAM-SHOCK solver of PAMCRASH is used to solve the birdstrike problem with large transient deformation. PAMCRASH is a collision simulation software package launched by ESI Group, which is widely used in the field of automobile collision. Among them, PAM-SHOCK is more suitable for high-speed impact environments for its powerful converge in large deformation calculations.

The stress contours of the four shapes of birds striking the flat blade are similar. Here, the cylindrical bird with hemispheres at both ends is taken as an example, as shown in Fig. 22.4. The high-stress circular area in the center is the impact area, and the stress contour perfectly shows the process of stress wave propagation from the central area to the surroundings. The flat blade has reached the ultimate strength under the impact of the four shapes of bird, so it is impossible to evaluate the impact damage level of different shapes of birds through the peak stress.

Figure 22.5 shows the impact process of birdstrike. Since the four shapes of bird have similar birdstrike processes, here we take a cylindrical bird with hemispheres at both ends as an example. Some particles of the bird hit the flat blade at 5 ms, and most of the bird particles hit the blade at 10 ms. By 15 ms, all the bird particles have

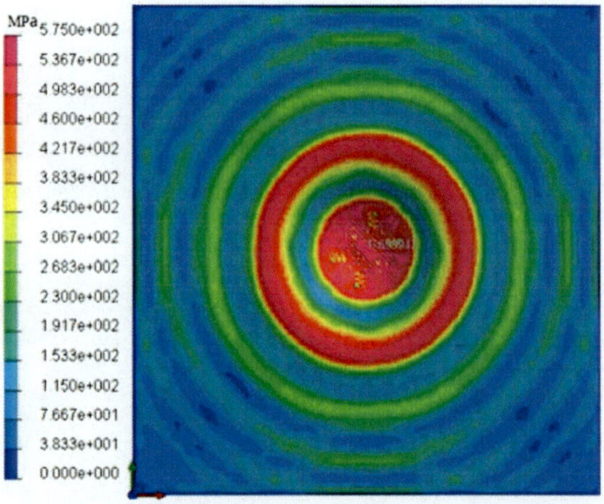

Fig. 22.4 Stress contour of the cylindrical bird with hemispheres at both ends hitting a flat blade

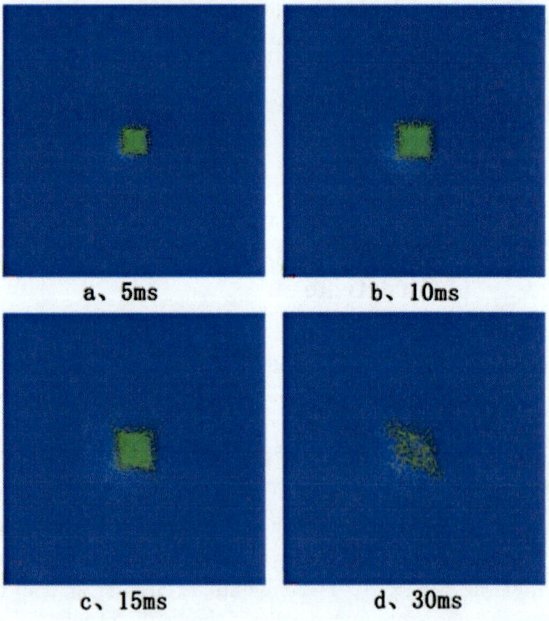

Fig. 22.5 The deformation process of the cylindrical bird with hemispheres at both ends

completed the impact, and sputtered around. Diffusion is particularly obvious in the bird deformation diagram at 30 ms.

Figure 22.6 shows the contact force history curve during the birdstrike through the master-slave contact between the bird and the flat blade. It indicates that the contact force of the spherical bird reaches its peak at 9 ms, which is significantly earlier than the other three shapes of bird. The peak contact force produced by the spherical bird is the largest, and it is significantly higher than other birds, reaching

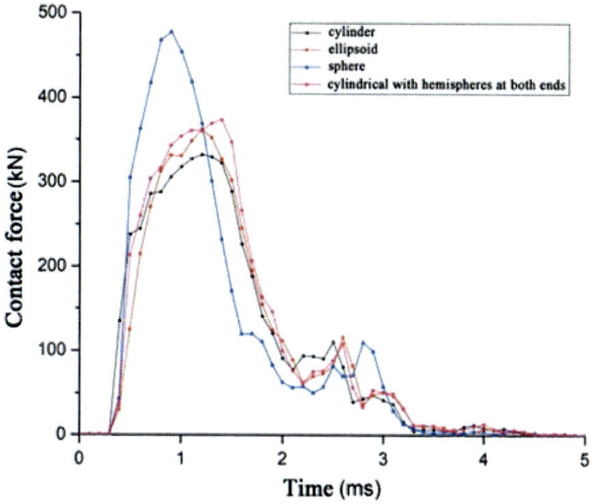

Fig. 22.6 Contact force history curves of different shapes of bird impacting on flat blade

477.44 kN. The spherical bird has the shortest length, so it can transmit all kinetic energy to the flat blade in the shortest time, which is the main reason for its maximum contact force peak. The contact force curves of the other three shapes of birds are not much different, and the contact force generated by the cylindrical bird is the smallest, which is 327.08 kN.

A specified node is set at the center point on the back of the flat blade, which represents the maximum displacement of the blade. Figure 22.7 shows the displacement history curves of the node under the birdstrike of the four birds. It indicates that the blade deformation caused by the spherical bird is the largest, reaching 81.09 mm, and the blade deformation caused by the ellipsoidal bird is the smallest, which is 69.88 mm.

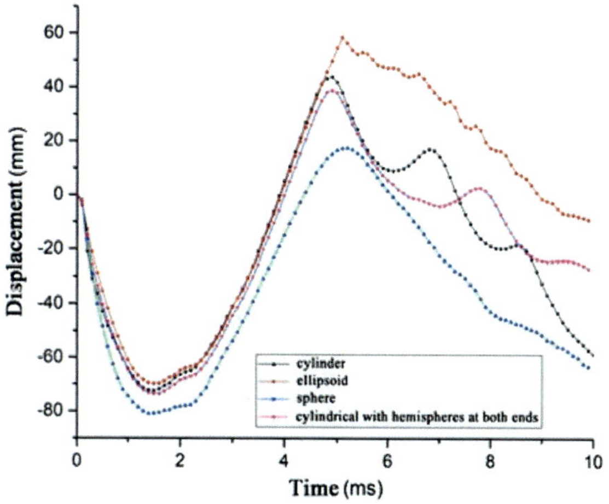

Fig. 22.7 Displacement history curves of different shapes of bird impacting on flat blade

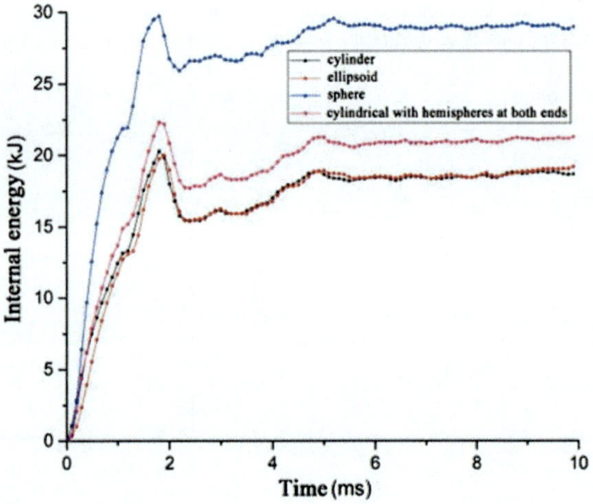

Fig. 22.8 Internal energy history curves of different shapes of bird impacting on flat blade

The damage of the flat blade can also be evaluated by the internal energy absorbed in birdstrike. Figure 22.8 shows the internal energy history curves of the flat blade. The flat blade absorbs largest energy under the impact of the spherical bird, reaching 29.72 kJ. The energy absorption of the flat plate caused by ellipsoidal bird is the smallest.

Table 22.3 shows the birdstrike results of the four shape birds. The spherical bird is the most destructive in terms of the maximum displacement of the flat plate, the peak contact force and the energy absorption of the flat blade. The destructive effect of the cylindrical bird with hemispheres at both ends ranks second. Cylindrical bird and ellipsoidal bird have different ranks in different evaluation indexes. It can be seen from Figs. 22.6, 22.7 and 22.8 that the contact force, energy absorption and deformation of the birdstrike caused by the spherical bird occur earliest. This is because the spherical bird has the smallest aspect ratio, so the kinetic energy of bird can act on the flat blade in a shortest time. This is why the spherical bird has the strongest destructive power.

Table 22.3 Impact response of four birdstrike models

Bird shape	Maximum displacement [mm]	Peak contact force [kN]	Maximum energy absorption[J]
Cylinder	72.50	327.08	20.27
Ellipsoids	69.88	361.08	19.98
Sphere	81.09	477.44	29.72
Cylindrical with hemispheres at both ends	73.62	373.73	22.33

22.4 Conclusions

In this paper, four shapes of bird models are established, which are cylinder, ellipsoid, sphere and cylinder with hemispheres at both ends. For the first time, the influence of bird shape on the birdstrike response of flat blades is fully discussed. The maximum displacement of the flat blade, the peak contact force and the peak energy absorption of the flat blade are used to evaluate the impact damage level for flat blade in the birdstrike. The simulation results show that spherical bird causes the most serious damage under the premise of the same mass and same velocity. Therefore, in the numerical study of birdstrike for engine, the use of spherical bird can make the research results more conservative and ensure the safety of the engine structure.

References

1. Zhang, Y.K., Li, Y.L.: Numerical study on bird mechanical model computer simulation. Comput. Simulat. **28**(9), 67–70 (2011). (in Chinese)
2. Guida, M., Marulo, F., Polito, T., Meo, M., Riccio, M.: Design and testing of a fiber-metal-laminate bird-strike-resistant leading edge. J. Aircr. **46**(6), 2121–2129 (2009)
3. Heimbs, S.: Computational methods for bird strike simulations: a review. Comput. Struct. **89**(23–24), 2093–2112 (2011)
4. Hu, D., Song, B., Wang, D., Chen, Z.: Experiment and numerical simulation of a full-scale helicopter composite cockpit structure subject to a bird strike. Compos. Struct. **149**, 385–397 (2016)
5. Diamantakos, I., Fotopoulos, K., Jamin, M., Eberhardt, A., Lampeas, G.: Investigation of bird strike events on composite wing panels. Fatigue Fract. Eng. Mater. Struct. **40**(10), 1538–1550 (2017)
6. Georgiadis, S., A. Gunnion, J., Thomson, R.S., Cartwright, B.K.: Bird-strike simulation for certification of the Boeing 787 composite moveable trailing edge. Compos. Struct. **86**(1–3), 258–268 (2008)
7. Zammit, A., Kim, M., Bayandor, J.: Bird-strike damage tolerance analysis of composite turbofan engines. In: 27th Congress of the International Council of the Aeronautical Sciences, Vol. 4, pp. 2824–2840, ICAS (2010)
8. Zhu, S.H., Tong, M.B.: Bird shape sensitivity to dynamic response of bird strike on aircraft windshield. J. Nanjing Univ. Aeronaut. Astronaut. **40**(4), 551–555 (2008). (in Chinese)
9. Kari, S., Gabrys, J., Lincks, D.: Birdstrike Analysis of radome and wing leading edge using LS-DYNA. In: 5th International LS-DYNA Users Conference 414, 87–91 (1998)
10. Zhu, B.B., Cai, J.: Study on impact energy of bird strike of transport aircraft based on MCMC method. Aeronaut. Comput. Techniq. **47**, 94–96,100 (2017). (in Chinese)
11. Zhang, D.H., Fei, Q.G., Liu, H.Y.: Effects of bird's striking orientation on bird impact analysis based on a realistic bird model. J. Vibrat. Shock **34**, 103–108 (2015). (in Chinese)
12. Liu, J., Li, Y., Guo, W., Shi, X., Peng, G., Ye, B.: Parameters inversion on bird constitutive model Part I: Study on experiment of bird striking on plate. Acta Aeronautica et Astronautica Sinica **32**(5), 802–811 (2011). (in Chinese)
13. Mccarthy, M. A., Xiao, J. Petrinic, R., N., Kamoulakos, A., Melito, V.: Modelling of bird strike on an aircraft wing leading edge made from fibre metal laminates—Part 1: Material modelling. Appl. Compos. Mater. **11**, 295–315 (2004)
14. Liu, J., Li, Y., Shi, X., Liu, Y.: Parameters inversion on bird constitutive model Part II: Study on model parameters inversion. Acta Aeronautica et Astronautica Sinica **32**(5), 812–821 (2011). (in Chinese)

15. McCarthy, M.A., Xiao, J.R., McCarthy, C.T., Kamoulakos, A., Ramos, J., Gallard, J.P., Melito, V.: Modelling of bird strike on an aircraft wing leading edge made from fibre metal laminates—Part 2: Modelling of impact with SPH bird model. Appl. Compos. Mater. **11**(5), 317–340 (2004)

Chapter 23
The Application of Time–Domain DQM to the Dynamically Forced Vibration of Simply Supported Plates

Fan Lin, Jianshe Peng, Shifeng Xue, and Liu Yang

Abstract The time–domain DQM is for discretizing space domain of continuous functions as well as time domain of that. In this paper, a study of the forced vibration of simply supported plates subjected to a transverse uniform load is presented on the basis of the time–domain DQM. Applying DQM both in space domain and time domain, the partial differential governing equations and initial–boundary conditions are discretized into linear algebraic equations which are solved easily to obtain the transverse deflection of simply supported plates subjected to a transverse load. The numerical results of two examples show that these numerical results of time–domain DQM are in good agreement with analytical solution and that the accuracy and efficiency of time–domain DQM are excellent under certain number of nodes. It is simple and easy for the mathematical principle of space–time DQM to implement in program, so it has a bright application prospect in the dynamic analysis field and is worthy of a further study.

Keywords Time–domain DQM · Forced vibration · Plate · Adj.R2

23.1 Introduction

The differential quadrature method (DQM) which was first developed by Bellman and Cacti in 1971 [1] has been extensively used for discretizing differential equations into algebraic ones. Compared to conventional numerical methods, DQM is relatively simple and has less computational effort in terms of achieving excellent accuracy of numerical solutions.

The time–domain DQM is for discretizing space domain of continuous functions as well as time domain of that and is utilized successfully for gaining the

F. Lin · S. Xue (✉)
College of Pipeline and Civil Engineering, China University of Petroleum (East China), Qingdao 266580, China
e-mail: sfeng@upc.edu.cn

J. Peng · L. Yang
School of Mechanical Engineering, Cheng Du University, Cheng Du 610106, China

J. Xu and K. M. Pandey (eds.), *Mechanical Engineering and Materials*, Mechanisms and Machine Science 100,
https://doi.org/10.1007/978-3-030-68303-0_23

numerical solutions of vibration analysis of structures. Kitipornchai et al. [2] used DQM to discretize space domain of partial differential motion equations for studying nonlinear vibration of edge cracked functionally graded Timoshenko beams. Similarly, Ke et al. [3] observed nonlinear free vibration of embedded double–walled carbon nanotubes based on nonlocal Timoshenko beam theory by the same way. Fung [4] performed DQM to discretize time domain of differential motion equations to solve first–order initial value problems. Furthermore, Peng et al. [5] applied the time–domain DQM to survey forced vibration analysis of beams. For the mechanical analysis of plate, Wang [6] used DQM to discretize two–dimensional space domain of partial differential motion equations for studying linear static analysis, elastoplastic buckling analysis, and linear buckling analysis of thin plates, ARANI and JAFARI [7] for Nonlinear vibration analysis of laminated composite Mindlin micro/nano–plates resting on orthotropic Pasternak medium, and Safarpour et al. [8] for static and free vibration analysis of graphene platelets. For the statically forced and free vibration of plates, however, there are DQM to only discretize two–dimensional space domain of partial differential motion equations and for the dynamically forced vibration no literature available on basis of the time–domain DQM.

In this paper, a study of the forced vibration of simply supported plates subjected to a transverse uniform dynamic load is presented on the basis of the time–domain DQM. Applying DQM both in space domain and time domain, the partial differential governing equations and initial–boundary conditions are discretized into linear algebraic equations for studying the forced vibration of simply supported plates subjected to a transverse uniform load.

23.2 Solution Technique

Equation (23.1) defines the governing equation [9] of simply supported plates subjected to a transverse uniform load q:

$$\nabla^2\nabla^2 w + \frac{\bar{m}}{D}\frac{\partial^2 w}{\partial t^2} = \frac{q}{D} \tag{23.1}$$

where $\nabla^2\nabla^2$ is double Laplacian Operator, w transverse deflection, $\bar{m}$ mass per unit volume, D bending stiffness, t time, and some of them are defined by Eqs. (23.2a–d):

$$\nabla^2\nabla^2 = \frac{\partial^4}{\partial x^4} + 2\frac{\partial^4}{\partial x^2 \partial y^2} + \frac{\partial^4}{\partial y^4},\ w = w(x,\ y,\ t),\ D = \frac{E\delta^3}{12(1-\mu^2)},$$
$$q = q(x, y, t) \quad \text{(23.2a–d)}$$

in which E is elastic modulus, δ thickness, and μ Poisson's ratio.

For simply supported plates, the boundary conditions are described by Eqs. (23.3a–d):

$$w = 0, \quad \frac{\partial^2 w}{\partial x^2} = 0 \text{ at } x = 0 \text{ and } L; \; w = 0, \quad \frac{\partial^2 w}{\partial y^2} = 0, \; y = 0 \text{ and } B \tag{23.3a–d}$$

where L and B are length and width, respectively.

On the other hand, the initial conditions are given by Eqs. (23.4a, b):

$$w = w_0, \quad \frac{\partial w}{\partial t} = v_0 \text{ when } t = 0 \tag{23.4a, b}$$

where w_0 and v_0 are initial deflection and velocity, respectively.

Setting dimensionless quantities $W = \frac{w}{\delta}$, $\eta = \frac{x}{L}$, $\zeta = \frac{y}{B}$, and $\tau = \frac{t}{\Gamma}$, Eq. (23.1) can be represented as Eq. (23.5):

$$\frac{\delta}{L^4}\frac{\partial^4 W}{\partial \eta^4} + 2\frac{\delta}{L^2 B^2}\frac{\partial^4 W}{\partial \eta^2 \partial \zeta^2} + \frac{\delta}{B^4}\frac{\partial^4 W}{\partial \zeta^4} + \frac{\bar{m}\delta}{D\Gamma^2}\frac{\partial^2 W}{\partial \tau^2} = \frac{q(\eta, \zeta, \tau)}{D} \tag{23.5}$$

and the boundary conditions are given in Eqs. (23.6a–d):

$$W = 0, \quad \frac{\partial^2 W}{\partial \eta^2} = 0 \text{ at } \eta = 0 \text{ and } 1; \; W = 0, \quad \frac{\partial^2 W}{\partial \zeta^2} = 0 \text{ at } \zeta = 0 \text{ and } 1 \tag{23.6a–d}$$

and the initial conditions as Eqs. (23.7a,b):

$$W = \frac{w_0}{\delta}, \quad \frac{\partial W}{\partial \tau} = \frac{v_0}{\delta} \text{ when } \tau = 0 \tag{23.7a,b}$$

Based on the discretization rules of DQM [10], the dimensionless transverse displacement field W and its *rth* partial derivatives can be approximated as Eqs. (23.8a–e):

$$W = \sum_{m=1}^{N_\eta}\sum_{n=1}^{N_\zeta}\sum_{p=1}^{N_\tau} \gamma_m(\eta)\chi_n(\zeta)\psi_p(\tau)W(\eta_m, \zeta_n, \tau_p),$$

$$\left.\frac{\partial^r W}{\partial \eta^r}\right|_{\substack{\eta=\eta_i\\ \zeta=\zeta_j\\ \tau=\tau_k}} = \sum_{m=1}^{N_\eta} C_{im}^{(r)} W(\eta_m, \zeta_j, \tau_k),$$

$$\left.\frac{\partial^r W}{\partial \zeta^r}\right|_{\substack{\eta=\eta_i\\ \zeta=\zeta_j\\ \tau=\tau_k}} = \sum_{n=1}^{N_\zeta} \bar{C}_{jn}^{(r)} W(\eta_i, \zeta_n, \tau_k), \quad \left.\frac{\partial^r W}{\partial \tau^r}\right|_{\substack{\eta=\eta_i\\ \zeta=\zeta_j\\ \tau=\tau_k}} = \sum_{p=1}^{N\tau} \underline{C}_{kp}^{(r)} W(\eta_i, \zeta_j, \tau_p),$$

$$\left.\frac{\partial^{r+s} W}{\partial \eta^r \partial \zeta^s}\right|_{\substack{\eta=\eta_i\\ \zeta=\zeta_j\\ \tau=\tau_k}} = \sum_{m=1}^{N_\eta}\sum_{n=1}^{N_\zeta} C_{im}^{(r)} \bar{C}_{jn}^{(s)} W(\eta_m, \zeta_n, \tau_k) \tag{23.8a–e}$$

where N_η, N_ζ, and N_τ are the number of nodes distributed along the x, y and t directions, respectively; $\gamma_m(\eta)$, $\chi_n(\zeta)$, and $\psi_p(\tau)$ are the Lagrange interpolation polynomials; $C^{(r)}_{im}$, $\bar{C}^{(r)}_{jn}$, and $\underset{-}{C}^{(r)}_{kp}$ are the weighting coefficients. The sampling points are obtained by using cosine pattern as Eq. (23.9):

$$\alpha_i = \frac{1}{2}\left\{1 - \cos\left[\frac{\pi(i-1)}{N-1}\right]\right\},\ i = 1, 2, \ldots, \beta \tag{23.9}$$

where α_i is equal to η_i for $\beta = N_\eta$, ζ_i for $\beta = N_\zeta$, and τ_i for $\beta = N_\tau$.

Substituting Eqs. (23.8a–e) into Eqs. (23.5–23.7a,b), the discretized algebraic equations governing the linear forced vibration of plates are written as Eq. (23.10):

$$\begin{aligned} &\frac{\delta}{L^4}\sum_{m=1}^{N_\eta} C^{(4)}_{im} W(\eta_m, \zeta_j, \tau_k) + 2\frac{\delta}{L^2P^2}\sum_{m=1}^{N_\eta}\sum_{n=1}^{N_\zeta} C^{(2)}_{im}\bar{C}^{(2)}_{jn} W(\eta_m, \zeta_n, \tau_k) \\ &+ \frac{\delta}{B^4}\sum_{n=1}^{N_\zeta}\bar{C}^{(4)}_{jn} W(\eta_i, \zeta_n, \tau_k) \\ &+ \frac{\bar{m}\delta}{D\Gamma^2}\sum_{p=1}^{N_\tau}\underset{-}{C}^{(2)}_{kp} W(\eta_i, \zeta_j, \tau_p) = \frac{q(\eta_i, \zeta_j, \tau_k)}{D} \end{aligned} \tag{23.10}$$

and the boundary conditions Eqs. (23.6a–d) for x direction as Eqs. (23.11a–d):

$$\begin{aligned} &W(\eta_l, \zeta_j, \tau_k) = 0,\ \sum_{m=1}^{N_\eta} C^{(2)}_{1m} W(\eta_m, \zeta_j, \tau_k) = 0,\ W(\eta_{N\eta}, \zeta_j, \tau_k) = 0, \\ &\sum_{m=1}^{N_\eta} C^{(2)}_{N_\eta m} W(\eta_m, \zeta_j, \tau_k) = 0 \end{aligned} \tag{23.11a–d}$$

and y direction as Eqs. (23.11e–h):

$$\begin{aligned} &W(\eta_i, \zeta_1, \tau_k) = 0,\ \sum_{n=1}^{N_\zeta}\bar{C}^{(2)}_{1n} W(\eta_i, \zeta_n, \tau_k) = 0,\ W(\eta_i, \zeta_{N\zeta}, \tau_k) = 0, \\ &\sum_{n=1}^{N_\zeta}\bar{C}^{(2)}_{N_\zeta n} W(\eta_i, \zeta_n, \tau_k) = 0 \end{aligned} \tag{23.11e–h}$$

and the initial conditions Eqs. (23.7a,b) as Eqs. (23.11i–j):

$$W(\eta_i, \zeta_j, \tau_1) = \frac{w_0}{\delta},\ \sum_{p=1}^{N_\tau}\underset{-}{C}^{(1)}_{p} W(\eta_i, \zeta_j, \tau_p) = \frac{v_0\Gamma}{\delta} \tag{23.11i–j}$$

where $i = 1, 2, \ldots, N_\eta$; $j = 1, 2, \ldots, N_\zeta$; $k = 1, 2, \ldots N_\tau$.

Obviously, there are $N_\eta \times N_\zeta \times N_\tau$ algebraic equations in the Eq. (23.10). Consequently, Eq. (23.10) can be rewritten in the block matrix form as Eq. (23.12):

$$[s] \cdot \{\Omega\} = \{F\} \tag{23.12}$$

where $[s]$ is the stiffness matrix, $[\Omega]$, that is to say, $[\Omega] = \left\{W(\eta_1, \zeta_1, \tau_1), W(\eta_1, \zeta_1, \tau_2), \ldots, W(\eta_{N_\eta}, \zeta_{N_\zeta}, \tau_{N_\tau - 1}), W(\eta_{N_\eta}, \zeta_{N_\zeta}, \tau_{N_\tau})\right\}$, is unknown dynamic displacement vector to be achieved next step, and $\{F\}$ is the generalized load vector coming from the right hand terms of Eq. (23.10).

Similarly, replacing these governing equations of Eq. (23.12) with the boundary conditions Eqs. (23.11a–h) and initial conditions Eqs. (23.11i–j) for the same nodes of space–time domain, we can get the following matrix Eq. (23.13) of simply supported plates subjected to a uniform transverse load

$$\left[\bar{S}\right] \cdot \{\Omega\} = \{\bar{F}\} \tag{23.13}$$

where $\left[\bar{S}\right]$ and $\{\bar{F}\}$ are the updated stiffness matrix and generalized load vector, respectively.

The dynamic displacement vector can be gained by solving Eq. (23.13) and the dimensional transverse displacement field in continuous space–time domain can be obtained by Eq. (23.14):

$$w = \delta \sum_{m=1}^{N_\eta} \sum_{n=1}^{N_\zeta} \sum_{p=1}^{N_\tau} \gamma_m(x) \chi_n(y) \psi_p(t) W(\eta_m, \zeta_n, \tau_p) \tag{23.14}$$

23.3 Numerical Results

To demonstrate the accuracy and efficiency of the adopted time–domain DQM, there are two examples for studying the forced vibration of simply supported plates subjected to a transverse uniform load. For the first one, we calculate the transverse deflection versus time response of a simply supported plate with $\bar{m} = 0.008\,\text{kg/cm}^3$, $E = 1.5 \times 10^7\,\text{N/cm}^2$, $L = 60\,\text{cm}$, $B = 60\,\text{cm}$, $\delta = 1\,\text{cm}$, $\mu = 0.333$. When it is subjected to a uniform suddenly applied load $q_1 = 0.3\,\text{N/cm}^2$, both the initial deflection and velocity are $w_0 = v_0 = 0$. For the second one, we load the same simply supported plate in the above example with a uniform cosine load $q_2 = 0.3 \times \cos(200t)\,\text{N/cm}^2$, and do the similar work.

Note that for the forced vibration of simply supported plates subjected to a transverse load, the Adj. R2 (so–called adjusted coefficient of determination) could be used for measuring the degree of the fitting curves, for examples, the degree of a w versus t curve (i.e., the estimated biomass) gained by DQM to fit the another (namely the observed biomass) gained by analytical solution, and described by following

Eqs. (23.15a, b) [11].

$$Adj.R^2 = 1 - \frac{n-1}{n-p}(1 - R^2),\ R^2 = 1 - \sum_{i=1}^{n}(w^i - \hat{w}^i)^2 / \sum_{i=1}^{n}(w^i - \bar{w})^2 \tag{23.15a, b}$$

in which w^i are values of the analytical solution of w at $t_i = (i-1)\Gamma/(n-1)$, $\hat{w}^i$ are that of the DQM, $\bar{w}$ is the average value of w^i, n is the sample size of w^i, for examples, $n = 301$ in this paper, p is the number of parameters, namely $N_\eta + N_\zeta + N_\tau$, and Γ is a period of vibration time t.

Figures 23.1 and 23.2 present w versus t curves obtained by DQM and analytical solution, where $DQM\alpha(\alpha = 1, 2, \ldots, 8)$ is the αth curve obtained by DQM, and three numbers of its followed parentheses are N_η, N_ζ, and N_τ, respectively; we make N_η equal to N_ζ and smaller than N_τ according to experience. Both in Figs. 23.1 and

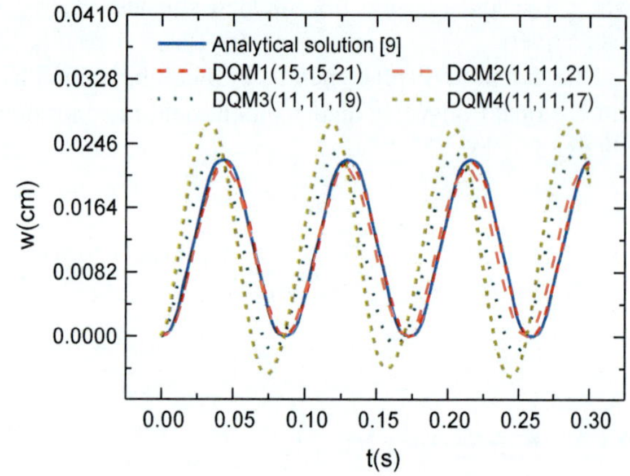

Fig. 23.1 Effect of the number of nodes $N_\eta \times N_\zeta \times N_\tau$ on the accuracy and efficiency of the proposed time–domain DQM for a simply supported plate subjected to a uniform suddenly applied load q_1

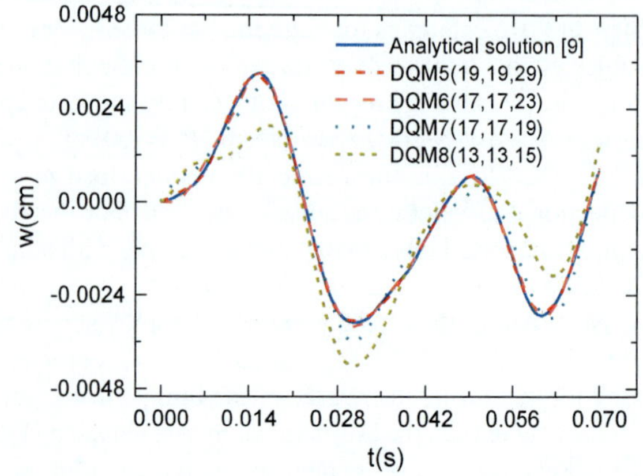

Fig. 23.2 Effect of the number of nodes $N_\eta \times N_\zeta \times N_\tau$ on the accuracy and efficiency of the proposed time–domain DQM for a simply supported plate subjected to a uniform cosine load q_2

Table 23.1 The degree of the fitting curves

	Curve name	$N_\eta + N_\zeta + N_\tau = p$	n	Adj. R^2_1	Adj. R^2_2
Figure 23.1	Analytical solution [9]	$25 + 25 + 0 = 50$	301	1.00000	0.99427
	DQM1	$15 + 15 + 21 = 51$	301	0.99391	1.00000
	DQM2	$11 + 11 + 21 = 43$	301	0.94689	0.95609
	DQM3	$11 + 11 + 19 = 41$	301	0.53077	0.50824
	DQM4	$11 + 11 + 17 = 39$	301	0.23734	0.20913
Figure 23.2	Analytical solution [9]	$33 + 33 + 0 = 66$	301	1.00000	0.99920
	DQM5	$19 + 19 + 29 = 67$	301	0.99920	1.00000
	DQM6	$17 + 17 + 23 = 57$	301	0.99781	0.99606
	DQM7	$17 + 17 + 19 = 53$	301	0.96621	0.96456
	DQM8	$13 + 13 + 15 = 41$	301	0.77113	0.77470

23.2, w versus t curve of DQM gradually become closer to that of the analytical solution as the number of nodes $N_\eta \times N_\zeta \times N_\tau$ increases. Especially, the curves DQM1 and DQM4 are the nearest and the farthest from curve of analytical solution in Fig. 23.1, respectively, while the curves DQM5 and DQM8 in Fig. 23.2.

Meanwhile, Table 23.1 gives the degree of the fitting curves that are shown in Figs. 23.1 and 23.2. For Adj. R2_1 both in Figs. 23.1 and 23.2, the w versus t curve obtained by analytical solution is the observed biomass, while for Adj. R2_2 in Figs. 23.1 the w versus t curve by DQM1 and in Fig. 23.2 by DQM5 are the observed biomass, respectively. From Eqs. (23.15a, b), we easily know that the degree of the observed biomass to fit itself is equal to 1.

For both Figs. 23.1 and 23.2, the Adj. R2_1 whose observed biomass is the w versus t curve obtained by analytical solution increases as the number of nodes $N_\eta \times N_\zeta \times N_\tau$ increases. Especially for Fig. 23.1, when $N_\eta \times N_\zeta \times N_\tau$ is smallest and equal to $11 \times 11 \times 17$, the DQM4 curve has the lowest Adj. R2_1 (0.23734) while the DQM1 curve has the highest Adj. R2 (0.99389) when $N_\eta \times N_\zeta \times N_\tau$ is biggest and equal to $15 \times 15 \times 21$. For Fig. 23.2, similarly, DQM5 and DQM8 have the same phenomena.

For both Figs. 23.1 and 23.2, looking as the observed biomass the w versus t curves obtained respectively by DQM1 and DQM5, these curves that have relatively large Adj. R2_1, such as DQM2(0.94689), DQM6(0.99781), and DQM7(0.96621), also have rather large Adj. R2_2, i.e., DQM2(0.95609), DQM6(0.99606), and DQM7 (0.96456). Without analytical solution, we can judge that whether the results are convergent or not on the basis of Adj. R2 whose observed biomass is that obtained only by DQM.

23.4 Conclusion

The time–domain DQM is for discretizing space domain of continuous functions as well as time domain of that. In this paper, a study of the forced vibration of simply supported plates subjected to a transverse uniform load is presented on the basis of the time–domain DQM. It is observed from results of two examples, the accuracy and efficiency of time–domain DQ method is excellent under certain number of nodes. It is simple and easy for the mathematical principle of space–time DQM to implement in program, so it has a bright application prospect in the dynamic analysis field and is worthy of a further study.

References

1. Bellman, R., Casti, J.: Differential quadrature and long-term integration. J. Math. Anal. Appl. **34**(2), 235–238 (1971)
2. Kitipornchai, S., Ke, L.L., Yang, J., Xiang, Y.: Nonlinear vibration of edge cracked functionally graded Timoshenko beams. J. Sound Vib. **324**(3–5), 962–982 (2009)
3. Ke, L.L., Xiang, Y., Yang, J., Kitipornchai, S.: Nonlinear free vibration of embedded double-walled carbon nanotubes based on nonlocal Timoshenko beam theory. Comput. Mater. Sci. **47**(2), 409–417 (2009)
4. Fung, T.C.: Solving initial value problems by differential quadrature method - Part 1: first-order equations. Int. J. Numer. Meth. Eng. **50**(6), 1411–1427 (2001)
5. Peng, J.S., Xie, G., Yang, L., Yuan, Y.Q.: A time-domain dq approach for vibration analysis of beams. Adv. Mater. Res. **631–632**, 957–961 (2013)
6. Wang, X.W.: Chapter 6-Static analysis of thin plate. Differential quadrature and differential quadrature based element methods, pp. 120–133 (2015)
7. Arani, A.G., Jafari, G.S.: Nonlinear vibration analysis of laminated composite Mindlin micro/nano-plates resting on orthotropic Pasternak medium using DQM. Appl. Math. Mech.-English Ed. **36**(8), 1033–1044 (2015)
8. Safarpour, M., Rahimi, A.R., Alibeigloo, A.: Static and free vibration analysis of graphene platelets reinforced composite truncated conical shell, cylindrical shell, and annular plate using theory of elasticity and DQM. Mech. Based Des. Struct. Mach. **48**(4), 496–524 (2020)
9. Xu, Z.L.: Theory of elasticity (the 2nd half). Higher Education Press, Beijing (2006)
10. Kuang, J.H., Chen, C.J.: Dynamic characteristics of shaped micro-actuators solved using the differential quadrature method. J. Micromech. Microeng. **14**(4), 647–655 (2004)
11. Saglam, S., Ozdemir, E., Ozkan, U. Y., Demirel, T. and Makineci, E.: Biomass estimation of aboveground tree components for Turkey oak (Quercus cerris L.) in south-eastern Turkey. Environ. Monitor. Assess. **192** (7), 418 (2020)